# ADAPTATION POLICY FRAMEWORKS FOR CLIMATE CHANGE: DEVELOPING STRATEGIES, POLICIES AND MEASURES

Adaptation is a process by which individuals, communities and countries seek to cope with the consequences of climate change. The process of adaptation is not new; the idea of incorporating future climate risk into policy-making is. While our understanding of climate change and its potential impacts has become clearer, the availability of practical guidance on adaptation has not kept pace. The development of the Adaptation Policy Framework (APF) is intended to help provide the rapidly evolving process of adaptation policy-making with a much-needed roadmap. Ultimately, the purpose of the APF is to support adaptation processes to protect – and enhance – human well-being in the face of climate change.

The Adaptation Policy Framework is built around four major principles that provide a basis from which integrated actions to adapt to climate change can be developed:

- Adaptation to short-term climate variability and extreme events serves as a starting point for reducing vulnerability to longer-term climate change;
- Adaptation occurs at different levels in society, including the local level;
- Adaptation policy and measures should be assessed in a development context; and
- The adaptation strategy and the stakeholder process by which it is implemented are equally important.

The APF can be used by countries to both evaluate and complement existing planning processes to address climate change adaptation. As an assessment, planning and implementation framework, it lays out an approach to climate change adaptation that supports sustainable development, rather than the other way around. The APF is about practice rather than theory; it starts with the information that developing countries already possess concerning vulnerable systems such as agriculture, water resources, public health, and disaster management, and aims to exploit existing synergies and intersecting themes in order to enable better informed policy-making.

This volume will be invaluable for everyone working on climate change adaptation and policy-making.

**Bo Lim** is the Senior Technical Advisor and Chief of the Capacity Development and Adaptation Group at the United Nations Development Programme (UNDP)-Global Environment Facility (GEF). Dr. Lim managed the GEF National Communications Support Programme for Climate Change at UNDP, that assisted 130 developing countries to meet their commitments under the United Nations Framework Convention for Climate Change.

**Erika Spanger-Siegfried** is an Associate Scientist with the Stockholm Environment Institute – Boston Center, where her work over the past five years has focused on the intersection of sustainable development and international policy, with special emphasis on climate change vulnerability and adaptation.

**Ian Burton,** a Scientist Emeritus with the Meteorological Service of Canada, and an Emeritus Professor, University of Toronto, is a specialist in natural hazards management, risk assessment, and adaptation to climate change. Dr. Burton now works as an independent scholar and consultant.

**Elizabeth L. Malone** is a Senior Research Scientist at Battelle Washington Operations working on policy-relevant social science research in global change issues. Dr. Malone's work has contributed to linkages among global environmental change, globalization, economic development, equity, and sustainability.

**Saleemul Huq** is the founding Executive Director of the Bangladesh Centre for Advanced Studies, the major non-government research and policy institute working on environment and development related issues in Bangladesh. Dr. Huq's environmental planning experience includes work on global environmental issues for numerous international agencies.

# Adaptation Policy Frameworks for Climate Change: Developing Strategies, Policies and Measures

**Edited by**

**Bo Lim**
**Erika Spanger-Siegfried**

**Co-authored by**

**Ian Burton**
**Elizabeth Malone**
**Saleemul Huq**

United Nations
Development
Programme

CAMBRIDGE
UNIVERSITY PRESS

# Contents

# Foreword

A key issue, especially for non-Annex I Parties, is how to develop national strategies for adaptation to climate change that are easy to integrate into sustainable development plans. Most national vulnerability and adaptation studies to date have focused on the selection of climate change scenarios and impact studies – an approach that has not always resulted in policy-relevant options for adaptation responses.

Through Swiss, Canadian and Dutch funding, the National Communications Support Unit of the United Nations Development Programme has developed *Adaptation Policy Frameworks for Climate Change: Developing Strategies, Policies and Measures,* hereafter referred to as the Adaptation Policy Framework (APF), that consists of a User's Guidebook and nine Technical Papers. This Framework provides a flexible approach that can be modified to meet the specific needs of countries in any region of the world. The main objective of the Guidebook and the Technical Papers is to assist and provide guidance to developing countries in identifying, prioritising, and shaping potential adaptation options into a coherent strategy that is consistent with their sustainable development and other national priorities. The Framework may also support the preparation of the National Communications of both Annex I and non-Annex I Parties.

The APF builds on several methods, including the 1994 *IPCC Technical Guidelines for Assessing Climate Change Impacts and Adaptations.* A key innovation is that it will work from *current* climate variability and extremes, and assess recent climate experiences. In other words, it is firmly grounded in the present, and it links the near-term to the medium- and longer-terms. Other innovations include developing an adaptation baseline and situating adaptation in the current policy context. The Framework will focus on adaptations and best practices that are known to reduce vulnerability in the most effective way.

The APF will assist Parties in mainstreaming the development of national strategies for adaptation in the sustainable development policy context. Other features include the involvement of stakeholders and public participation at the community level, and the integration of adaptation measures with natural hazard reduction and disaster prevention programmes. All of these elements are being developed in the dual contexts of capacity building and the need to strengthen adaptive capacity.

The APF has been developed for implementation of Global Environment Facility (GEF) and other initiatives, including regional projects and national efforts to respond to the challenge of climate change.

José Romero
Senior Scientific Officer
Conventions Section, International Affairs Division
Swiss Agency for Environment, Forests and Landscape
Berne, Switzerland

Jean-Bernard Dubois
Deputy Head
Division of Natural Resources and Environment
Swiss Agency for Development and Cooperation
Berne, Switzerland

Frank Pinto
Executive Co-ordinator
UNDP Global Environment Facility
New York, United States

September 2004

# Executive Summary

Climate change impacts can affect all sectors and levels of society. In the past few years, reducing vulnerability to climate change has become an urgent issue for the world's developing countries. Not only do these countries lack the means to cope with climate hazards, but their economies also tend to have greater dependence on climate-sensitive sectors, such as agriculture, water, and coastal zones. For these countries, climate change adaptation remains at the forefront of any sustainable development policy agenda.

Adaptation is a process by which individuals, communities and countries seek to cope with the consequences of climate change, including variability. The process of adaptation is not new; throughout history, people have been adapting to changing conditions, including natural long-term changes in climate. What is innovative is the idea of incorporating future climate risk into policy-making. Although our understanding of climate change and its potential impacts has become clearer, the availability of practical guidance on adaptation to climate change has not kept pace.

The development of the Adaptation Policy Framework (APF) was motivated because the rapidly evolving process of adaptation policy making has lacked a clear roadmap. The APF seeks to address this gap by offering a flexible approach through which users can clarify their own priority issues and implement responsive adaptation strategies, policies and measures.

The United Nations Development Programme – Global Environment Facility (UNDP-GEF), with support from the Swiss, Canadian and Dutch governments, developed the APF as an innovative set of guidance for the development and implementation of adaptation strategies. The APF aims to help countries as they integrate adaptation concerns into the broader goals of national development. Ultimately, the purpose of the APF is to support adaptation processes to protect and, when possible, enhance human well-being in the face of climate change, including variability.

## The United Nations Development Programme Vision

Looking ahead, the United Nations Development Programme (UNDP) envisages that the guidance embodied in the APF could help launch engagement across broad segments of society on how to advance sustainable development in the face of climate risks.

At the broadest level, this could lead to harmonisation of adaptation with a country's additional, often more pressing, development priorities such as poverty alleviation, food security enhancement, and disaster management.

At a more operational level, the UNDP believes that the following realignments can take place as dialogue around adaptation unfolds in the years ahead:

- Initiation of a process to reverse trends that increase maladaptation and raise the risks for human populations and natural systems;
- Reassessment of current plans for increasing the robustness of infrastructure designs and long-term investments;
- Improvement of societal awareness and preparedness for future climate change, from policy-makers to local communities;
- Increased understanding of the factors that enhance or threaten the adaptability of vulnerable populations and natural systems; and
- A new focus on assessing the flexibility and resilience of social and managed natural systems.

## Principles of the Adaptation Policy Framework

The APF is structured around four major principles that provide a basis from which actions to adapt to climate change can be developed. Embedded in these principles are features that distinguish the APF from previous guidance.

- *Adaptation to short-term climate variability and extreme events is included as a basis for reducing vulnerability to longer-term climate change.* As users seek to prepare for near-, medium- and longer-term adaptation, the APF helps them to firmly ground their decisions in the priorities of the present.
- *Adaptation policy and measures are assessed in a developmental context.* By making policy the centrepiece of adaptation, the APF shifts the focus away from individual adaptation projects as a response to climate change, and toward a fundamental integration of adaptation into key policy and planning processes.
- *Adaptation occurs at different levels in society, including the local level.* The APF combines national policy-making with a proactive "bottom-up" risk management approach. It enables the user to hone in on and respond to key adaptation priorities, whether at the national or village scale.
- *Both the strategy and the process by which adaptation is implemented are equally important.* The APF places a strong emphasis on the broad engagement of stakeholders. Stakeholders are seen as instrumental in driving each stage of the adaptation process.

The APF's strong emphasis on flexibility underpins each of these principles. In the APF, users will find a comprehensive review of the available analytical techniques, as well as clear encouragement to use only those techniques that meet their unique needs.

The APF recognises the value of building on what is already known, utilising synergies and intersecting themes to enable more informed and effective policy-making, and to guide adaptation. At its heart, the APF is about practice rather than theory. For any country or community using the APF, the starting point is the information that already exists on vulnerable systems such as agriculture, water resources, public health, and disaster management.

**The Adaptation Policy Framework Process**

The primary use of the APF is to guide studies, projects, planning and policy exercises (collectively referred to hereafter as "projects") toward the identification of appropriate adaptation strategies, policies and measures. Depending on the level of knowledge about the vulnerable system, the particular APF process used can vary widely from one project to the next.

The APF is comprised of five Components:

**Component 1: Scoping and designing an adaptation project** involves ensuring that a project – whatever its scale or scope – is well-integrated into the national policy planning and development process. This is the most vital stage of the APF process. The purpose is to put in place an effective project plan so that adaptation strategies, policies and measures can be implemented.

**Component 2: Assessing current vulnerability** involves responding to several questions, such as: Where does a society stand today with respect to vulnerability to climate risks? What factors determine a society's current vulnerability? How successful are the efforts to adapt to current climate risks?

**Component 3: Assessing future climate risks** focuses on the development of scenarios of future climate, vulnerability, and socio-economic and environmental trends as a basis for considering future climate risks.

**Component 4: Formulating an adaptation strategy** in response to current vulnerability and future climate risks involves the identification and selection of a set of adaptation policy options and measures, and the formulation of these options into a cohesive, integrated strategy.

**Component 5: Continuing the adaptation process** involves implementing, monitoring, evaluating, improving and sustaining the initiatives launched by the adaptation project.

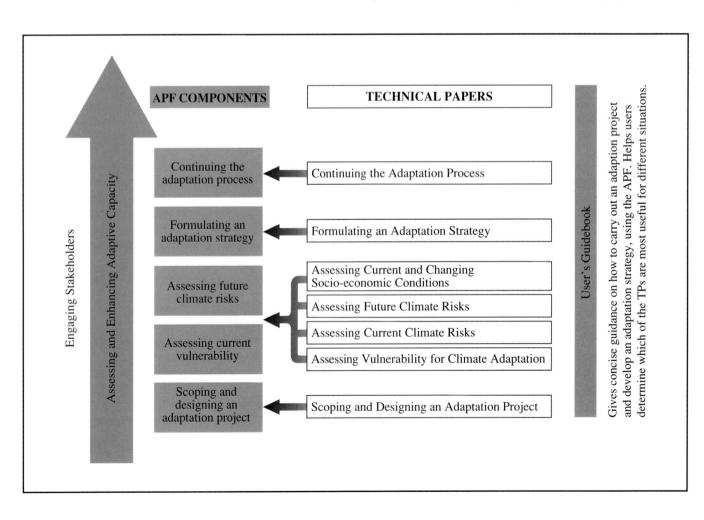

---

Implementing the APF will invariably be characterised by:

- Careful application of the scoping and design process;
- Strong stakeholder engagement;
- Assessing and enhancing adaptive capacity;
- Analysis of adaptation to cope with current and future climate change; and
- A programme to monitor, evaluate and improve the impact of the adaptation activity.

---

These components are supplemented by two cross-cutting processes: 1) Engaging stakeholders in the adaptation process, and 2) Assessing and enhancing adaptive capacity.

**Engaging stakeholders in the adaptation process** is seen as essential to each APF component, and is ultimately crucial to the successful implementation of an adaptation strategy. Engaging stakeholders requires an active and sustained dialogue among affected individuals and groups.

**Assessing and enhancing adaptive capacity** involves catalysing change management processes so that societies can better cope with climate change, including variability.

Users will approach the APF with a wide range of needs. For some, addressing all components of the APF process will be the most strategic, perhaps more resource-intensive route. Others may already have significant information on current vulnerability, but not on future climate risks, and may choose to fill gaps in information by investing heavily in only one or two of the components. The APF accommodates either of these approaches, and a wide range of other uses.

Implementing the APF does not require an abundance of high quality data, or extensive expertise in computer-based models. Rather, it relies upon a thoughtful assessment and a robust stakeholder process. While not costless in terms of time and resources, the APF process is readily manageable if correctly applied.

**Resources**

The APF is supported by a series of nine Technical Papers (TPs). Each TP explores a specific component of the APF and provides detailed guidance. The papers are:

1. *Scoping and Designing an Adaptation Project*
2. *Engaging Stakeholders in the Adaptation Process*
3. *Assessing Vulnerability for Climate Adaptation*
4. *Assessing Current Climate Risks*
5. *Assessing Future Climate Risks*
6. *Assessing Current and Changing Socio-economic Conditions*

7. *Assessing and Enhancing Adaptive Capacity*
8. *Formulating an Adaptation Strategy*
9. *Continuing the Adaptation Process*

The TPs are aimed at both the scientific community and actual practitioners of adaptation. Many of these papers feature annexes of resources, toolkits, and other guidance tools. Users who are more interested in a general understanding of the APF are encouraged to refer to the APF User's Guidebook, which outlines the APF process.

# Section I

# User's Guidebook

Section Co-ordinators:
BO LIM (UNDP), ELIZABETH MALONE (USA)

# User's Guidebook

ERIKA SPANGER-SIEGFRIED[1] AND BILL DOUGHERTY[1]

Contributing Authors
*Tom Downing[2], Molly Hellmuth[3], Udo Hoeggel[4], Andreas Klaey[4], and Kate Lonsdale[2]*

Reviewers
*Ayite-Lo N. Ajavon[5], Boni Biangini[6], Yamil Bonduki[7], Henk Bosch[8], Nick Brooks[9], James B. Chimphamba[10], Kristie L. Ebi[11], Ermira Fida[12], Pascal Girot[13], Mamadou Honadia[14], Saleemul Huq[15], Roger Jones[16], Emilio L. La Rovere[17], Elizabeth L. Malone[18], Taito Nakalevu[19], Isabelle Niang-Diop[20], Nicole North[21], Rosa Perez[22], Olga Pilifosova[23], Eduardo Reyes[24], Andy Reisinger[25], Othmar Schwank[21], Barry Smit[26], Jessica Troni[27], and Gary Yohe[28]*

[1] Stockholm Environment Institute, Boston, United States

[2] Stockholm Environment Institute, Oxford, United Kingdom

[3] UNEP Collaborating Center on Energy and Environment, Roskilde, Denmark

[4] Centre for Development and Environment, University of Berne, Berne, Switzerland

[5] University of Lomé, Lomé, Togo

[6] Global Environment Facility, Washington DC, United States

[7] United Nations Development Programme – Global Environment Facility, New York, United States

[8] Government Support Group for Energy and Environment, The Hague, The Netherlands

[9] Tyndall Centre for Climate Change Research, University of East Anglia, Norwich, United Kingdom

[10] Department of Geography and Earth Sciences, University of Malawi, Zomba, Malawi

[11] Exponent, Alexandria, United States

[12] National Environmental Agency, Tirana, Albania

[13] Bureau of Development Policy – Latin American Region SURF, United Nations Development Programme, San Jose, Costa Rica

[14] Secretariat permanent du CONAGESE, Ougadougou, Burkina Faso

[15] International Institute for Environment and Development, London, United Kingdom

[16] Commonwealth Scientific & Industrial Research Organisation, Atmospheric Research, Aspendale, Australia

[17] Centre for Integrated Studies on Climate Change and the Environment, Rio de Janeiro, Brazil

[18] Pacific Northwest National Laboratory, Washington, DC, United States

[19] South Pacific Regional Environment Programme, Apia, Samoa

[20] Department of Geology, Faculty of Science, University Cheikh Anta Diop, Dakar, Senegal

[21] INFRAS, Zurich, Switzerland

[22] Philippine Atmospheric, Geophysical and Astronomical Services Administration, Manila, Philippines

[23] United Nations Framework Convention on Climate Change, Bonn, Germany

[24] Autoridad Nacional de Ambiente, Panamá City, Panamá

[25] Climate Change Office, Ministry for the Environment, New Zealand

[26] University of Guelph, Guelph, Canada

[27] Department for International Development, Oxford, United Kingdom

[28] Wesleyan University, Middletown, United States

# CONTENTS

## Introduction

This User's Guidebook summarises guidance prepared by the United Nations Development Programme (UNDP) for the development and implementation of climate change adaptation strategies. The UNDP developed this Guidebook in collaboration with leading experts from around the globe.

### *Why was this Guidebook written?*

- The Guidebook was motivated by the lack of practical guidance on adaptation to climate change. While a substantial amount of literature exists regarding climate change impacts, information on adaptation policy and strategies is limited.
- The Guidebook explains how to use the Adaptation Policy Framework (APF). It is designed to provide user-friendly guidance on the most appropriate adaptation approaches and tools, customised to a country's unique national circumstances.
- For most countries, climate change adaptation is a new undertaking. In general, when examining adaptation strategies, numerous conceptual, technical and operational challenges arise. Developed as part of the APF, the Guidebook is a quick reference that addresses the above challenges in the nine Technical Papers (TPs) that comprise the Framework.

### *What are the Guidebook's objectives?*

- The Guidebook reviews key concepts, methods, and case studies to formulate adaptation strategies and measures – emphasising readability. Graphics, topic text boxes, and markers are used to present particularly important issues. The TPs are cited throughout this document for technical background. In this way, the Guidebook helps users understand their options for carrying out an adaptation project, as well as the range of technical and other resources available.
- Countries exhibit a great range in the types of vulnerability and adaptation projects they have undertaken, the role of stakeholders in development planning, and the degree of technical capacity. The APF can support the adaptation process at any point in this range depending on local constraints, resources, and opportunities.
- The Guidebook's major objective is to assist in the process of incorporating adaptation concerns into local, sector-specific, and national development planning processes.

### *Who should read this Guidebook?*

- Although anyone can use this Guidebook – policy-makers, the academic community, project developers, and local stakeholders – it is primarily designed for technical analysts, climate project coordinators and developers, and climate change policy makers.
- The Guidebook will also be useful to stakeholders interested in sustainable development. It can promote dialogue among local communities, policy-makers, the private sector and the general public regarding adaptation to climate change in general, and the prospects for incorporating adaptation into national development priorities.

In short, this Guidebook articulates the APF's flexible approach to the design and implementation of climate change adaptation activities.

## How to use this Guidebook

Throughout this Guidebook, **adaptation** is used to describe a process by which strategies to moderate and cope with the consequences of climate change, including variability, are developed and implemented. The APF and its set of TPs have been developed to provide guidance to all climate adaptation efforts – from the national to the local scale.

Globally, countries are already adapting to current climatic events on different levels (national, provincial, and/or local) and over various time frames (short- to long-term). Adaptation planning occurs primarily through government policy making. When unplanned, adaptation tends to be triggered by unexpected changes in natural or human systems.

Developing an adaptation strategy for future climate change requires a set of key objectives. At the broadest level, these should fit within a nation's development priorities (e.g., poverty alleviation, food security enhancement, action plans under multilateral environmental agreements). At an operational level, there are at least five important objectives:

- Initiation of a process to reverse trends that increase maladaptation and raise the risks for human populations and natural systems;
- Reassessment of current plans for increasing the robustness of infrastructure designs and long-term investments;
- Improvement of societal awareness and preparedness for future climate change, from policy-makers to local communities;
- Increased understanding of the factors that enhance or threaten the adaptability of vulnerable populations and natural systems; and
- A new focus on assessing the flexibility and resilience of social and managed natural systems.

Developing an adaptation strategy that can respond to these objectives requires a vision that balances the need to reduce climate change impacts with the constraints of national policy-making processes. Whatever adaptation options and measures emerge, packaging these decisions into an effective adaptation strategy will require increased policy coherence across economic sectors, societal levels and time frames.

## Approach

The APF is designed as a roadmap rather than a cookbook. For users who want more details on analytical issues, this information is available in the references and citations in each TP of the APF.

In essence, the Guidebook is an interface between its users and the technical information in the accompanying papers (and, by extension, the general literature on climate vulnerability and adaptation). This Guidebook does *not* replace the APF Technical Papers; rather, it is part of a package of material to orient users to the key Components of the APF.

The Guidebook aims to strike a balance between encouraging flexibility in designing adaptation plans and providing concrete recommendations. Because each country's needs and resources are different, this Guidebook avoids presenting a step-by-step list. Instead, it outlines the basic parameters of the APF process and its major Components, while identifying the strategic issues and policy decisions involved.

This point deserves special emphasis. The APF is a flexible process that project teams use to formulate and implement their climate change adaptation strategies. It can be applied at various levels – e.g., policy development, project formulation, and multi-sectoral studies. Given the APF's flexibility, considerable effort is devoted in the Guidebook to helping users identify an appropriate entry point, key outputs, appropriate methods and tools, and the extent of analysis. Throughout the discussion, the underlying APF principle is that all adaptation activities should be compatible with a country's broader development context.

## Target audience

Users do not need any prior knowledge of climate vulnerability and adaptation techniques. The APF will valuable to anyone who wants to know more about climate change adaptation, including those charged with making policy or developing projects.

## Structure of the Guidebook

The Guidebook starts with an overview of the APF, including the relationship between the seven APF Components and the nine Technical Papers, the range of options for using the APF, and important adaptation concepts.

The Guidebook devotes a section to each of the Components, addressing key concepts and tasks, as well as the challenges of carrying out these activities. To orient users to the principal themes of each APF Component, the **purpose**, **process**, and desired **outputs** are indicated at the beginning. TP references have been inscribed in the text throughout the Guidebook to facilitate easy reference to the TPs and to specific sections where the reader can find technical guidance.

Each Component section concludes with **Key Issues** and a **Checklist**. These highlight the major issues, decisions, and interim products that need to be addressed and/or developed within each Component.

## The Adaptation Policy Framework

### What is the Adaptation Policy Framework?

The APF is structured around four major principles:

- Adaptation to short-term climate variability and extreme events serves as a starting point for reducing vulnerability to longer-term climate change.

- Adaptation policies and measures are best assessed in a developmental context.

- Adaptation occurs at different levels in society, including the local level.

- The adaptation strategy and the process by which it is implemented are equally important.

*Think of the APF as a structured approach to formulating and implementing adaptation strategies, policies, and measures to ensure human development in the face of climate variability and change. The APF links climate change adaptation to sustainable development and global environmental issues.*

To address climate change impacts, countries add adaptation policies and measures to existing planning processes, including assessment, project development, implementation and monitoring. As a framework, it lays out an approach to climate change adaptation that enhances sustainable development, rather than the other way around. It also facilitates the process of identifying, characterising, and promoting "win-win" adaptation options.

The APF is about practice rather than theory. It starts with the information – which countries already possess – on vulnerable systems such as agriculture, water resources, public health, and disaster management. This information can be used to initiate a shift in the way risk, vulnerability and climate change are viewed. The APF builds on what is already known rather than "reinventing the wheel". By making use of existing synergies and intersecting themes, this approach can ultimately lead to a more informed policy-making process.

### Intended Adaptation Policy Framework outputs and outcomes

The APF is capable of providing a variety of outputs, depending on how it is applied. While specific outputs depend on particular needs and goals, in general, a completed APF process leads to a clarification of adaptation strategies, policies and measures, implementation plan, and enhanced adaptive capacity.

A particular use of the APF depends on the desired outputs. Several major outputs are envisioned, as follows:

- **Policy development:** The APF can be used to identify policy options to reduce climate change impacts, either through measures that enhance society's resilience or actions that expand the range of coping strategies. This policy focus may be directed at certain aspects of a national development strategy, at specific geographic areas, or at important sectors of the national economy (e.g., agriculture, forestry, water resources, transportation, coastal zone management, public health, ecosystem management and risk management).
- **Integrated assessments:** Adaptation in one sector often has consequences for another. For example, reduction of the impacts of drought can improve nutrition levels and overall public health. For this reason, the APF has been designed to facilitate a process of integrated assessment, including a consultation process in which links between sectors can be identified and assessed. Such assessments can also offer valuable input to National Communications under the United Nations Framework Convention on Climate Change (UNFCCC).

- **Project formulation:** The APF process can be used for formulating adaptation projects, or for exploring the potential to add adaptation considerations to other types of projects. These projects can focus on any population scale, from the village to the national level.

A well-implemented APF initiative can catalyse a policy process that extends well beyond the project's lifetime. During the process of implementing an adaptation project, public awareness should be raised, individual, community, sectoral and national capacities enhanced, and policy processes established or modified.

Ideally, an "adaptation community" will be created – one that is capable of supporting the new adaptation process. At the end of the effort, both the team and the stakeholders should have a better understanding of the key strengths and vulnerabilities of their priority system, with respect to climate change.

### *The Adaptation Policy Framework Components*

Figure 1 illustrates the APF process. Five basic Components (the shaded boxes) are linked by two cross-cutting Components (represented by the arrow (adaptive capacity) and

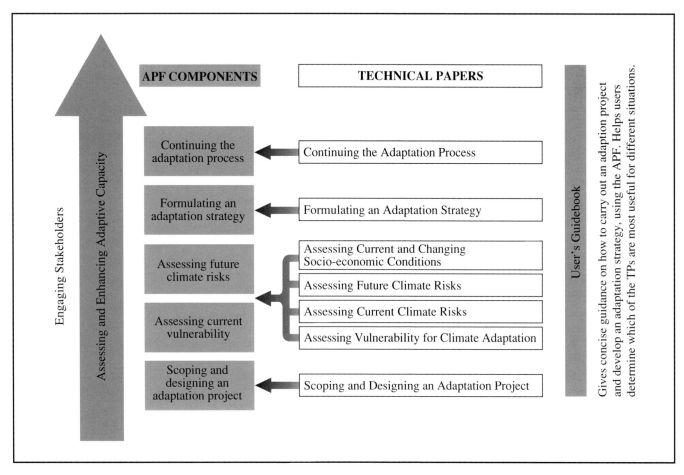

**Figure 1:** Outline of the Adaptation Policy Framework process

the larger frame (the stakeholder context) within which all Components are played out). Details regarding the technical underpinnings of the APF are provided by the nine **TPs**.

Since specific adaptation measures are usually implemented at various levels, the APF is intended to be accessible to technical analysts, the private sector, the general public and other stake-holders. In particular, the APF process emphasises both stake-holder engagement and the need to mobilise local action to increase adaptive capacity.

Each Component of the APF is briefly summarised below.

- *Scoping and designing an adaptation project* involves ensuring that the project is well designed and can be integrated into the national policy process.
- *Assessing current vulnerability* involves an assess-ment of the present situation. It addresses the ques-tions: "Where does society stand today with respect to vulnerability to climate risks?" "What factors deter-mine its current vulnerability?" And "How successful are its efforts to adapt to current climate risks?"
- *Assessing future climate risks* involves developing scenarios of future climate, vulnerability, and socio-economic and environmental trends as a basis for con-sidering future climate risks.
- *Formulating an adaptation strategy* involves the creation of a set of adaptation policy options and measures in response to current vulnerability and future climate risks.
- *Continuing the adaptation process* involves imple-menting, monitoring, evaluating, and sustaining the initiatives started by the adaptation project.
- *Engaging stakeholders in the adaptation process* is crucial to the successful implementation of adapta-tion. This cross-cutting Component involves creating and sustaining an active dialogue among affected indi-viduals and groups.
- *Assessing and enhancing adaptive capacity,* another cross-cutting Component, involves the integration of activities to better cope with climate change, including variability, into national capacity strengthening efforts.

Each of the above Components has its logic and purpose. However, the APF is sufficiently flexible to permit projects to use only one or two Components, or to apply modified versions of the Components. Decisions about how to use the APF will depend on a country's prior adaptation work, needs, goals, and resources (see *Getting started* and *Scoping and designing an adaptation project*).

### The Adaptation Policy Framework Technical Papers

As mentioned earlier, the APF is supported by a series of nine TPs, each of which explores a specific aspect of the APF and provides detailed guidance on one or more of the APF Components. Each TP also contains annexes with additional information on methodologies and tools.

- **Technical Paper 1:** *Scoping and designing an adaptation project* focuses on the first Component of the APF. It is a general guide to all of the tasks and activities involved in formulating and implementing adaptation.
- **Technical Paper 2:** *Engaging stakeholders in the adaptation process* focuses on the role of stakeholders in identifying appropriate adaptation strategies. This TP outlines a cross-cutting Component with implica-tions for each of the other APF Components.
- **Technical Paper 3:** *Assessing vulnerability for climate adaptation* focuses on methods and tools for a vulnera-bility assessment for climate adaptation. This paper outlines the vulnerability-based approach to adaptation.
- **Technical Paper 4:** *Assessing current climate risks* outlines a conceptual framework for assessing current climate risks using the natural hazards-based and the vulnerability-based approaches. This TP emphasises the (natural) hazards-based approach to adaptation.
- **Technical Paper 5:** *Assessing future climate risks* describes risk assessment techniques for determining climate risks and adaptation needs under a changing climate. This TP also emphasises the (natural) haz-ards-based approach to adaptation.
- **Technical Paper 6:** *Assessing current and changing socio-economic conditions* presents how to charac-terise socio-economic conditions and how they relate to vulnerability and climate analyses. This TP outlines the policy-based approach to adaptation while sup-porting other approaches.
- **Technical Paper 7:** *Assessing and enhancing adap-tive capacity* discusses how to assess and enhance the capacity of human systems to cope with climate change, including variability. TP7 outlines the second cross-cutting Component and has implications for each of the APF Components. This TP describes the adaptive capacity approach to adaptation while sup-porting other approaches as well.
- **Technical Paper 8:** *Formulating an adaptation strat-egy* focuses on how to formulate a strategy that responds effectively to a system's key vulnerabilities and to the project's unique policy context and nation-al development goals.
- **Technical Paper 9:** *Continuing the adaptation process* focuses on the processes of barrier removal, incorporating adaptation into the development process, and improving implemented adaptation activ-ity over time, through monitoring and evaluation.

### Getting started

Applying the APF – and its associated methods and tools – ini-tially depends on the nature of the output desired (e.g., policy development, integrated assessment, or project formulation). Once this is established, APF users should identify specific approaches, methods, and tools that are appropriate, consider-ing the resources available.

Given the range of potential uses of the APF, it is important that users evaluate project priorities, desired outcomes, and resources. They should address several aspects of an APF including:

- **Approach:** A variety of conceptual frameworks or approaches can be used when applying the APF. In fact, each of the four different methods – (natural) hazards-based, vulnerability-based, policy-based, and adaptive capacity-based – emphasises a different aspect or Component of the adaptation process.
- **Coverage:** Uses of the APF can vary considerably in the level of coverage. For example, one country's strategic focus may address all of its geographic areas and major sectors for some long-term planning period. Another's might be highly localised geographically (e.g., coastlines) and in terms of sectors (e.g., fisheries). An example of the latter would be an APF project to address the vulnerability of fishing communities to frequent storm surges and future sea level rise.
- **Methods and tools:** The methods and tools used will depend on the level of complexity and/or comprehensiveness of the effort. Within each of the four primary approaches indicated above, a variety of analytical tools are available. Some of these can be highly quantitative (e.g., agent-based simulation modelling, multi-criteria analysis, scenario analysis), while others are more suitable for qualitative assessments (e.g., stakeholder consultations, focus groups).
- **Components:** The specific tasks that are carried out will depend on the particular Components of the APF that are most relevant to a country's situation. For example, some countries have existing robust vulnerability assessments. In others, no one has ever explored the process of formulating and implementing an adaptation strategy.

Applying the APF does not necessarily require an abundance of high-quality data, or extensive expertise in computer-based models. It is possible to use the APF to conduct a project in entirely qualitative terms. Applying the APF requires thoughtful assessment of adaptation to climate change, a robust stakeholder process – and what would be considered manageable costs in terms of time and funding. For some countries, addressing all five basic and both cross-cutting Components of the APF process will be the most strategic, but more resource-intensive, option. Other countries may, e.g., already have significant information on current, but not future vulnerability. These countries may choose to fill gaps in information by focusing on the latter Components. In short, there are a number of options.

That said, implementing the APF will invariably be characterised by:

- careful application of the scoping and design process
- a strong stakeholder engagement process
- assessment and enhancement of adaptive capacity
- analysis of adaptation to cope with current and future climate change; and

- a programme to monitor and evaluate the impact of adaptation.

### Implementing the Adaptation Policy Framework

The following sections provide guidance on how to implement an APF process. As users proceed, it is important to remember that the detailed technical guidance is found in the accompanying TPs.

Each section corresponds to an APF Component; within it, the different key tasks are outlined. It is important to emphasise that the APF guidance should be tailored by users and adapted to local circumstances so that it is: (a) modified to national goals, resources, and expected outcomes; (b) limited to time and resource constraints; (c) as substantive as possible; and (d) designed to meet applicable standards and/or criteria of key national (i.e., sectoral ministries), multilateral (e.g., GEF), and bilateral (e.g., industrialised country donors) organisations.

It is possible that a reasonable outcome for a given project would be to carry out a subset of the tasks, or to modify the tasks to better fit existing resources and constraints.

The Guidebook helps users navigate the decisions that must be made to implement the APF process effectively. They include: (1) an appropriate project approach; (2) prioritising Components and tasks; (3) specific methods and technical resources; and (4) plans for implementation, awareness building and continuation of the process.

Users will note a more detailed description of the tasks in the *Scoping and designing an adaptation project* Component, particularly in comparison with those in subsequent Components. This Component is emphasised somewhat since getting the APF process off on the right track – the aim of Component 1 – is the most important aspect of the entire APF process.

### Scoping and designing an adaptation project

| Key TPs: 1 and 2 | This section introduces the process of scoping and designing adaptation projects – the focus of the first Component of the APF process. The main purpose is to establish an effective project plan, so that APF users can design adaptation strategies, policies and measures.

The **process** is illustrated by the flow chart in Figure 2, and includes four major tasks:

1. scope project and define objectives;
2. establish the project team;
3. review and synthesise existing information on vulnerability and adaptation; and
4. design the adaptation project.

The expected **output** is a detailed implementation plan, including clearly stated objectives, activities and outcomes.

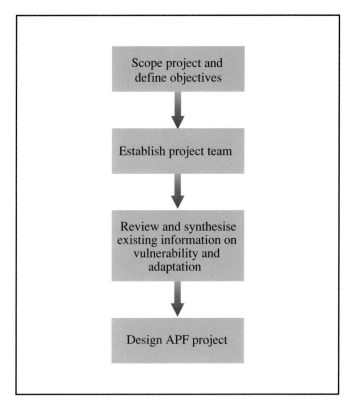

**Figure 2:** Scoping and designing an adaptation project

*Think of scoping and designing an adaptation project as essentially a small-scale exploration of all of the APF Components to get a sense of the "big picture".*

Since the full adaptation team will likely not have been assembled at this point, only the core team members will be participating. Although stakeholder input is valuable for determining project priorities and strategies, it is probably preferable to only include a small group at this juncture.

### Task 1: Scope project and define objectives

| TP1 Section 1.4 | APF studies are intended to identify strategies, policies, and measures that will have far-reaching and long-lasting effects. To achieve this, the APF process starts by scoping the key elements to consider in (a) existing development policies and priorities, and (b) adaptation needs and constraints. The set of activities outlined below is intended to guide the initial project team through the process of identifying these key considerations.

### Establish the stakeholder process

| TP1 Section 1.4.1, Box 1-1; TP2 Section 2.6.1 | To establish adaptation priorities, needs and constraints, engaging stakeholders is essential. The core team members should initiate an inclusive stakeholder dialogue process – one that accommodates a range of diverse view-

points. This type of stakeholder input can help to ensure that the project will respond to critical adaptation priorities.

The stakeholder process may need to be iterative: the initial project team consulting with a small group of stakeholders to inform the initial development of the project objectives; the full project team engaging a broad group of stakeholders throughout the project. It is important to note that this activity is a subset of activities outlined under the cross-cutting Component, *Engaging stakeholders in the adaptation process.*

### Prioritise key systems

| TP1 Section 1.4.1; TP6 Section 6.4.1 | Countries have a range of vulnerabilities to climate change from drought risk to an increased burden of vector-borne diseases. Users will need to narrow the focus of their project to a strategic subset of adaptation priorities. In principle, priority should be placed on systems where there is both high vulnerability and a high likelihood of significant impacts from climate hazards.

| TP1 Section 1.4.1, Annex A.1.1; TP3 Section 3.4.2 | Some adaptation projects will start with a clear choice of priority system. For others, users can develop a list (ranked or unranked) of who is vulnerable, to what, where and to what extent. The information, although somewhat general at this stage, should be adequate to make the necessary comparisons and prioritisations. In addition to the list, a qualitative description of the reasons underlying the choice of priorities can be helpful.

Adaptation priorities can be identified using existing vulnerability assessments, consultations with people likely to be affected, the advice and needs of decision makers, scientific experts, etc. However, to be legitimate in the public eye, the prioritisation process should include some form of stakeholder input.

### Review policy process

| TP1 Section 1.4.1 | The major goal of reviewing policy processes is to recognise how adaptive capacity can be developed. Understanding national, sectoral and local policy-making processes is essential for assessing how an adaptation strategy might be implemented through these processes.

Output for this activity might include a brief overview of:

- relationship between key policy processes and climate change adaptation;
- potential for integrating adaptation concerns into policy agendas; and
- ways to improve existing linkages for policy coherence and to strengthen commitment to adaptation.

It will be especially useful to identify situations within the policy process where adaptation recommendations may be difficult to implement or sustain. Once identified, approaches may then be developed to manage these barriers.

## Determine project objectives and outcomes

Framing the project objectives and expected outcomes is a critical step. This process will determine whether the project is responsive to the needs of stakeholders and policy makers.

This process should result in a set of concise objectives and a corresponding set of expected outcomes that are achievable within the scope of the project. The process of setting objectives can be accomplished using facilitated stakeholder fora, expert opinion and input from policy makers.

As an aid to future monitoring and evaluation (M&E) efforts, the APF project team should also develop criteria to evaluate the APF's success at this point. These can help to judge the degree to which the expected outcomes have been achieved.

## Develop communication plan

| TP1 Section 1.4.1 | The results of the adaptation process will be most useful if they are shared with key stakeholders, decision makers and the general public. Therefore, it is important to produce a communication plan that is tailored to the needs of target audiences. The communication strategy should be designed in such a way that its effectiveness can be monitored and evaluated, and it can be adjusted and modified on the basis of such an evaluation.

## Task 2: Establish project team

| TP1 Section 1.3 | Effective adaptation requires a team that closely reflects the needs and objectives of the project. In selecting the team, the goal will be to develop an interdisciplinary panel that both represents a range of sectors and scales of society and is capable of responding to each of the project's priorities. Ideally, team members should be able to commit for the duration of the APF process.

## Task 3: Review and synthesise existing information on vulnerability and adaptation

| TP1 Section 1.4.2 | In some countries, work on vulnerability and adaptation may have already been carried out. Task 3 involves identifying such resources and distilling the most important information for input to the development of a **project baseline**.

Baselines are used to sketch the current situation and to give researchers a snapshot against which to view change. It is against this baseline that the effectiveness of adaptation action can later be assessed. A well-defined project baseline should outline the current level of vulnerability and adaptation in the system of interest. Though this is a preliminary baseline to be refined in later tasks, it is a critical APF task, as the project baseline serves as the point of departure for the entire process.

## Develop indicators

| TP1 Section 1.4.3; TP6 Section 6.4.2 | Current vulnerability is often assessed through the use of indicators – quantitative, qualitative, or both. These are used to describe various characteristics of vulnerable systems. The approach chosen (e.g., policy-based, vulnerability-based, etc.) will dictate which indicators will be relevant.

After review of the available information, indicators may be identified that can be used to sketch the baseline for the priority system(s). This baseline will include current vulnerability and level of adaptation. The baseline should be described as fully and clearly as possible. Ideally, indicators chosen to describe the project baseline will also be used in the project monitoring and evaluation process.

## Review and synthesise existing information

APF users can gain insight from previous studies, expert opinion and the policy context into the vulnerability of key systems. Examples include national development plans, Poverty Reduction Strategy Papers, environmental sustainability plans and natural hazards assessments. These and other existing sources should be identified and explored, and information can be extracted for use in developing the baseline. The focus of this effort should be on both key concerns of the priority system(s) – e.g., the history of drought and crop failure in a region – and the relationship between risk and the priority system(s) – e.g., the impact of drought on smallholders.

| TP1 Section 1.4.4 | Users can refer to expert opinion, analogue or historical studies, and/or modelling to understand available information. Existing sources of information include: (a) studies/projects that have focused on climate change-related impacts (e.g., previous vulnerability and impact studies); (b) studies/projects that have been carried out that may not have an explicit focus on climate change (e.g., national action plans under the Desertification Convention), but are nonetheless highly relevant; and (c) the existing policy context for coping with current climate risks and variability. This material may provide information necessary for constructing a baseline.

| TP1 Section 1.4.3 | Where specific actions or programmes have been implemented to address the threat of climate-related hazards, there may be an extensive national literature to draw from. In addition, APF users may review studies, policies and measures that were designed to address other problems (e.g., disasters, poverty, resource management, biodiversity conservation), as such information often contains examples of very relevant forms of adaptation.

### Task 4: Design the adaptation project

The completion of the previous tasks will determine a certain pathway for the APF process. For example, in countries in which substantial prior work on current and future vulnerability assessment has been done, the APF could be used to formulate an adaptation strategy and provide advice on continuing the process. In this case, less emphasis would be placed on the Components dealing with current and future vulnerability assessment, and the project's unique needs would define its pathway. If none emerges, the user will have to identify and revisit the particular activity that requires clarification. The rest of Task 4 provides guidance on defining specific characteristics of the proposed pathway.

### Select approaches and methods

The Objective of Task 4 is to select an approach that both fits with the scope of the adaptation project, and is compatible with the available resources. In choosing an approach, the user begins to put clear parameters on priority steps and methods (see Box 1 for approaches recommended for use in an APF process).

---

**Box 1: Recommended approaches to Adaptation Policy Framework studies**

- **(Natural) hazards-based approach:** Analyse possible outcomes from a specific climate hazard
- **Vulnerability-based approach:** Determine the likelihood that current or desired vulnerability may be affected by future climate hazards
- **Adaptive-capacity approach:** Analyse the barriers to adaptation and propose how they can be overcome.
- **Policy-based approach:** Investigate the efficacy of an existing or proposed policy in light of a changing exposure or sensitivity

---

These approaches are complementary rather than mutually exclusive. The vulnerability-based and adaptive capacity approaches resemble two sides of a coin on which the climate hazard approach could be overlaid.

If there is an approach already in use – for instance in development planning – then it may make sense to adopt it. Alternatively, if existing reviews and plans are not available or suitable, the team will need to develop its own approach. Stakeholder-led exercises and the scoping activities outlined in Task 1 can be very useful for this decision-making process. Users will need to make this choice carefully, with an eye to the effect it will have on the nature of the APF process.

| TP1 Sections 1.4.3 and 1.4.4 | The choice of approach has direct implications for the level of effort associated with data acquisition, modelling, and |

other aspects of adaptation. For instance, if a (natural) hazards-based approach has been selected, significant effort will need to be devoted toward the assessment of current and future climate risks, which will influence the choice of methods for these tasks. If a policy-based approach has been selected, more resources may need to be focused on understanding socio-economic aspects of current vulnerability and developing socio-economic scenarios. The selection of methods will often flow from the selection of approach, as is discussed in each of the corresponding TPs.

### Develop synthesis plan

| TP1 Section 1.4.4 | Any adaptation project, regardless of approach, will require a careful synthesis in |

order to be useful to the overall goal of the adaptation process. At this early stage, users are encouraged to outline a preliminary plan for synthesising results and for providing input to the identification of adaptation options and recommendations.

### Develop monitoring and evaluation strategy

| TP1 Section 1.4.4; TP9 Section 9.4 | The key outputs of an adaptation project are the strategy, the policies and/or the measures for reducing vulnerability and increasing |

ing adaptive capacity in the priority system(s). How effectively these recommendations are implemented needs to be monitored and periodically evaluated. APF users should develop indicators for each element of the strategy, policy and measure to assess their effectiveness. Having a strategy in place can help to ensure that indicators are developed to enable effective M&E at a later stage.

### Develop terms of reference

Finally, terms of reference should be developed that clearly describe the project objectives and expected outcomes, the respective project activities, the stakeholders involved in the project, the budget, timelines, etc. This activity is integral to any project planning process. Many techniques are available to accomplish this. Perhaps the most useful is the logical framework analysis approach. Users may also find it useful to consult with additional stakeholders and the general public. Their input can help refine or reframe the policy context or the project objectives, if necessary. Wide dissemination of the terms of reference will help ensure that the process remains open and transparent.

### Key issues

The tasks above raise a number of institutional, analytical, and operational issues. This section reiterates the key issues and outlines some overarching considerations.

*Project linkages:* Most likely, there are ongoing and/or planned projects within the user's country that are highly relevant to adaptation. These projects may have complementarities and synergies that could (a) increase the strategic value of the adaptation process, (b) enable more detailed assessments, (c) increase the impact of results, and (d) increase the efficiency of available funding.

*Engaging stakeholders:* APF studies may require two stages of stakeholder dialogue. The first is a pilot stage, in which the initial project team consults with a small stakeholder group in order to develop appropriate priorities and objectives, and then identifies additional stakeholders. The second is a longer-term process that engages a larger stakeholder group. This process is sustained for the duration of the project.

*Method selection:* The methods used will vary significantly from one adaptation project to the next. Methods chosen should:

- respond to project goals;
- respect the constraints of project resources;
- have political support; and,
- if applicable, be sufficiently credible for potential donors.

*Uncertainties:* There are several key points in the project design process for addressing uncertainty. By taking the time to understand and articulate uncertainties and assumptions (e.g., with regard to developing indicators and baselines), users can plan a project that both minimises and accounts for uncertainties.

*Policy process:* The project team should start their project with a clear understanding of the policy processes they wish to inform. The team must identify potential obstacles within the policy process that may make it difficult to implement or sustain adaptation policies and measures. Examples include a particular inertia of the policy process, vested interests of groups or individuals, and unclear priorities.

---

**Checklist**

This checklist is a quick reference to the activities in the *Scoping and Designing an Adaptation Project* Component of the APF. Before proceeding to the next APF Component, users may want to consider whether they have:

- Defined **priority systems** and project boundaries?
- Established a plan for identifying and engaging **stakeholders**?
- Determined project **objectives** and desired **outcomes**?
- Developed a **plan for communication** of results to stakeholders and decision makers?
- Selected the adaptation project **team**?
- Identified, assembled, and reviewed **pertinent information**?
- Selected an **approach** (e.g., from the four recommended)?
- Analysed the national **policy-making process and barriers** in the context of adaptation?
- Prepared **terms of reference** for the overall project?

---

## Assessing current vulnerability

One of the APF's key innovations is that it begins with an emphasis on current climate conditions since, for many countries, adaptation to current climate risk is the most immediate adaptation task.

Key TPs: 3, 4, 6 and 7   The second Component of the APF addresses two key aspects of current conditions –vulnerability to current climate, and the scope and effectiveness of existing adaptation measures. Starting with the current conditions helps ensure that any resulting policies and measures are based on current experience.

*Think of this APF Component as a process that helps to distinctly define current vulnerability and adaptation in the context of the priority system.*

The main **purpose** of assessing current vulnerability and adaptation is to understand the characteristics of climate-related vulnerability in the priority system and the scope of the system's adaptive responses. Specifically, APF users must address three key questions:

1. What is the status of national development policies and plans with respect to the vulnerability of priority system(s) to current climate risks?
2. Which factors determine the vulnerability of those priority systems?
3. How successful are current adaptation approaches?

Component 2 of the APF is an early point at which adaptation projects can take very different pathways, depending on project priorities.

The **process** includes four major tasks to assess:

1. Climate risks and potential impacts;
2. Socio-economic conditions;
3. Adaptation experience (including policies and measures) and adaptive capacity; and
4. Vulnerability (to both socio-economic conditions and climate).

Rather than being sequential, these tasks are interactive.

The expected **output** is a comprehensive assessment of the priority system's vulnerability to current climate and the adaptation options it uses.

### Task 1: Assess climate risks and potential impacts

Under this task, users acquire an understanding of current climate risks. This understanding provides a basis for formulating adaptation strategies to manage future climate risks.

The assessment of current risks can be either qualitative, quantitative or a combination of the two. In its most comprehensive form, Task 1 entails: characterising climate variability, extremes and hazards; assessing impacts; developing risk assessment criteria; and assessing current climate risks.

| TP4 Sections 4.1-4.4 | The two key elements for Task 1 are a conceptual model of the system, and an understanding of the hazards and vulnerabilities in order to prioritise risk (Component 1).

| TP4 Section 4.4.6; TP6 Section 6.4.5; TP7 Section 7.4.2 | How current climate risks are assessed depends on the approach selected (i.e., (natural) hazards-based approach, vulnerability-based approach, policy-based approach, or the adaptive-capacity approach). With any of these approaches, qualitative and quantitative methods are available for assessing risk. The methods will depend on a number of factors, including the level and quality of information needed by stakeholders (Box 2).

### Task 2: Assess socio-economic conditions

| TP6 Section 6.4.3 | The purpose of this task is to assess current socio-economic conditions within the priority system. The task output is a concise description of current conditions affecting vulnerability and risk. This description can also be used later in the project to develop socio-economic scenarios to inform projections of future vulnerability and climate risk.

A comprehensive assessment of current socio-economic conditions can include: (1) clarifying system boundaries, (2) developing system indicators, (3) describing socio-economic conditions today, and (4) analysing critical characteristics. This assessment can be a detailed process, or it can be qualitative

---

**Box 2: Key points for assessing current vulnerability and climate risks**

- **Approach:** Climate vulnerability and risk arise from interactions between climate and society. There are different ways of assessing these. They can be approached from their social aspects, through vulnerability-based assessment; from their climatic aspect, through natural hazards-based assessment; or through complementary approaches that combine elements of both.
- **Quantitative versus qualitative methods:** A key decision is how quantitative or qualitative the climate risk assessment should be. In Component 1, Task 3, preliminary methods were selected which users can now begin to apply. It may be useful to revisit decisions on methods and adjust as needed, as understanding of the specific project direction grows.

---

input to other tasks, such as the vulnerability assessment. It is likely to involve working with stakeholders to determine the most appropriate socio-economic indicators (e.g., qualitative, quantitative or mixed), and assembling descriptions (e.g., data-rich or qualitative) on current socio-economic conditions. These descriptions should include demographic, economic, natural resources, governance/development and cultural aspects of current conditions.

### Task 3: Assess adaptation experience

To be effective and acceptable to stakeholders, adaptation measures must be consistent with past experience, current behaviour and future expectations. Characterising this collective adaptation experience is essential.

The purpose of this task is to evaluate the success of the priority system's current adaptation (baseline). This baseline is a description of the recent and current adaptation experience, including policies and measures currently in place, as well as an assessment of current adaptive capacity. (This should not be confused with a "project baseline" discussed under Component 1.)

Assessing adaptation experience involves two main processes. First, thorough scoping and synthesis of information on existing policies and measures relevant to adaptation in the priority system(s). Second, an assessment of the system's capacity to adapt to current hazards – i.e., how well have these policies and measures worked? Both autonomous and planned adaptations should be explored.

### Task 4: Assess vulnerability

| TP3 Sections 3.1-3.4 | The purpose of this activity is to identify and characterise the priority system's sensitivity to climate hazards. The primary output of this task will be a rich description of current vulnerability – both socio-economic and climate. This description can build on the outputs of the previous three tasks. Task 4 can also provide key input to the assessment of future climate risks by describing potential future vulnerabilities.

Vulnerability assessment can involve a detailed synthesis of the assessments in the preceding tasks (e.g., climate risks, socio-economic conditions). It can be a simple synthesis of pre-existing vulnerability assessments. Or it can be something in between. Vulnerability assessment can be a stand-alone process or input to an assessment of current climate risks.

This current vulnerability assessment can be used later in the adaptation project to describe potential future vulnerabilities, to compare vulnerability under different socio-economic and climatic futures, or to identify key options for adaptation. All of these activities intersect in important ways with efforts to enhance adaptive capacity.

In its most complete form, assessing current vulnerability can clarify definitions and analysis questions; define key vulnerable groups (priority systems); define exposure to climate risk (using socio-economic indicators); and assess current vulnerability (the conjunction of climate hazards and socio-economic conditions).

### *Key issues*

The tasks above raise a number of institutional, analytical and operational issues. This section reiterates key issues and outlines new, overarching considerations.

*Current climate:* Several types of statistical data can be used to describe current climate (e.g., mean, standard deviations, the frequency of extreme events). Some stakeholders may be keen to define variability in a more people-centred way that ties processes within the climate system to changes in readily observed patterns.

*Understanding climate risk:* Risk refers to the combination of: the magnitude of a climatic event; the likelihood of that event; and the consequences of that event. Stakeholders should understand these elements of risk early in the APF process.

*Defining vulnerability:* Stakeholders are likely to have different definitions of vulnerability. For clarity in communication, the project team should agree on a definition to use with stakeholders throughout the APF process.

*Types of baselines:* Two types of baselines should be developed for most adaptation projects. The first is the project baseline, described under Component 1. The second, the adaptation baseline, describes adaptations to current climate. These baselines can provide input to future reference scenarios, which are touched on in the following section.

---

**Checklist**

This checklist is a quick reference to the activities in the *Assessing Current Vulnerability* Component of the APF. Before proceeding to the next APF Component, users may want to consider whether they have:

- **Characterised** climate variability, extremes, and hazards?
- Described **socio-economic conditions** affecting current vulnerability and risk?
- Conducted an assessment of the **adaptation baseline**?
- Identified and characterised the **vulnerability** of the priority system to current climate hazards?

---

### Assessing future climate risks

*Key TPs: 3, 5, 6 and 7* While improved adaptation to current climate risks is important, it is not sufficient to deal with all of the possible future risks of climate change. To understand these risks, APF users should take into account future scenarios of climate change, vulnerability to climate impacts, and socio-economic dynamics. This section outlines the process of assessing future climate risks in the priority system(s).

Another APF innovation is its expanded view of the analytical techniques for assessing future climate risks. The conventional approach has been to develop a climate change scenario by perturbing a baseline climate scenario, use a modelling system to assess the impact of the perturbation, and evaluate adaptation options to mitigate those impacts. The types of assessments and their analytical needs have multiplied since that method was first formulated. Today, more robust assessment techniques are available.

***Think of this APF Component as a process that most closely resembles the analyses that have been conducted as part of earlier climate change vulnerability assessments, in which the focus was on future climate trends.***

With the APF approach, many different analytical techniques can be used to assess future climate risk. These range from qualitative analysis (e.g., partitioning the outcomes into low, medium and high risk) to highly sophisticated quantitative techniques (probabilities calculated using statistical and/or modelling techniques).

*TP6 Section 6.4.6* The assessment of future climate risks involves examining intersections between trends (e.g., climate, natural resources, socio-economic conditions) and factors that influence the development of adaptive responses (i.e., barriers and opportunities). Figure 3 illustrates these intersections. Against the background of climate trends are socio-economic trends, adaptation barriers, and environmental trends. The points of overlap represent impacts caused by future climate change. These impacts may be diminished or increased, depending upon a system's level of adaptation or adaptive capacity.

The main **purpose** of this APF Component is to characterise future climate risks in a priority system so that adaptation policies and measures can be designed to reduce the system's exposure to future climate hazards.

This **process** includes four major tasks to characterise:

1. Climate trends, risks and opportunities;
2. Socio-economic trends, risks and opportunities;
3. Natural resource and environment trends; and
4. Adaptation barriers and opportunities.

The **output** will be a series of potential scenarios, outlining future climate change and vulnerability, socio-economic conditions, and trends in natural resource and environmental management.

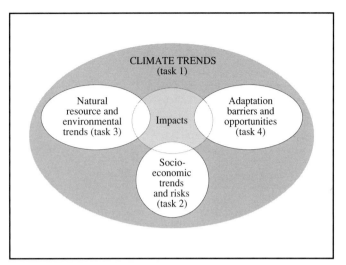

**Figure 3:** Conceptualisation of tasks for Component 3

For these tasks, users will continue to apply the approach that was adopted in the previous Component, *Assessing Current Vulnerability*. As discussed earlier, two major approaches to risk assessment are the natural hazards-based and the vulnerability-based approaches; others can also be used. The choice of one approach over the other has important implications for the nature of the tasks outlined below.

***Task 1: Characterise climate trends, risks and opportunities***

TP6 Sections 6.4.4, 6.4.6
TP7 Sections 7.3.5, 7.4.3

The purpose of this task is to describe potential future climate risks and opportunities associated with it. Generally, the output of Task 1 will consist of two elements – a set of future climate change scenarios and an analysis of associated risk.

A comprehensive characterisation of climate trends, risks and opportunities can require: clarification of the priority system's sensitivities to climate change; construction of appropriate planning horizons; development of climate change scenarios; linkage of scenarios to impact models (with input from socio-economic scenarios); and risk analysis. To implement these sub-tasks, a range of methods and options is available.

***Task 2: Characterise socio-economic trends, risks and opportunities***

TP5 Section 5.4;
Annex A.5.1

In order to design adaptation strategies for the unknown hazards of future climate change, it is useful to construct possible accounts of what the future might be like – i.e., in what kind of future world (or in what kind of priority system) will adaptation be taking place?

The purpose of this task is to develop and describe prospective socio-economic conditions for the priority system. Characterising

future socio-economic conditions involves building on an assessment of current conditions. There are two primary tasks involved. The first is to develop alternative "storylines" of the future for an appropriate time period (e.g., between 20 and 50 years into the future). The second is to make projections about how certain socio-economic conditions will change in the future under these alternative storylines.

The output will be a series of qualitative and/or quantitative scenarios. When integrated with additional trends, this series can include baselines without new adaptation (i.e., the adaptation baseline or reference scenario), and two or three scenarios incorporating additional adaptation policies and measures. These scenarios can then be used as input to projections of future vulnerability and climate risk. This can be done by applying various climate scenarios to each of the socio-economic scenarios and assessing future vulnerability and risk.

To develop socio-economic scenarios, users can build their own, or use/adapt existing ones. This can be a detailed, quantitative process, or a more qualitative one. Either way, the process will likely involve working with stakeholders to determine the most appropriate storylines and scenarios for the priority system(s).

***Task 3: Characterise natural resource and environmental trends***

The growth in consumption of natural resources raises important issues regarding vulnerability to future climate risks. There are many current examples of serious environmental degradation caused by the exploitation of fossil fuels, mineral, and other resources. Since climate impacts are likely to be exacerbated as environmental degradation increases, an assessment of natural resource management trends can provide essential input to assessments of the risks associated with future climate change. Such an assessment links the communities who may be vulnerable to climate change impacts with the potential sources of their vulnerability.

Environmental scenarios may need to be developed in which important feedbacks exacerbate climate risks, where environmental conditions influence adaptive capacity, or where environmental management options can be used to assess adaptation. Environmental scenarios can be developed from models via socio-economic storylines or as regular changes in conditions designed to assess sensitivity. Such scenarios include land-use/land cover change, ozone depletion, ultraviolet exposure and water resource scenarios. More information on the construction of these scenarios can be found in Chapter 3 of the IPCC Third Assessment Report.

TP6 Section 6.4.6
TP7 Section 7.4.3
TP9 Section 9.3

The purpose of this task is to develop and describe prospective natural resource management conditions in the priority system(s). This will typically require that storylines be integrated into the socio-economic assessment described in the previous task. The output will be a series of qualitative and/or quantitative scenarios.

## Task 4: Characterise adaptation barriers and opportunities

Aspects of current development and environmental policy are essential for assessing potential barriers to adaptation. Especially important are recent or planned state reforms for economic development (e.g., privatisation and liberalisation of trade). Policies and programmes related to the priority system should be evaluated for their potential to support effective adaptation to climate change, in the context of sustainable development.

The purpose of this task is to identify aspects of national decision-making processes that either pose potential barriers to incorporating adaptation into development planning, or that provide important opportunities to build adaptive capacity. This will typically require that the institutional, environmental, and participatory aspects of the planning and policy-making process be well understood, and that the pathways for implementing policy (e.g., laws, standards, regulations) are assessed regarding their roles, effectiveness and institutional linkages.

### Key issues

The tasks above raise a number of institutional, analytical and operational issues. This section reiterates key issues and outlines new, overarching considerations.

*Dealing with uncertainty:* Since uncertainty permeates climate change assessments, projects have relied on specialised methods, such as the development and use of climate scenarios. But the uncertainty in predicting future climate is one reason why the APF recommends that adaptation assessment be anchored with an understanding of current climate risk. This helps to provide a road map from known territory into uncertain futures.

*Scenario development:* APF users should verify that the scenarios incorporated in their assessments are based on an internally consistent set of assumptions about driving forces and key relationships (e.g., between socio-economic conditions, natural resource management, and policy-making processes).

---

**Checklist**

This checklist is a quick reference to the activities involved in the *Assessing Future Climate Risks* Component of the APF. Before proceeding to the next APF Component, users may want to consider whether they have:

- **Characterised** climate trends, risks and opportunities?
- Described scenarios of **socio-economic (and environmental) conditions?**
- Addressed **uncertainties** in the choice of methods and tools for trend predictions?
- Laid a basis for its incorporation into **risk management strategies,** and planning under uncertainty?

---

## Formulating an adaptation strategy

*Key TPs: 8; TP2 Section 2.6.4; TP7 Section 7.4.4*

An adaptation strategy for a country refers to a broad plan of action for addressing impacts of climate change. The APF was developed to provide guidance on adaptation assessment.

Operationally, the formulation of an adaptation strategy can pose a big challenge. It means situating the climate change issue in a policy world that is full of competing priorities, interest groups, short attention spans, election-driven priorities and a host of potential unpredictable events. Ultimately, whatever options and measures the project team proposes to reduce the priority system's vulnerability to climate risks, packaging those decisions into an adaptation strategy will require overcoming practical constraints.

*Think of the process of formulating an adaptation strategy as trying to identify a suite of policies and measures that are extensions of the previous Components and fit within the priority system's unique policy-making process.*

Clearly, significant momentum has occurred in recent decades through international participation in multilateral environmental agreements (e.g., the United Nations Convention on Desertification, United Nations Framework Convention on Climate Change) that could be effectively leveraged. Although this effort was not directly motivated by climate change adaptation, their objectives overlap. The adaptation strategy development process should build on such experience.

The **purpose** of this APF Component is to integrate all of the preceding APF work into a well-defined strategy to direct adaptation action.

As Figure-4 illustrates, the **process** will generally include four major tasks:

1. Synthesise previous steps/studies on potential adaptation options;
2. Identify and formulate adaptation options;
3. Prioritise and select adaptation options; and
4. Formulate an adaptation strategy.

The **outputs** will be the adaptation strategy itself, including recommendations for planning policies and specific measures.

To help ensure broad-based endorsement and effective implementation of the resulting adaptation strategy, the full group of stakeholders should be involved in this Component. The formulation of an adaptation strategy should proceed in conjunction with the guidance discussed in the *Assessing and Enhancing Adaptive Capacity* Component.

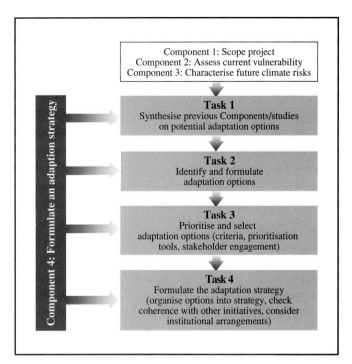

**Figure 4:** Tasks in formulating an adaptation strategy

### Task 1: Synthesise previous Components/studies on potential adaptation options

The main goal of Task 1 is to take stock of what has emerged so far in the APF process. Once the assessments of current and future climate risks have been completed, the results can be synthesised. The output will be a preliminary, non-prioritised list of potential adaptation options.

*TP8 Section 8.4.2* From these preceding efforts (especially, Component 2, *Assessing current vulnerability*), APF users will have identified adaptation options currently in place. In addition to collating potential options in a list, the project team should also provide a brief assessment of these experiences – i.e., what worked and why? Of course, an adaptation strategy should also respond to an analysis of future climate risks. Suggestions for options can be obtained from the previous APF Component as well as from studies in the literature from countries facing similar adaptation challenges.

### Task 2: Identify and formulate adaptation options

*TP8 Section 8.4.3* The main goals of Task 2 are to characterise adaptation options in terms of their costs, impacts, and potential barriers, and to develop criteria for prioritising options.

The development of criteria should be a stakeholder-driven process. To ensure that criteria reflect the needs of the priority

system, stakeholder input is critical. The criteria developed will be used to prioritise measures and policies. They can also act as indicators of the project's longer-term success in achieving the adaptation objectives.

An example of a set of criteria is provided by the National Adaptation Programmes of Action (NAPA) Guidelines[1]. As the NAPA and APF are highly complementary processes, these criteria may be of use:

1. *Expected level of damage* as an indication of the benefits to be gained by preventing or mitigating damage;

2. *Poverty reduction* as an indication of enhanced adaptive capacity;

3. *Synergies with multilateral environmental agreements* as an indicator of cost savings and/or additional benefit;

4. *Cost effectiveness* (or just costs). Even in cases where existing criteria are used, they should be adjusted as needed.

### Task 3: Prioritise and select adaptation options

*TP8 Section 8.4.4; TP8 Annex 8.1* The goal of Task 3 is to identify priorities from the array of possible adaptation policy options and measures. Using selected criteria and prioritisation methods, the output will be a ranked list of adaptation options.

This task involves selecting and applying prioritisation methods. In view of the diversity of climate change adaptation options, probably more than one method may be needed to review all choices. To decide which should be used in the prioritisation process, users should carefully consider the available methods (e.g., cost benefit analysis, cost effectiveness analysis, multi-criteria analysis, expert judgment). Some methods require higher levels of data and resource inputs (in terms of time and skills of stakeholders).

---

### Task 4: Formulate the adaptation strategy

*TP8 Section 8.4.4; TP8 Annex 8.1* The goal of Task 4 is to assemble priority adaptation options into a cohesive strategy. The output will be a strategy document that outlines an alternative package of policy mixes and measures, implementation plans (who, where, with what resources), time frames (when) and operational issues (what types of institutional support).

This task will generally involve the following activities:

---

[1] Annex C, Global Environment Facility Operational Guidelines for Expedited Financing for the Preparation of National Adaptation Programmes of Action by Least Developed Countries (April 2002).

1. Drafting the adaptation strategy;
2. Reviewing the coherence of the strategy with existing strategies;
3. Scoping issues related to strategy implementation (e.g., barriers and barrier removal plans); and
4. Finalising the strategy.

Stakeholder support may be the single most important factor in determining whether the adaptation strategy is successfully implemented. For this reason, broad stakeholder input to the strategy development process is critical.

### Key issues

The tasks above raise a number of institutional, analytical and operational issues. This section reiterates key issues and outlines new, overarching considerations.

*Approach for formulating adaptation strategy:* The choice of approach depends on the dynamics of the stakeholder process that has unfolded. If this process has been dominated by high-level policy-makers and technical analysts, the top-down approach is likely to work best. If stakeholder engagement has been broad and inclusive, a bottom-up or hybrid approach may serve best.

*Designing the implementation strategy:* The formulation of the adaptation strategy is not the end of the APF process. The strategy then needs to be implemented and sustained. Given this, the strategy should be designed with the specific needs of the implementation process in mind. The strategy should be coherent and fit the dictates of the policy process. The adaptation strategy is thus a "living" document – a continuous process, flexible enough to integrate new elements, including the climate "surprises" that will certainly occur in the future.

---

### Checklist

This checklist is a quick reference to the activities involved in the *Formulating an adaptation strategy* Component of the APF. Before proceeding to the next APF Component, users may want to consider whether they have:

- **Taken stock** of what has emerged thus far in the APF process?
- Characterised **adaptation options** in terms of their costs, impacts, and barriers?
- Created a **ranked set** of adaptation interventions?
- Prepared the adaptation **strategy document** that outlines measures, implementation plans, time frames and operational issues?

---

### Continuing the adaptation process

| *Key TPs: 8; 9;* |
| *TP2 Section 2.6.5;* |
| *TP7 Section 7.4.5* |

To be effective over the long term, the adaptation process should lay the groundwork for similar efforts in the future in ways that support overarching national development objectives. To do so, the adaptation strategy must be integrated with processes to update plans, policies and programmes.

Effectively incorporating adaptation into a country's development planning is a challenging endeavour. It requires cross-sectoral cooperation, an interdisciplinary approach and considerable political will. Monitoring implemented adaptation strategies is also demanding. It requires both an ongoing commitment to monitoring and evaluation (M&E) and a high-level government response to addressing barriers that are impeding the strategy.

***Think of this APF Component as the start of a long-term process of adaptation, begun by your project.***

The **purpose** of this APF Component is to implement and sustain the adaptation strategy, policies and measures, through:

- effective integration with existing processes and plans;
- strong institutional support;
- M&E processes;
- responsive mechanisms for adjusting the adaptation process; and
- creative mainstreaming strategies.

The **process** will generally involve three major tasks:

1. Incorporate adaptation policies and measures into development plans;
2. Implement the adaptation strategy and institutionalise follow-ups; and
3. Review, monitor and evaluate the effectiveness of policies, measures and projects.

There is no single **output** from this APF Component. Instead, it is the starting point of what will hopefully be a sustained adaptation process. The set of tasks seeks to initiate new adaptation action through policies and measures, to promote a supportive institutional structure, and to launch iterative feedback loops designed to improve the process over time.

As in other APF Components, stakeholders play an essential role. For example, integrating the strategy with existing development plans will require the close involvement of selected government stakeholders. They represent key parts of the institutional framework and can help to build and sustain the necessary M&E process. Engaging stakeholders for this longer-term activity should not be an afterthought. Instead, users are encouraged to plan carefully for this phase in the stakeholder engagement strategy (discussed in the next section and in Component 1).

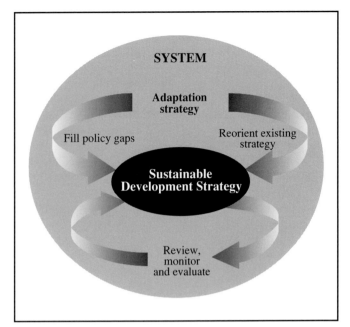

**Figure 5:** Conceptualisation of continuing the adaptation process

Figure 5 illustrates the activities and feedback loops in this APF Component. The underlying concept is that there are two approaches to continuing the adaptation process. On one hand, countries can re-orient existing policies and practices to make them more responsive to the increased vulnerability caused by increased variability and change (right top arrow). Disaster management practices are a good example of this phenomenon.

Alternatively, countries can choose to address policy gaps regarding climate risks, while also enhancing the resilience of the priority system (left top arrow). These interventions remove existing barriers to the adoption of policies that are sensitive to the impacts of climate change, including variability.

### Task 1: Incorporate adaptation policies and measures into development plans

| *TP8 Section 8.4.5* |
| *TP9 Sections 9.4.2, 9.4.3* |

An adaptation strategy needs to be incorporated with key development policies, processes and plans. The strategy is likely to have significant co-benefits in terms of improving resource management, enhancing capacity development, reducing poverty, and reducing vulnerability to a variety of current stresses. Integrating the adaptation strategy – e.g., by "piggy-backing" it onto related plans and activities – can make its implementation more efficient. In fact, given the competitive nature of policy-making, a fledgling policy such as climate change adaptation may be unlikely to succeed if it is not integrated with other more familiar and established issues.

The goal of Task 1 is to effectively incorporate the adaptation strategy, policies and measures into relevant existing processes and plans. The output may include a detailed plan for integra-

tion and an ongoing process through which strategy is actively integrated.

The integration plan should pay particular attention to potential barriers, including capacity. Users will need to be open and creative about addressing the challenges. Institutional inertia and ongoing policy debates, for example, can thwart the integration process. Understanding these and formulating strategies to overcome them will significantly improve chances of success.

Incorporating adaptation policies and measures refers to their formal integration into national and/or regional development process and budgets. The idea is to make the adaptation strategy a basic Component of existing national development plans. To start with, it is recommended that users establish common ground between the adaptation strategy and existing policy processes. This will enable an assessment of how the adaptation strategy complements – or even advances – the system's broader objectives of poverty reduction and sustainable development. The following are likely places to incorporate Components of a climate change adaptation strategy:

• environmental management plans (particularly when they incorporate environmental impact assessments);
• national conservation strategies;
• disaster preparedness and/or management plans; and
• sustainable development plans for specific sectors (e.g., agriculture, forestry, transportation, fisheries).

### Task 2: Implement the adaptation strategy and institutionalise follow-ups

| *TP9 Sections 9.4.2, 9.4.3* |

The goal of Task 2 is to transform outputs – in particular, the adaptation strategy – into an ongoing adaptation process. The output is likely to include a well-documented implementation plan, including details on establishing an institutional support structure. The less tangible – but more important – output will be the adaptation process itself.

This task will generally involve the following activities:

1. Assembling the resources for implementation of the adaptation strategy (e.g., staff, facilities, funds);
2. Launching management and oversight structures for each aspect of implementation (e.g., local teams, national managers, advisory groups);
3. Initiating the implementation activities; and
4. Formalising an institutional structure for follow-up and support.

Activity 3 might also include the launch of policy integration meetings (e.g., on integrating adaptation with activities of the national poverty reduction strategy), new sectoral strategies (e.g., to provide improved support to water harvesting activities) and/or specific adaptation projects.

### Task 3: Review, monitor and evaluate the effectiveness of policies, measures and projects

TP9 Section 9.4.1 The goal of Task 3 is to enable the necessary M&E so that the adaptation process can be sustained and improved over time: an M&E system that identifies what aspects of the adaptation process are working, are not working, and why, and that provides mechanisms for adjusting the adaptation process as needed. The output will be a detailed M&E plan.

This task will generally involve scoping and planning for M&E and launching the M&E framework. To do so, users will want to consider key M&E options and approaches, such as participatory monitoring and evaluation (i.e., "learning by doing").

### Key issues

The tasks above raise a number of institutional, analytical and operational issues. This section reiterates key issues and outlines new, overarching considerations.

*Confronting barriers to adaptation M&E:* Barriers may exist to implementing adaptation strategies, policies and measures. These may be due to resource constraints and governance issues. These barriers need to be openly confronted and possible solutions explored.

*Linking indicators:* The APF provides insight into who will be adapting to what, how they will be adapting, and why. Answers to these questions define, not only the scope of prospective adaptation, but also the basis for monitoring and evaluation.

---

**Checklist**

This checklist is a quick reference to the activities involved in the *Continuing the adaptation process* Component of the APF. Users may want to consider whether they have:

- Developed a detailed plan for effective **incorporation** of the adaptation strategy into national development plans?
- Prepared an adaptation **implementation plan** and identified how to institutionalise follow-up?
- Assembled a strategy for **reviewing, monitoring, and evaluating** adaptation impacts?

---

### Engaging stakeholders

Key TPs: 1, 2; TP2 Section 2.6 The APF is an explicitly stakeholder-driven approach to climate change adaptation projects. Engaging stakeholders is a universal activity that cuts across all APF Components. Stakeholders can contribute significantly to understanding current vulnerability and adaptation and to identifying the necessary adaptation measures. At the same time, their involvement in a project can educate stakeholders about the risks associated with climate change, and encourage them to support the adaptation process. Done well, this process of engagement can assist the implementation of adaptation policies and the formation of an adaptation community. More important, it can provide the momentum to carry the adaptation process forward.

*Stakeholder involvement at different levels and stages is crucial to successful adaptation.*

The **purpose** of this cross-cutting Component is to ensure that key stakeholders are fundamentally engaged in the adaptation project. Here "key stakeholders" refers to both those affected by climate change and those positioned most effectively to advance adaptation.

This **process** will include three major tasks:

1. Identify stakeholders;
2. Clarify the roles of stakeholders; and
3. Manage the dialogue process.

The **output** should be an active, inclusive stakeholder dialogue that is developed and sustained over the course of the project.

### Links with Adaptation Policy Framework Components

In broad terms, stakeholder-related tasks and participation levels should be closely linked with the APF Components, as outlined below.

- *Scoping and designing an adaptation project:* Key tasks should focus on reviewing existing policies, identifying stakeholders, and clarifying stakeholder interests and roles. At this stage of the process, stakeholder participation should be rather limited and focused on a subset of the stakeholder group.
- *Assessing current vulnerability:* Key tasks should focus on developing a common understanding, identifying successful coping strategies, and providing equitable access to information. Participation should be extended to representatives of the most vulnerable groups, technical specialists, and policy makers at the appropriate levels (local, regional, and/or national).
- *Assessing future climate risks:* Key tasks should focus on defining planning horizons and other parameters, and stakeholder perceptions (e.g., with regard to future

scenarios). Participation for this Component should be the same as for the previous Component.

*Formulating an adaptation strategy:* Key tasks should focus on assessing and prioritising adaptation options. Participation should be very broad for this Component.

*Continuing the adaptation process:* Key tasks should focus on fostering stakeholder action on, and support of, adaptation activities. As with the previous Component, participation should be as broad as possible – essentially all stakeholders should be involved.

Conducting stakeholder tasks and facilitating their participation does not in itself guarantee equity, fairness or acceptance. The process must be carefully designed, implemented and managed. Some stakeholders will be centrally involved throughout the process. Others may play more specialised roles. The aim is to create a stakeholder engagement process that leads to open dialogue, mutual learning, and consensus decisions.

To achieve this, each of the three major task areas is summarised below.

### *Identify key stakeholders*

Major stakeholders include the most vulnerable groups and those who have a role in influencing climate change adaptation in the priority system.

| TP1 Section 1.4.1; TP2 Sections 2.5, 2.6 | Identifying stakeholders is a key task in the initial APF Component (i.e., *Scoping and designing an adaptation project).* The |

selected stakeholders should not simply be those known to be involved in these issues, such as government representatives, non-governmental organisation (NGO) volunteers, and academics. Every effort should be made to include other individuals, particularly the highly vulnerable in society, so that they are represented in the adaptation process.

A simple but effective way to identify stakeholders is as follows:

1.  Conduct initial scoping of stakeholders and identify a core group;
2.  Ask this core group to suggest other stakeholders; and
3.  Ask this larger group to ask whom they consider to be relevant stakeholders until no new names are identified.

Those invited to participate should have the capacity to influence the adaptation process, or be part of a group that will be directly affected by a predicted climatic impact and willing to participate in the process. Preliminary outputs of this process are a thorough scoping of the key stakeholders to engage in each of the five APF Components, and a plan for soliciting their participation.

### *Clarify stakeholder roles*

| TP2 Section 2.5, Box 2.4, Annex A.2.2 | Stakeholders involved throughout the APF process should have suitable and productive roles. Stakeholder roles can be |

defined in a number of ways. One useful option is to organise roles according to stakeholder type and their influence/potential. This enables: a clarification of appropriate general roles for each stakeholder type; suggestions of more specific roles for particularly pivotal stakeholders, to the advantage of the project; and clarification of expected level of contribution for each APF Component.

Each of the APF Components implies a different project role for the stakeholders involved. A plan should be developed that outlines what specific stakeholder activities will be useful in each Component; which stakeholders are best suited to carrying out these activities; and what methods will be used to engage these stakeholders in these activities. Attention to these issues will help ensure that expectations regarding stakeholder contributions are consistent with the demands of the particular APF Component.

### *Manage the dialogue process*

Stakeholders, particularly those whose livelihoods are directly affected by the impacts of current climate variability, will often have a rich experience of, and knowledge about, what kind of adaptation is practical. However, these people may face greater logistical challenges to participating in the project. Furthermore, they may distrust or feel uncomfortable with the process. Since the involvement of these groups is essential, it may be necessary to devote more effort or support to effectively manage the sustained involvement of these groups.

Stakeholder dialogues need to be transparent in order to be effective. There are many techniques available to accomplish this, including techniques to: explore expectations and build trust; promote discussion and scope issues; conduct participatory analysis; and evaluate the process. These techniques should be applied flexibly and in response to project needs and those of stakeholders. At a strategic management level, this will require a broad-based plan for effective stakeholder communication and a plan to help stakeholders sustain the adaptation process after completion of the adaptation project.

### *Key issues*

The discussion above raises a number of institutional, analytical and operational issues. This section reiterates key issues and outlines new, overarching considerations.

*Communicating project outputs:* Stakeholders may be the most effective resource for communicating project outputs. A key aspect of the stakeholder engagement plan should be a strategy for communicating with stakeholders and the broader groups they represent. Stakeholders themselves can help to construct this communication strategy. Key stakeholders can provide guidance (prior to and during the project) on how best to communicate with certain groups. This guidance can be used to develop and improve the project's stakeholder process, and as input to a larger strategy for communicating project outputs.

*Ensuring representation and inclusiveness:* The aim in creating a stakeholder dialogue is to enable open exchange and foster mutual (team and stakeholder) learning. Through listening to the views and experience of other people involved in the process, stakeholders can build a shared understanding of the issues. Priority areas for action can emerge that take account of everyone's perceptions. This process can build mutual understanding and trust between the groups and individuals involved. A substantial literature on working with stakeholders exists, including engagement approaches and principles of effective engagement.

## Assessing and enhancing adaptive capacity

| *Key TP: 7* | Identifying ways to increase adaptive capacity is a universal activity that cuts across all APF Components. The focus should be on adaptive capacity that is directly relevant to climate change including variability.

| *TP7 Section 7.4* | One of the APF's innovations is that it urges countries to view adaptive capacity as a policy change process, and stakeholders as change agents. Furthermore, it treats adaptive capacity as a multidisciplinary approach to respond to different dimensions of climate change, e.g., temporal (current and future); strategic (policy and governance implementation); or operative (assessment determinants and indicators).

> *Adaptive capacity is the property of a system to adjust its characteristics or behaviour, in order to expand its coping range under existing climate variability, or future climate conditions.*

The main **purpose** of this cross-cutting Component is to provide guidance on how adaptive capacity can be assessed and enhanced.

The **process** includes three major tasks:

1. Assessing current adaptive capacity;
2. Identifying the constraints of adaptive capacity; and
3. Developing actions to enhance adaptive capacity.

The expected **outputs** of this cross-cutting Component should be an assessment of current adaptive capacity in the priority system(s), and a strategy for enhancing adaptive capacity in response to project results.

### Links with Adaptation Policy Framework Components

Consider the assessment of adaptive capacity as a set of questions that can be introduced and explored during the stakeholder dialogue process within each of the five APF Components. In effect, the discussion of adaptive capacity should be integrated into the broader stakeholder engagement process as early in the project timeline as possible. A set of exploratory questions to help to highlight key linkages between adaptive capacity and each APF Component is outlined below.

- *Scoping and designing an adaptation project:* What baseline capacity is evident from existing development and poverty studies and from recent country experience?
- *Assessing current vulnerability:* What adaptive capacity already exists to reduce current vulnerability to familiar hazards? Among which vulnerable groups and systems?
- *Assessing future climate risks:* What additional capacity is needed for vulnerable population groups, regions and sectors to adapt to future climate hazards? Given existing socio-economic and environmental trends, how would systems and population groups cope with increasing frequency and severity of existing hazards, or with new hazards?
- *Formulating an adaptation strategy:* How can national capacity be enhanced to promote autonomous adaptation? What barriers confront the implementation of adaptation strategies? What strategies can be formulated to encourage people to be more receptive to and positive about adaptation? Who needs to adapt?
- *Continuing the adaptation process:* What types of monitoring and evaluation protocols can help to continuously involve stakeholders in the adaptation process?

### Assess current adaptive capacity

To assess adaptive capacity, indicators may be developed. Adaptive capacity indicators are more difficult to identify than, e.g., risk indicators. However, with care, users can develop a set of indicators that will be applicable to the priority system(s) under consideration. Determinants and indicators of adaptive capacity may be identified by, e.g., posing a set of targeted questions to the range of stakeholders.

| *TP1*<br>*TP7 Sections 7.3.7, 7.4* | Based on the set of indicators and determinants that have been developed, a qualitative assessment of adaptive capacity in the priority system can be carried out. Adaptive capacity can be generic (i.e., a population's ability to cope with a range of climatic, environmental, economic or other stresses) or specific (i.e., capacity to cope with specific current climate). The output of this activity will consist of the identification of the level to which adaptive measures have already been implemented, and the implications of promoting future adaptive capacity. This task is relevant to Component 2, *Assessing current vulnerability.*

### *Identify barriers to, and opportunities for, developing adaptive capacity*

It is important to identify the existing barriers to implementation of adaptive measures, as well as the particular opportunities and strengths that may facilitate the introduction of adaptive measures. Generally, the output of this activity will consist of a description of local, regional, and national policy and governance roles for enhancing adaptive capacity. This task is relevant to the assessment of future climate risks and the characterisation of adaptation barriers and opportunities in the formulation of an adaptation strategy (Components 3 and 4).

### *Develop strategies to integrate adaptive capacity into adaptation*

| TP7 Section 7.4 | The purpose of this task is to develop strategies to enhance both generic and specific adaptive capacity, facilitate anticipatory adaptation, and promote enabling environments for autonomous adaptation. There are seven steps in this process that cut across each APF Component and, in particular, Component 4, *Formulating an adaptation strategy*. The output for this activity will consist of a set of policy and governance initiatives that will enhance adaptive capacity.

### *Key issues*

The discussion above raises a number of institutional, analytical and operational issues. This section reiterates key issues and outlines new, overarching considerations.

| TP7 Section 7.4 | *Governance strategies for enhancing adaptive capacity:* These refer to multiple ways (e.g., institutional, regulatory, educational) that governments respond to elements in society to implement adaptation policies.

*Policy strategies for enhancing adaptive capacity:* These strategies may be viewed as the fiscal, legislative and other instruments for addressing climate change. Policy strategies encompass a range of options including, e.g.:

| TP8 Box 8.2 |
- Changes in taxation or regulatory regimes;
- Redistribution or reallocation of resources; and
- Support for research agencies and public information projects.

# Section II

# Technical Papers

# Preface

With support from UNDP-GEF, a team of the world's leading experts developed the nine APF Technical Papers (TPs) presented in this section. Each of the TPs offers progressive step-by-step guidance, tools, examples, and tips. The complete series is intended to expand the existing body of guidance for climate change adaptation assessment, planning and implementation.

## Why were these Technical Papers written?

The authors of these TPs were brought together to respond to the need of national analysts and planners for guidance on adaptation. As the array of Initial National Communications submitted to the United Nations Framework Convention on Climate Change illustrates, effective consideration of adaptation options has been difficult for many countries. Out of necessity, the adaptation policy-making process has been evolving quickly. By contrast, the coverage of technical guidance has, up to now, not kept pace. Adaptation Policy Framework (APF) authors have developed the TPs to help meet user demand for effective adaptation.

## What are the objectives of these Technical Papers?

Today, the need for clear understanding of climate vulnerability and adaptation exists at many levels, from the local decision-making process, to national development planning. At both ends of the decision-making spectrum and at the multitude points in-between, technical resources are needed to effectively guide adaptation inquiries through the set of questions, and to the most effective solutions. The TPs provide these resources, offering a range of routes that users can take to consider, understand and respond to unique adaptation needs. The objective of these papers is to offer – both individually, and as a coherent set – guidance that is at the same time accessible to the average user and rich enough in technical detail to support comprehensive technical assessments and strategy development.

## Who should read these Technical Papers?

The TPs have been developed for both practitioners and the scientific community. Certain TPs, and the APF process more generally, are geared towards feeding into the policy process.

Different TPs will be most useful to readers at different stages of an adaptation project. As Figure 1 of the User's Guidebook outlines, the TPs correspond closely to specific APF Components, and can be drawn upon most intensively as a project proceeds through that Component. Still, APF users will be well-served to review all TPs prior to project design. This can help to ensure that the project team is aware of the range of key considerations and accounts for these in their project design.

Not all APF users will need to use all the TPs. The question of which TPs to rely upon most heavily depends almost entirely on the goals of the adaptation project, the information and resources available at the outset, and the specific project approach that is consequently selected (e.g., vulnerability-based, hazards-based, policy-based, or adaptive-capacity; see User's Guidebook and TP1). To elaborate, two examples are provided here. It is important to note that countless additional examples exist of how the TPs can be used. *These are intended only to illustrate how two hypothetical projects might use the TPs.*

*Example 1:* A hypothetical project that is building off of an existing, comprehensive climate change vulnerability assessment, such as from an in-depth Initial National Communication, may not need to spend a great deal time on APF Components 2 and 3, i.e., *Assessing current vulnerability* and *Assessing future climate risks.* Consequently, this project team may choose not to use the TPs (3-6) that most directly support these Components.

*Example 2:* By contrast, a hypothetical project that requires a thorough assessment of current and future climate risk may choose to spend significant effort on Components 2 and 3 and its related TPs. However, depending on the approach the team selects for their project, they may carry out Components 2 and 3 by relying more heavily on one particular subset of TPs. For instance, the team may focus on TP3 if using the vulnerability-based approach, TP4 if using the hazards-based approach, TP6 if using the policy-based approach, or TP7 if using the adaptive-capacity approach. For each of these approaches, additional TPs would be drawn upon for a complete assessment. The use of APF resources would depend on the primary objective of the project.

Despite the range of potential TP uses, there will be some consistency in their use across projects. Specifically, all adaptation projects that use the APF will want to draw upon TP1 for a customised project design, on TP2 for successful stakeholder engagement, on TP7 for a rich understanding of adaptive capacity, on TP8 for effective adaptation strategy development, and on TP9 for insights on taking the adaptation process forward.

# 1

# Scoping and Designing an Adaptation Project

KRISTIE L. EBI[1], BO LIM[2], AND YVETTE AGUILAR[3]

Contributing Authors

*Ian Burton[4], Gretchen deBoer[5], Bill Dougherty[6], Saleemul Huq[7], Erika Spanger-Siegfried[6], and Kate Lonsdale[8]*

Reviewers

*Mozaharul Alam[9], Suruchi Bhawal[10], Henk Bosch[11], Mousse Cissé[12], Kees Dorland[13], Mohamed El Raey[14], Roger Jones[15], Ulka Kelkar[10], Liza Leclerc[16], Mohan Munasinghe[17], Stephen M. Mwakifwamba[18], Atiq Rahman[9], Samir Safi[19], Emilio Sempris[20], Barry Smit[21], Juan-Pedro Searle Solar[22], Henry David Venema[23], and Thomas J. Wilbanks[24]*

[1]  Exponent, Alexandria, United States

[2]  United Nations Development Programme-Global Environment Facility, New York, United States

[3]  Environment and Natural Resources Ministry, San Salvador, El Salvador

[4]  University of Toronto, Toronto, Canada

[5]  Canadian International Development Agency, Quebec, Canada

[6]  Stockholm Environment Institute, Boston, United States

[7]  International Institute for Environment & Development, London, United Kingdom

[8]  Stockholm Environment Institute, Oxford, United Kingdom

[9]  Bangladesh Centre for Advanced Studies, Dhaka, Bangladesh

[10] The Energy and Resources Institute, New Delhi, India

[11] Government Support Group for Energy and Environment, The Hague, The Netherlands

[12] ENDA Tiers Monde, Dakar, Senegal

[13] Institute for Environmental Studies, Amsterdam, Netherlands

[14] University of Alexandria, Alexandria, Egypt

[15] Commonwealth Scientific & Industrial Research Organisation, Atmospheric Research, Aspendale, Australia

[16] Canadian International Development Agency, Quebec, Canada

[17] Munasinghe Institute for Development, Colombo, Sri Lanka

[18] The Centre for Energy, Environment, Science & Technology, Dar Es Salaam, Tanzania

[19] Lebanese University, Beirut, Lebanon

[20] Water Center for the Humid Tropics of Latin America and the Caribbean, Panama City, Panama

[21] University of Guelph, Guelph, Canada

[22] Comisión Nacional Del Medio Ambiente, Santiago, Chile

[23] International Institute for Sustainable Development, Winnipeg, Canada

[24] Oak Ridge National Laboratory, Oak Ridge, United States

# CONTENTS

## 1.1. Introduction

Using the Adaptation Policy Framework (APF), this Technical Paper (TP) will assist project teams in designing projects to develop and implement adaptation strategies, policies and measures that can ensure human development in the face of climate change, including variability. The APF provides a basis by which countries can evaluate and modify existing planning processes and practices to address climate change impacts. To do so, this TP walks the reader through a series of recommended tasks, preparing them for the hands-on work of project scoping and design.

In the pages that follow, equal importance is placed on designing an adaptation project and on launching an adaptation policy process that will extend beyond the project lifetime. During the process of conducting an adaptation project, public awareness is raised, individual, community, sectoral and national capacities are enhanced, and policy processes are established or modified. At the end of the project, the team will have a better understanding of the resilience and vulnerabilities of priority systems with respect to climate change, including variability.

Stakeholder involvement throughout the project should promote equity in decision-making, a thorough and transparent exchange of information and viewpoints, agreement on key objectives and a general consensus on recommended measures and policies. Ideally, an adaptation community that is capable of supporting future adaptation activities will be created by the end of the project.

An adaptation project can result in a variety of outputs, including sectoral and integrated policy analysis and implementation. A typical adaptation project will identify adaptation strategies, policies, and measures aimed at different levels of society for different spatial and temporal scales.

The APF can be used to develop and incorporate adaptation concerns into national, sector-specific and local development planning processes. Because adaptation in one sector often has consequences for another, the APF has been designed to facilitate a process of integrated assessment throug a consultation process in which links between sectors are identified and assessed. The APF also may be applied to add adaptation Components onto on-going projects. By reviewing the scoping and design process outlined here and in the User's Guidebook, readers should be equipped to develop an adaptation project that suits their particular needs.

## 1.2. Relationship with the Adaptation Policy Framework as a whole

TP1 provides guidance on the first Component of the APF process: *Scoping and designing an adaptation project* (Figure

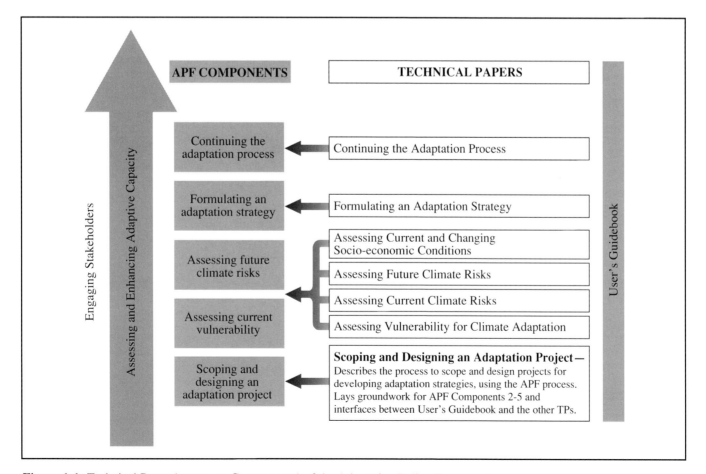

**Figure 1-1:** Technical Paper 1 supports Component 1 of the Adaptation Policy Framework

1-1), and lays the groundwork for Components 2 through 5. As such, TP1 can be used as initial guidance for launching an adaptation project. Understanding the methods described in TPs 2 (*Engaging Stakeholders in the Adaptation Process*), 7 (*Measuring and Enhancing Adaptive Capacity*), 8 (*Formulating an Adaptation Strategy*), and 9 (*Continuing the Adaptation Process*) will be helpful in designing the adaptation project. Depending on its objectives, an adaptation project can include additional Components, such as assessment of current vulnerability in the priority system (TP3), assessment of current and future climate risks (TPs 4 and 5), and assessment of the relevant socio-economic conditions and prospects (TP6) and of adaptation itself (TP 7-9).

## 1.3. Key concepts

Brief definitions of concepts used throughout the APF are provided in the User's Guidebook and in this paper. Many of these concepts are discussed in greater detail in other TPs (e.g., stakeholders in TP2, vulnerability in TP3, (natural) hazards-based approach and risk in TPs 3-5, methodological approaches and baselines in TPs 3-7, indicators in TP6, systems and adaptive capacity in TP7, strategies, policies and measures in TP8, monitoring, evaluation and mainstreaming adaptation in TP9). For ease of reference, concepts central to this TP are outlined here.

*Adaptation* is a process by which strategies to moderate, cope with and take advantage of the consequences of climate events are enhanced, developed and implemented.

An *adaptation baseline* is a comprehensive description of adaptations that are in place to cope with current climate. The baseline may be both qualitative and quantitative, but should be operationally defined with a limited set of parameters (indicators).

An *adaptation community* is the network of stakeholders that takes shape over the course of an adaptation project and persists following the project's completion; its goals are to implement, support and improve adaptation strategies, policies and measures.

An *adaptation project* for developing and implementing adaptation strategies, policies and measures may be designed and carried out, using some or all of the concepts of the APF.

*Adaptive capacity* is the property of a system to adjust its characteristics or behaviour, in order to expand its coping range under existing climate variability, or future climate conditions.

Different adaptations will have a variety of priorities and needs. A project *approach* is selected to respond to these unique needs. The four major approaches discussed in the APF are outlined here:

1.   With the hazards-based approach, a project assesses current climate vulnerability or risk in the priority system

(TP4), and uses climate scenarios to estimate changes in vulnerability or risk over time and space (TP5).

2.   With the vulnerability-based approach (TP3), a project focuses on the characterisation of a priority system's vulnerability and assesses how likely critical thresholds of vulnerability are to be exceeded under climate change. Current vulnerability is seen as a reflection of both development conditions and sensitivity to current climate. The vulnerability-based approach can be used to feed into a larger climate risk assessment (TPs 3-5).

3.   With the adaptive-capacity approach, a project assesses a system with respect to its current adaptive capacity, and proposes ways in which adaptive capacity can be increased so that the system is better able cope with climate change including variability (TP7).

4.   With the policy-based approach, a project tests a new policy being framed to see whether it is robust under climate change, or tests an existing policy to see whether it manages anticipated risk under climate change (TP6).

In the APF, *baselines* have two primary uses:

1.   First, there is the ***project baseline***. This is a description of where the project is starting from – who is vulnerable to what, and what is currently being done to reduce that vulnerability. Project baselines are generally focused on the priority system, and are thus site-specific and limited to the duration of the project. Depending on the approach used in an adaptation project, a project baseline will be characterised by a set of quantitative or qualitative indicators, and may take the form of, e.g., a vulnerability baseline (TP3), a climate risk baseline (TPs 4 and 5), an adaptive capacity baseline (TP7), or an adaptation baseline (TP6). Project baselines can later be used in the monitoring and evaluation process to measure change (in, e.g., vulnerability, adaptive capacity, climate risk) in the priority system, and the effectiveness of adaptation strategies, policies and measures.

2.   Second, depending on their project needs and design, APF users may choose to develop ***reference scenarios*** that represent future conditions in the priority system in the absence of climate adaptation. Scenarios may also be developed in which various adaptation measures are applied. Both reference scenarios and scenarios may be compared with baselines to evaluate the implications of various adaptation strategies, policies and measures. Scenarios differ from project baselines in that they deal with the longer term and are used for informing policy decisions concerned with various development pathways at the strategic planning level.

An *indicator* is a quantitative or qualitative parameter that provides a simple and reliable basis for assessing change. In the context of the APF, a set of indicators is used to characterise an adaptation phenomenon, to construct a baseline, and to measure and assess changes in the priority system.

***Policies and measures*** address the need for climate adaptation in distinct, but sometimes overlapping ways. Policies typically refer to instruments that governments can use to change economic structures and individual behaviours. Measures are usually specific actions, such as planting different crops.

A ***priority system*** is the focus of an adaptation project. It is a system that is characterised by high vulnerability to different climate hazards, as well as being strategically important at local and/or national levels. Socio-economic and biophysical criteria are often used to select priority systems by a given stakeholder group, and to set system parameters (indicators) for a given project.

***Stakeholders*** are those who have interests in a particular decision, either as individuals or as representatives of a group. This includes people who influence a decision, or can influence it, as well as those affected by it. A stakeholder analysis often involves institutional mapping.

A ***strategy*** is a broad plan of action that is implemented through policies and measures. Strategies can be comprehensive or targeted.

A ***system*** may be a region, a community, a household, an economic sector, a business, a population group, etc., that is exposed to specific climate hazards.

### 1.4. Guidance on scoping and designing an adaptation project

The process of scoping and designing an adaptation project will involve a number of related activities. In general, these can be grouped into the following four task areas:

- Scope project and define objectives
- Establish a project team

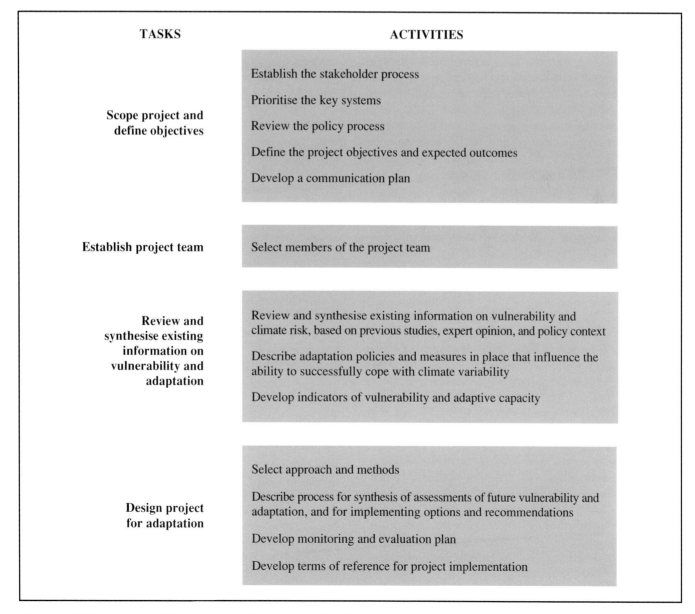

| TASKS | ACTIVITIES |
|---|---|
| **Scope project and define objectives** | Establish the stakeholder process<br><br>Prioritise the key systems<br><br>Review the policy process<br><br>Define the project objectives and expected outcomes<br><br>Develop a communication plan |
| **Establish project team** | Select members of the project team |
| **Review and synthesise existing information on vulnerability and adaptation** | Review and synthesise existing information on vulnerability and climate risk, based on previous studies, expert opinion, and policy context<br><br>Describe adaptation policies and measures in place that influence the ability to successfully cope with climate variability<br><br>Develop indicators of vulnerability and adaptive capacity |
| **Design project for adaptation** | Select approach and methods<br><br>Describe process for synthesis of assessments of future vulnerability and adaptation, and for implementing options and recommendations<br><br>Develop monitoring and evaluation plan<br><br>Develop terms of reference for project implementation |

**Figure 1-2:** Tasks and activities to scope and design an adaptation project

- Review and synthesise existing information on vulnerability and adaptation
- Design adaptation project

In Figure 1-2, these tasks are presented as a linear process, but they will likely be carried out concurrently, and with significant feedback among them. Annex A.1.1 provides questions and tables that may be used by the project team as they work through these tasks.

### 1.4.1. Scope project and define objectives

Effective adaptation projects will have long-lasting benefits for a given country. Building on the principles and "lessons learned" from prior experience from related disciplines will help to make climate change adaptation successful. This longer-term perspective is critical for addressing climate change impacts because of its decadal time horizon.

Specifically, the project should carefully take into account existing development plans in order to identify linkages between adaptation to climate change and other priorities. This approach recognises the importance of understanding the drivers of vulnerability at different levels – whether national, regional, sectoral or local. Strategies, policies, and measures identified should be consistent with national development plans (e.g., to meet Millennium Development Goals) while offering the co-benefits of reducing exposure to a range of future climate hazards and conditions.

The project team needs to identify ongoing and/or planned projects within the country that have relevance to the adaptation project. These projects may be complementary, and possibly synergistic. Together they may increase the strategic value of

the adaptation process, enable more focused assessments, increase the policy impact of the results, and increase the efficiency of the available funding. Given that adaptation is not a stand-alone issue, leveraging is essential.

Key activities in this scoping exercise include:

- Establishing the stakeholder process
- Prioritising the key systems
- Reviewing the policy process
- Defining the project objectives and expected outcomes
- Developing a communication plan

*Establish the stakeholder process*

As an initial step, the project team will need to establish a process for generating stakeholder input to the design, implementation, and conduct of an adaptation project. Two stages of stakeholder involvement may be required. At the initial stages of project scoping, the stakeholder group will probably be small, in order to enable the quick development of priorities and objectives, and to identify additional stakeholders. Following initial stakeholder scoping activities, the full project team and a broad, diverse group of stakeholders are engaged for the project duration. In most situations, it is necessary to increase the interest and commitment of government organisations beyond those directly involved in the United Nations Framework Convention on Climate Change (UNFCCC).

A stakeholder process inclusive of a wide range of viewpoints is needed to facilitate a shared understanding of the issues, including the fact that adaptation strategies, policies and measures may result in winners and losers. In addition, the stake-

---

**Box 1-1: Questions to aid description of stakeholders**

- Who is affected by climate change, including variability, in the priority system?

- Who in the priority system are the potential leaders in the government, research communities, and civil society (e.g., non-governmental organisations (NGOs), associations, local communities)? Who is responsible for facilitating and implementing policies and measures for adaptation?

- Who controls the largest financial contributions for sectoral lending, or direct foreign investment?

- Who is actively working in the priority system on relevant issues (e.g., disaster management, poverty alleviation, forest management or community development)?

- Who is concerned with the priority system and the project results? Possibilities include national or local government, scientists, technology suppliers, economists, universities, private companies, NGOs, co-operatives, trade unions, communities and women and youth movements.

- Which stakeholders are responsible for formal and informal dissemination of knowledge? Is there a media presence?

- Which stakeholders are likely to be affected by the implementation of adaptation policies and measures in the priority system?

holder community may offer data, analytic capabilities, insights, and understanding of relevant problems that can contribute directly to the adaptation project. Given the valuable role of stakeholders in the project, a documentation log should be kept of key decisions agreed to by stakeholders in order to retain institutional memory after the project lifetime.

TP2 describes participatory techniques and tools to identify the key stakeholders, define their roles and responsibilities, and engage them in determining the best methods to encourage effective communication. Box 1-1 can be used to identify stakeholders; Figure 1-3 places these stakeholder groups in the local, regional and global context.

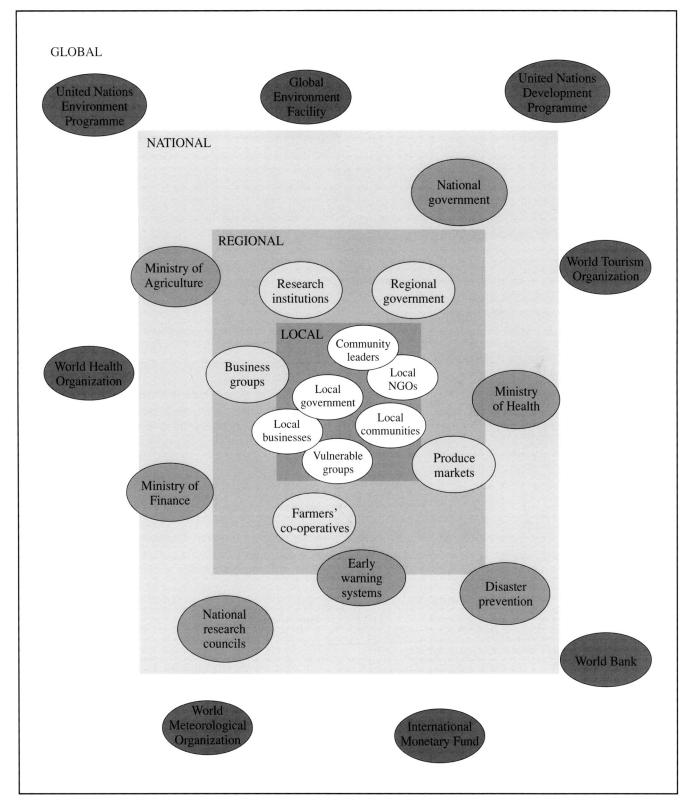

**Figure 1-3:** Stakeholder groups and institutional mapping

*Prioritise the key systems*

Countries have multiple vulnerabilities to climate change, including variability, from drought risk to an increased burden of vector-borne diseases. A particular adaptation project will be selected based on prioritisation of who is vulnerable, to what, where, and to what extent within the priority system (TP3, Section 3.4.2). Although the information on priorities will be general at this stage, it should be sufficient to make the necessary comparisons. The questions in Annex A.1 may be helpful in the prioritisation process; these questions can be modified as needed.

Adaptation priorities can be identified from existing vulnerability assessments, from stakeholders likely to be affected, or from the advice and needs of decision-makers and scientific experts. To be genuinely valuable for adaptation, and legitimate in the public eye, the prioritisation process should include extensive stakeholder input. Prioritisation can be determined along various dimensions such as a particular sector, region or climate hazard. Prioritisation should recognise that sectors are often interdependent, e.g., both human health and agriculture are dependent on water resources. Prioritisation should consider relevant factors such as geographical location, time horizon, level of governance, current and future climate vulnerability, current and future socio-economic conditions, integration across systems, etc. In principle, priority should be placed on systems where there is both high vulnerability and a high likelihood of significant potential impacts from climate change, including variability. For example, specific climate hazards, such as major floods or droughts within a specific sector, could be the focus of the project. Finally, there should be consideration of how the climate risk(s) in the priority system could interact with development patterns and plans.

*Review the policy process*

Understanding national, sectoral and local policy-making processes is essential for determining how to design and implement an adaptation strategy, policy, or measure. In support of their commitments under the UNFCCC, almost all governments have prepared national climate change reports. In such countries, the APF process can build upon the national vulnerability and adaptation assessments that were conducted as part of National Communications and/or National Adaptation Programmes of Action. In many countries, institutional structures were also established through the UNFCCC process, but these externally-funded structures are often weakly integrated into the national policy-making process.

The structure of relevant decision-making processes – whether national and sectoral policy-making processes, or a community's social choice mechanism – needs to be identified to understand how an adaptation strategy, policy, or measure can be implemented through these processes. The basic questions to be answered include the following:

- Which level of decision-making is most relevant to the adaptation process? E.g., central government, municipal, and/or local community level.
- At each level, how can strategies, policies, and measures recommended by an adaptation project be included in the decision-making agenda?
- If a project is being carried out at the community level, how can the results provide input to the national policy-making process?
- How might the policy processes initiated during the project be sustained beyond the project lifetime?

It is important to identify situations where adaptation recommendations may be difficult to implement or sustain in order to develop approaches to manage these situations. Examples include a particular inertia of the policy process, and vested interests of groups or individuals. (See TP6, Section 6.4.4, for guidance on a more in-depth characterisation of current conditions within the policy process.)

The output for this activity could be a brief report that summarises the relationship of the key policy processes to adaptation to climate change, including variability, the potential for integrating adaptation concerns into these processes and the methods by which adaptation can be incorporated into existing processes.

*Define the project objectives and expected outcomes*

Framing the project objectives and expected outcomes is critical to developing a project that will be informative and responsive to the needs of stakeholders and policy-makers. The project objective should state what the project is specifically intended to enable in the priority system, both during and after the project lifetime. The objective must be achievable within the project constraints, such as available funding. The process of setting objectives can be accomplished using facilitated stakeholder fora, expert opinion, and direction from policy makers. A number of tools can assist in creating a consensus on the central and sub-objectives of the project (TP2). Tools such as the political science-based "x goal-tree", for instance, can help the user to map the central goal of a project with the goals and interests of the stakeholders involved.

The basic objectives of an adaptation project might be to:

- Increase the robustness of infrastructure designs and long-term investments
- Increase the flexibility and resilience of managed natural systems
- Enhance the adaptive capacity of vulnerable groups
- Reverse trends that increase vulnerability
- Improve societal awareness and preparedness for future climate change
- Integrate adaptation in national and sectoral planning

Although the above objectives will differ from project to project, all would require information generated by vulnerability and adaptation assessments described in the APF. If the objective is to develop guidelines for including adaptation in national and sectoral planning, then the project will need to inform major project-planning or policy-making processes about the risks and opportunities associated with climate change.

To facilitate future monitoring and evaluation of the project output, consideration should be given to developing evaluation criteria during this task. Clear criteria will help in evaluating whether or not the desired outcome(s) was achieved.

*Develop a communication plan*

The project will only be effective if the results are effectively communicated with key stakeholders, decision-makers and the public. Therefore, it is important to produce a communi-

cation plan that is closely tailored to the needs of the target audiences, rather than the needs of the information generator. Communication should be adjusted and modified as required, based on monitoring of its effectiveness.

Key questions to be considered in the development of a communication strategy include: Who is responsible for the communication process? Which are the key audiences? How will the impact of communication be evaluated?

A national workshop could be held to present the results and to solicit feedback from stakeholders and decision-makers on key areas for further action. In addition to a project report, the team may produce a review of adaptation options, a summary of findings for stakeholders, and a technical report for the scientific community. (See TP9 for ideas on taking the adaptation process forward, which can be reflected in a project's communication plan.)

*Table 1-1:* Identifying adaptation project focus according to scale of implementation

|  | **Hazards-based approach** | **Vulnerability-based approach** | **Adaptive-capacity approach** | **Policy-based approach** |
|---|---|---|---|---|
|  | **Increasing resilience to severe flooding and future climate risks** | **Improving access to new markets and supporting livelihood diversification under future climate** | **Improving awareness in and the resilience of the business community to climate change, including variability** | **Reducing vulnerability to storm surges and sea level rise induced by climate change** |
| **National** | How can national meteorological services be changed to better monitor the evolution of future hazards? | How will recent changes in world markets affect aquaculture in Bangladesh (already at risk of inundation from sea level rise) under future climate? | Which business sectors will be most affected by climate change and why? What awareness raising is needed, and for whom? What fora should be involved? | What incentives or disincentives should be used to discourage the development of coastal zones vulnerable to sea level rise and storm surges induced by climate change? |
| **Regional** | How can flood early warning systems be made more effective under future climate for hard-to-reach communities? | How can access to new markets required by livelihood diversification activities be facilitated to moderate future climate? | How can regional businesses most effectively support livelihoods identified as being vulnerable to climate change, including variability? | Realignment or retreat? How to decide which areas are protected and which will become submerged under future climate? |
| **Local** | What techniques are most appropriate for effective local-level disaster preparedness planning under future climate? | How can credit schemes best support livelihood diversification in rural areas to reduce climate risks? | Which participatory visioning processes are most appropriate to identify threats and potential opportunities resulting from scenarios of climate change for members of local trade associations and businesses? | What stakeholder-led projects are most appropriate for investigating ways to mitigate flood damages in an urban area under future climate? |

Table 1-1 provides examples of how projects for a given sector might change in focus depending on the scale at which they are being implemented for each of the approaches described in the APF.

### 1.4.2.   Establish project team

The composition of an adaptation project team should be motivated by the project needs and goals. The interdisciplinary team could represent a range of sectors, and will likely include individuals with experience in vulnerability and adaptation assessments, climate science, and socio-economic research, as well as representatives of relevant stakeholders (including NGOs and potentially affected communities). It is essential to include practitioners with management expertise in the key issues of the priority systems. Other project team members may be drawn from universities and other research institutions, government agencies, non-government organisations, or private enterprises. The team members should commit to making a significant contribution over the course of the project.

### 1.4.3.   Review and synthesise existing information on vulnerability and adaptation

In this task the project team will identify and synthesise prior work on vulnerability and adaptation that is relevant to the priority system. This work may have been conducted within the team's country or in another country with relevant circumstances. Synthesis of this information will be used to develop a project baseline (below). It is against this baseline that future vulnerability and adaptation options can be considered and against which future progress toward adaptation goals can be viewed. A well-defined baseline should describe the current level of vulnerability and the adaptation measures in place to reduce that vulnerability. Key activities involved in this task include:

- Review and synthesise existing information on current vulnerability and climate risk, based on previous studies, expert opinion, and policy context
- Describe adaptation policies and measures in place that influence the ability to successfully cope with climate variability, including the effectiveness of those policies and measures
- Develop indicators of vulnerability and adaptive capacity

*Review and synthesise existing information on vulnerability and climate risk, based on previous studies, expert opinion, and policy context*

In cases where specific policies and measures have been implemented to address the impacts of climate change, including variability, on vulnerable systems, there may be extensive national literature upon which to draw. Existing data, information and analyses may be found in case studies, academic literature, publications by development practitioners, consultation with experts, community knowledge, and in policies and measures designed to

address other issues within the priority system. In addition, expert opinion and the policy context may offer information on the vulnerability of the priority system. Examples include national development plans, Poverty Reduction Strategy Papers and natural hazards assessments. It is important to let the scope and objectives of the project determine the relevance of this information. The synthesis should identify key factors of concern in the priority system, and outline what is known about the relationship between the risk and the priority system. Existing information on current socio-economic conditions that affect vulnerability to climate variability should also be evaluated.

Synthesis of available information can be based on expert opinion, analogue or historical studies, and/or modelling. The synthesis should outline the extent of knowledge on the key factors of concern, and the certainty and nature of the relationship between the risk and the system under study.

*Describe adaptation policies and measures in place that influence the ability to successfully cope with climate variability, including the effectiveness of those policies and measures*

Understanding the adaptations in place to cope with current climate risks is necessary to inform the development of adaptations to manage future climate risks. The output from this activity would be a preliminary adaptation baseline that describes the policies and measures in place to reduce vulnerability. It would involve identifying the autonomous and planned adaptations currently implemented to address climate risks in the priority system, including the level at which these have been implemented (national, regional and community level), their effectiveness and any barriers to their implementation. Also, it will identify institutions that can support implemented adaptation policies and measures. This evaluation will facilitate understanding what worked in the past, how policies and measures in place could be improved, and what strategies, policies and measures might be needed in the future. The project team should take a broad perspective and include relevant policies and measures that were designed to address other problems. TP4 provides guidance on conducting an assessment of adaptive responses to historic climate risks, and on developing the relationship between current climate risks and adaptive responses that can be used to calculate future climate risks. TP8 helps the user to define adaptation strategies, policies and measures relevant to the climate risks in the system.

*Develop indicators of vulnerability and adaptive capacity*

The information generated from the activities above can be used to summarise the existing vulnerability of and adaptations in place for the priority system (the project baseline). The project baseline describes where the project is starting from – who is vulnerable to what, what is currently being done to reduce that vulnerability, etc. In essence, the project baseline describes how well adapted the system is to current climatic conditions. With a sound understanding of where the project is starting from, a more accu-

---

**Box 1-2: Note on indicators**

Desirable indicators fulfil three criteria: (1) summarise or otherwise simplify relevant information; (2) make the phenomena of interest visible or perceptible; and (3) quantify, measure and communicate relevant information. They may be qualitative, quantitative or both. If quantitative scenarios of the future relevant to climate change vulnerability and adaptive capacity are desired, the process involves choosing indicators, collecting or locating appropriate data, and estimating future values for those proxies (see TP6 for more information on using indicators).

---

rate assessment can later be made of success of the project. Indicators chosen to describe the baseline should be used during project monitoring and evaluation whenever possible (TP9).

The APF discusses four main approaches to baseline development as outlined in section 1.3, *Key Concepts*. Projects will develop and rely on the baseline that corresponds to their project approach.

- TP3 describes methods to construct a *vulnerability baseline*, including the development of a set of vulnerability indicators for the project. The current vulnerability of the priority system may be quantified through the development of exposure-response relationships, or through the development of indicators to describe various aspects of the conditions of the system.
- TP4 Figure 4-2 provides a flow chart for assessing the current *climate risk baseline*.
- TP6 outlines the development of an *adaptation policy baseline*. This is a comprehensive description of adaptation-relevant policies that are in place to cope with current climate. TP6 also explains how to construct baseline indicators of socio-economic conditions that can drive vulnerability, risk and adaptive capacity.
- TP7 describes an approach for selecting indicators for defining the *adaptive capacity baseline*.

Additional guidance on choosing indicators is provided in Annex A.1.1.

### 1.4.4. Design project for adaptation

Key activities in this task include:

- Select approach and methods to:
  - Assess future vulnerability and adaptation
  - Characterise future climate-related risks
  - Assess future socio-economic conditions and prospects
  - Assess capacity to adapt
  - Characterise uncertainties
- Describe process for synthesising assessments of future vulnerability and adaptation, and for implementing options and recommendations
- Develop monitoring and evaluation plan
- Develop terms of reference for project implementation

The output of this task is a detailed project document.

*Select approach and methods*

The selection of an approach to and methods for acquiring the information needed for the project is addressed in some detail in TPs 3 through 6. The methods selected should be appropriate to the goals of the project, compatible with the potential constraints of available resources and sufficiently credible. Preference should be given to methods that build the national capacity for policy-making.

It may be appropriate to adopt an approach that is already in use, such as in development planning. Otherwise, the team will need to develop its own approach. Approaches recommended for adaptation projects are outlined in Section 1.3., Key concepts, and include:

- hazards-based approach (i.e., analyse possible outcomes from a specific climate hazard);
- vulnerability-based approach (i.e., determine the likelihood that current vulnerability may be affected by future climate hazards);
- policy-based approach (i.e., investigate the efficacy of an existing or proposed policy in light of a changing climate exposure or sensitivity); and
- adaptive-capacity approach (i.e., focus on increasing adaptive capacity and removing barriers to adaptation).

See Section 1.3 in this paper; TP4, Sections 4.4 and 4.4.2; and TP5 Section 5.4.1 for additional guidance on selecting approaches.

Stakeholder-led exercises discussed earlier may be useful when selecting methods (TP2, Annex A.2.2). The project team needs to select methods that provide sufficient information to enable stakeholders to make policy and investment decisions. Credibility of the assessment is, of course, extremely important to the policy-making and stakeholder communities. The assessment may lose its value if the methods are not appropriate to the objectives of the project, or if insufficient time and resources are spent to ensure reliability of results. In addition, methods should be internationally comparable and acceptable to facilitate the comparison of results among areas with similar vulnerabilities. Communicating results in a way that the underlying assumptions and the degree of uncertainty are understood can help to establish transparency.

The amount of information required for a specific project and the techniques and tools for obtaining that information will differ widely. For example, depending on whether the objective is to prepare for migrating disease vectors, to rationalise crop selection and tillage methods, or to find new employment for flood victims, the inputs may be significantly different. Methods and level of effort will change with the level of complexity or comprehensiveness of the adaptation project goal and objectives. For example, a comprehensive national adaptation strategy to cover all geographic areas and all sectors for the next 50 years will need more research support and different approaches than a five-year plan to help coastal fisheries to adapt to sea level rise and increasing storm surges. (See TP3 Annex A.3.3 for an illustration of the variation in methods and depth of analysis that are needed for different types of projects.)

For any proposed method, the benefits should justify any new data collection efforts. Practical considerations include the research skills needed, data availability, the cost and the length of time required to carry out the analysis, and computational requirements. In some cases, computer-based models may be available "off the shelf". In this case, the data requirements of the model and the availability of modelling skills in the project team are factors to be considered. Most often, the project team will have to assemble a variety of methods appropriate to their situation. The choice of methods therefore has to consider the relevant criteria and make a balanced judgment in terms of the trade-offs amongst them. Some methods will be precluded by constraints, such as the lack of financial resources, the lack of long-term data sets, the capacity for implementation and the time required to obtain results. The project team should weigh the practical considerations against both the project objectives and the need for credible research.

Uncertainties need to be addressed in the process of project design. Time should be taken to understand and clearly articulate uncertainties and assumptions, and to minimise them in project design.

*Describe process for the synthesis of assessments of future vulnerability and adaptation, and for implementing options and recommendations*

Synthesising the information generated by the project can bring together and make sense of the various results in order to recommend policies and measures for the priority system. During this phase of the project, the team might develop an outline for the synthesis of results that is structured to facilitate the identification and implementation of adaptation options. Methods for synthesis of assessments and generation of options and recommendations are discussed in some detail in TP8.

*Develop monitoring and evaluating plan*

An adaptation project should provide realistic recommendations for implementing strategies, policies and measures for reducing vulnerability and increasing adaptive capacity in the

priority system. It is only possible to evaluate the effectiveness of these measures if monitoring and evaluating plans are incorporated into the project design. The development of monitoring and evaluation plans is discussed in detail in TP9. The initial monitoring and evaluation plan needs to describe how evaluation results will be fed back to the management process, and how these plans could contribute to the establishment of a long-term monitoring and evaluation capability in the country. For each project recommendation, indicators of success should be developed to facilitate assessment of effectiveness.

Barriers may exist for the deployment or the evaluation of adaptation strategies, policies, and measures such as resource constraints, lack of ability to deploy available resources or an unwillingness to do so. These barriers need to be identified and possible solutions explored.

*Develop terms of reference for project implementation*

The terms of reference for the project should clearly describe the project objectives and expected outcomes, the specific project activities, the stakeholders involved in the project, the budget, due dates, etc. A logical framework analysis (logframe) matrix of activities, describing objectives, activities, and outputs, may be useful for organisation. The tasks and activities necessary for accomplishing the project objectives should be detailed. The process of developing the terms of reference may include consultation with additional stakeholders and the general public to refine or reframe the policy context or the project objectives. Wide dissemination of the terms of reference will help ensure that the process of conducting the project is open and transparent.

## 1.5.  Conclusions

As an output of APF Component 1 (and TP1), adaptation project teams will generally prepare a project proposal, with a detailed implementation plan including clear statements of objectives, activities and outcomes. Teams can use the checklist below to verify the comprehensiveness of their plans. (Each bullet below has been explored in the preceding guidance in Section 1.4.)

Has the team:

- Defined priority systems and project boundaries?
- Established a plan for identifying and engaging stakeholders?
- Determined project objectives and expected outcomes?
- Developed a plan for communicating results to stakeholders and decision-makers?
- Selected the project team?
- Identified, assembled, reviewed and synthesised pertinent information?
- Described the project baseline?
- Selected indicators?
- Selected an approach and the methods to be used?
- Described a process for synthesising assessments of

vulnerability and adaptation, and for implementing options and recommendations, if appropriate?

- Developed a strategy for assessing, monitoring and evaluating project effectiveness, including a preliminary strategy to overcome barriers to implementation of recommended adaptation measures?
- Analysed the national policy-making process in the context of adaptation?
- Prepared terms of reference for the overall project?

The main purpose of developing a detailed implementation plan for an adaptation project – one with clear statements of objective, activities, and outcomes – is to ensure that the project will ultimately result in the identification and implementation of effective adaptation strategies, policies and measures. In essence, this is a small-scale exploration of all the APF Components relevant to the priority system(s) in order to better design and implement the project. Conceptualising and defining the process at this stage, in a manner that is consistent with the APF principles, can greatly facilitate the implementation of the adaptation project.

# ANNEX

### Annex A.1.1. Questions to aid prioritisation of key systems

These questions are categorised under human, economic, and physical vulnerability to enable the project team to explore a range of vulnerabilities that can affect a single system. These questions are for organisational purposes only and can be modified as needed.

*Human vulnerability (sample system – smallholders):*

- Are there vulnerable groups within the system? Which groups?
- What is their key vulnerability (e.g., crop failure due to drought)?
- Historically, what is the typical impact on these groups (e.g., food shortage and malnutrition)?
- Historically, what is the magnitude of the impact (e.g., 250,000 people affected over two years)?
- Historically, have lives been lost because of this impact? How many?
- What has been done to mitigate this impact? How effective were these measures?
- What is the current level of risk?

*Economic vulnerability (sample system – water resources):*

- Is the system closely linked to the economy?
- What are the key links (e.g., crop irrigation, agricultural livelihoods, industrial processes)?
- What is the vulnerability associated with these links (e.g., reduced productivity or lost crops through drought)?
- Historically, what is the typical impact (e.g., drop in sorghum production, reduction of the workforce)?
- Historically, what is the magnitude of the impact (e.g., over five-year period, two of five regions were affected, a 10% drop in sorghum export, a 5% increase in unemployment)?
- What has been done to mitigate this impact? How effective were these measures?
- What is the current level of risk?

*Physical vulnerability (sample system – coastal region):*

- Is the system physically vulnerable (e.g., to coastal land loss or infrastructure damage)?
- What is the specific vulnerability (e.g., infrastructure damage through coastal inundation)?
- Historically, what is the magnitude of the impact (e.g., in a 1997 event, 20% of coastal structures in District X were damaged)?

# 2

# Engaging Stakeholders in the Adaptation Process

CECILIA CONDE[1] AND KATE LONSDALE[2]

**Contributing Authors**
*Anthony Nyong[3] and Yvette Aguilar[4]*

**Reviewers**
*Mozaharul Alam[5], Suruchi Bhawal[6], Henk Bosch[7], Mousse Cissé[8], Kees Dorland[9], Mohamed El Raey[10], Carlos Gay[11], Roger Jones[12], Ulka Kelkar[6], Maria Carmen Lemos[13], Erda Lin[14], Maynard Lugenja[15], Shiming Ma[14], Ana Rosa Moreno[16], Mohan Munasinghe[17], Atiq Rahman[5], Samir Safi[18], Barry Smit[19], and Juan-Pedro Searle Solar[20]*

[1] Center for Atmospheric Sciences, Universidad Nacional Autónoma de México, Mexico DF, Mexico
[2] Stockholm Environment Institute, Oxford, United Kingdom
[3] University of Jos, Jos, Nigeria
[4] Environment and Natural Resources Ministry, San Salvador, El Salvador
[5] Bangladesh Centre for Advanced Studies, Dhaka, Bangladesh
[6] The Energy and Resources Institute, New Delhi, India
[7] Government Support Group for Energy and Environment, The Hague, The Netherlands
[8] ENDA Tiers Monde, Dakar, Senegal
[9] Institute for Environmental Studies, Amsterdam, Netherlands
[10] University of Alexandria, Alexandria, Egypt
[11] Center for Atmospheric Sciences, Universidad Nacional Autónoma de México, Mexico DF, Mexico
[12] Commonwealth Scientific & Industrial Research Organisation, Atmospheric Research, Aspendale, Australia
[13] University of Michigan, Ann Arbor, Michigan, United States
[14] Institute of Agro-Meteorology, Chinese Academy of Agricultural Sciences, Beijing, P.R. China
[15] The Centre for Energy, Environment, Science & Technology, Dar Es Salaam, Tanzania
[16] The United States-Mexico Foundation for Science, Mexico DF, Mexico
[17] Munasinghe Institute for Development, Colombo, Sri Lanka
[18] Lebanese University, Faculty of Sciences II, Beirut, Lebanon
[19] University of Guelph, Guelph, Canada
[20] Comisión Nacional Del Medio Ambiente, Santiago, Chile

# CONTENTS

## 2.1. Introduction

Adaptation is a process by which strategies to moderate, cope with and take advantage of the consequences of climatic events are enhanced, developed and implemented. Adaptation occurs through public policy-making and decisions made by stakeholders, i.e., individuals, groups, organisations (governmental agencies or non-governmental organisations (NGOs)) and their networks. Relevant stakeholders need to be brought together to identify the most appropriate forms of adaptation. Analysing the capacity of stakeholders to cope with and adapt to climatic events is fundamental to characterising current and possible future vulnerability. Understanding the role of stakeholders in the decision-making process will assist in the implementation of adaptation policies. In short, stakeholders are central to the adaptation process.

Many countries have already undertaken what are called the first generation impact, vulnerability and adaptation (V&A) studies. Some countries have also undertaken more in-depth projects aimed at preventing or ameliorating climate impacts and risks. The Adaptation Policy Framework (APF) seeks to support new V&A studies, as well as a range of other adaptation-related inquiries. In doing so, it emphasises the importance of a more stakeholder-driven approach. Stakeholders are fundamental to the process of adaptation, as it is they who will comprise the "adaptation community" that is required to sustain the process.

Each of the five Components of the APF involves stakeholders in a number of ways. The composition of the stakeholder group may change as the types of activities change. The involvement of stakeholders will be essential throughout in: designing the project, determining the analytical approach to be used, evaluating candidate policies and measures, continuing the process and communicating results of the efforts. This Technical Paper (TP) gives guidance on how and why to engage stakeholders at each of these points. It aims to assist the user in designing a stakeholder involvement strategy and engaging different stakeholders in such a way that their basis for interaction is strengthened and broadened. The second and third sections outline, respectively, the relationship of this TP to the larger APF and the definition of stakeholders. The fourth section explores why the engagement of stakeholders is so valuable to an adaptation project. Section 2.5 sketches general approaches to engaging stakeholders, while section 2.6 provides specific guidance on engaging stakeholders in each Component of the APF. The TP concludes with key reflections on the stakeholder engagement process.

## 2.2. Relationship with the Adaptation Policy Framework as a whole

A distinguishing feature of the APF is that it is stakeholder-driven. As such, this TP relates to all five Components of the APF (Figure 2-1). TP2 suggests an overall strategy and specific

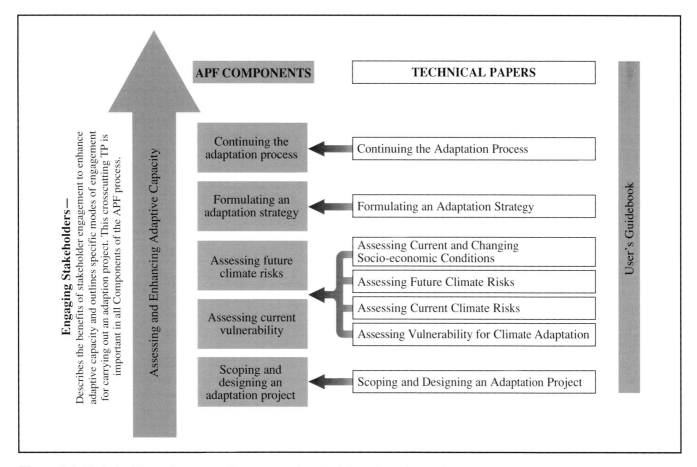

**Figure 2-1:** Technical Paper 2 supports Components 1 to 5 of the Adaptation Policy Framework

techniques for engaging stakeholders at each of these stages. Further, since stakeholders represent the primary source of adaptive capacity, this TP is closely aligned with the other cross-cutting paper (TP7), which is concerned with assessing and enhancing adaptive capacity.

The participants in the stakeholder process, the types of participation and the outcomes are discussed in the remaining sections of this TP.

## 2.3.    Key concepts

The term "stakeholder" in climate change studies refers to policy makers, scientists, administrators, communities, and managers in the economic sectors most at risk. In this context, stakeholders can be brought together from both public and private enterprises to develop a joint understanding of the issues and to create adaptations.

---

**Box 2-1: Stakeholder analysis in a community-based forest and wildlife resources management project in northern Mozambique**

The Mecuburi Forest Reserve was included as a pilot area of the Mozambique government project, "Support for Community Forestry and Wildlife Management (1997 – 2002)". The two project objectives were to:

1.    Improve the standard of living in rural communities through increased access to forest and wildlife products for household use and marketing; and generate income from employment, small industries and hunting fees.
2.    Protect and manage the resource base of forestry, wildlife, agriculture and animal husbandry through local communities in a rational way.

Table 2-1 outlines the outputs of the project's stakeholder analysis.

*Table 2-1: Stakeholder Analysis in Mecuburi Pilot Project Area*

| Stakeholder | Stake | Comments |
|---|---|---|
| Farmers living inside Mecuburi Forest Reserve | Arable land, spare arable land, basic needs for survival, cultural value of the forest | High migration indices due to the civil war that ended in 1992; some farmers "own" additional land outside the reserve |
| Farmers living next to Mecuburi Forest Reserve | Construction material, bush meat, cultural value of the forest | Not very interested in the proper utilisation of the resources in the reserve |
| Cotton and tobacco merchants | Cotton and tobacco produced by farmers living in the reserve | Promote cotton and tobacco cropping through credit schemes supplying basic inputs (e.g., technical advice) |
| Merchants dealing in construction material | Construction material (e.g., poles, bamboo, rope, thatch) in the forest reserve | These materials are often extracted illegally |
| Professional hunters | Wildlife for sport hunting and meat | Most hunt illegally, or in collusion with corrupt police officials |
| Commercial logging companies | Commercial timber (e.g., umbila, panga panga, chanfuta) growing inside the forest reserve | Often illegally extend their concession areas on adjacent public land to include the forest reserve |
| Local government/ administrative structures | Rural development, revenue for the local authority | Unlawfully superimpose authority in conservation area |
| Provincial Forest and Wildlife Services | Conservation, programme implementation, revenue | Caught in the paradoxical dilemma of having the duties of the police at certain times and of the extensionist at others |

*Source:* Presentation prepared by Patrick Mushove for the workshop "Climate Change, Vulnerability and Adaptation: AIACC Development Workshop", Third World Academy of Sciences, Trieste, Italy, 3-14 June, 2002

The definition of stakeholders used here is "those who have interests in a particular decision, either as individuals or as representatives of a group. This includes people who influence a decision, or can influence it, as well as those affected by it" (Hemmati, 2002).

## 2.4. Why engage stakeholders?

Stakeholders are individuals or groups who have the current and past experience of coping with, and adapting to, climate variability and extremes. The principal resource for responding to climate change impacts is people themselves, and their knowledge and expertise. Through an ongoing process of negotiation, they can assess the viability of adaptive measures. Together, the research community and stakeholders can develop adaptive strategies by combining scientific or factual information with local knowledge and experience of change and responses over time too. Box 2-1 describes an example of the importance of stakeholders' involvement, outlining individual stakeholders, their stake and the observed particularities of each group. This example corresponds roughly to Components 1 and 2 of the APF.

Stakeholders, at different levels and stages, are crucial to the success of an adaptation project. Through listening to the views of others, stakeholders can build a shared understanding of the issues. Priority areas for action emerge that take account of everyone's perceptions. This process requires time to build trust between the groups and individuals involved, and can be empowering, as solutions are worked out collaboratively (Box 2-2). If each participant is seen as having a valid view, a stakeholder process can encourage longer-term capacity development by developing pathways for co-ordinated action. Adaptive capacity is developed if people have time to strengthen networks, knowledge, resources and the willingness to find solu-

tions. However, the process must be carefully designed and implemented, as stakeholder participation does not in itself guarantee equity, fairness or eventual buy-in.

## 2.5. Approaches for stakeholder engagement

There are a great number of approaches to stakeholder engagement, and no single formula for success. Rather, there are combinations of tools and techniques that will be well-suited to a given situation. The choice of which to use depends on the complexity of the issues to be discussed and the purpose of the engagement, both of which will be determined in the initial steps of the project where a careful evaluation of the time and resources available should be performed.

Stakeholder engagement approaches vary from quite passive interactions, where the stakeholders provide information, to "self-mobilisation", where the stakeholders themselves initiate and design the process. The different levels of participation can be illustrated using the "ladder of participation" outlined in Figure 2-2. Engagement closer to self-mobilisation is not necessarily better because it is more participatory. Different levels of participation will be appropriate for different stages of the project and given the experience of the research team. However, it is important that the stakeholders understand how they are being involved, how the information they provide will be used and whether they have any power to influence decisions.

It is also important to consider the scope of the issues that stakeholders will participate in defining and solving (Thomas, 1996). When designing the engagement, it is important to take into account the stage at which the engagement is occurring in terms of the policy-making process, what decisions have already been taken and what positions are already fixed. It may be that the engagement, though very participatory in itself, is

---

**Box 2-2: Benefits of stakeholder engagement** (adapted from Twigg, 1999)

- Participatory initiatives are more likely to be sustainable because they build on local capacity and knowledge, and because the participants have "ownership" of any decisions made and are thus more likely to comply with them. Participatory initiatives are thus more likely to be compatible with long-term development plans.
- Working closely with local communities through stakeholder engagement can help decision-makers gain greater insight into the communities they serve, enabling them to work more effectively and produce better results. In turn, the communities can learn how the decision-making process works and how they can influence it effectively.
- The process of working and achieving things together can strengthen communities and build adaptive capacity through developing awareness of the issues within the community, as well as finding ways to address them. It can reinforce local organisations, and build up confidence, skills and the capacity to cooperate. In this way it increases people's potential for reducing their vulnerability. This, in turn, empowers people and enables them to tackle other challenges, individually and collectively.
- Stakeholder participation in planning, through priority-setting and voicing preferences, as well as in implementation, accords with people's right to participate in decisions that affect their lives. Processes of engagement can improve the likelihood of equity in decision-making and provide solutions for conflict situations.
- Engaging stakeholders may take longer than conventional, externally-driven processes, but may be more cost-effective in the long term; a stakeholder process is more likely to be sustainable because the process allows the ideas to be tried, tested and refined before adoption.

An additional level of participation can be added – that of **Catalysing change**, where community members influence other groups to initiate change.

**Self-mobilisation.** Stakeholders take the initiative. They may contact external organisations for advice and resources but ultimately they maintain the control. Likely outcome for stakeholders: very strong sense of ownership and independence.

**Interactive participation.** Joint analysis and joint action planning. The stakeholders themselves take control and have a common goal to achieve. Likely outcome for stakeholders: strong sense of shared ownership, long-term implementation structures.

**Functional participation.** Enlisting help in meeting the pre-determined objectives of a wider plan/programme. Stakeholders tend to be dependent on external resources and organisations. Likely outcome for stakeholders: can enable implementation of sound intentions, as long as support is available.

**Participation by consultation.** Asking for views on proposals and amending them to take these views into account. May keep participants informed of the results but ultimately, no real share in the decision-making.

**Participation in giving information.** People are involved in interviews or questionnaire based "extractive" research. No opportunity is given to influence the process or to contribute to or even see the final results. Likely outcome for stakeholders: generates information but that is all.

**Figure 2-2:** Ladder of participation (adapted from Pretty, 1994)

not effective because the scope is too constrained and there is no opportunity for developing creative solutions.

### 2.6.    Guidance for stakeholder engagement

In this section, actions for developing a stakeholder engagement strategy are outlined based on the five Components of the APF. For each of these Components, the project team may wish to review several participatory techniques and, with the facilitator's input, decide which they feel comfortable using (see examples in Annex A.2.1).

#### 2.6.1.    *Component 1: Scoping and designing an adaptation project*

*Who is involved?*

The scope of the project will be determined by the project team (TP1). This project team will propose the scope of research (e.g., region, sector, vulnerable group) based on the results of previous

studies and on the advice and needs of decision-makers and experts. The results of this first stage should be made widely available to NGOs and other interested groups for comments. This helps to ensure transparency and build trust in the process.

*Tasks in Component 1*

As outlined in TP1 (*Scoping and Designing an Adaptation Project*), in the first stage of the APF, the project team performs a brief review of the current national policies for climate change (e.g., United Nations Framework Convention on Climate Change ((UNFCCC)) National Communications), for development and for the environment (e.g., the conventions on biodiversity and desertification) as a way to identify national priorities and the institutions that could be engaged in the project. In this review process, the project team can start to build up a directory of national and international entities (e.g., experts, agencies, NGOs and project managers) whose work is related to adaptation and who could be a source of information and support. It is important to include key people at an early stage of the project. The relevant national and regional governmental decision-makers should be encouraged to read and comment on these initial reports. Being

---

**Box 2-3: Guidelines for effective engagement**

*Clarity*

Clarify the objectives and goals of the engagement and evaluate the appropriateness of the techniques. Work towards agreement on defining the problem, acknowledging differences in people's perception. Be realistic about what can be achieved given the constraints of time and money, the available expertise and the political realities. Communicate clearly in all phases of the engagement; this strategy should include access to and presentation of all relevant information. Short-term interests inevitably take over when resources are scarce.

*Understanding of related processes*

Be clear about how the engagement fits in with official decision-making processes. Will the engagement process feed into and inform these other processes effectively? It is important to identify people, groups and structures that can provide support to achieve any actions identified through the engagement process.

*Management of information*

Having access to information is a form of power. Some groups will need to be persuaded of the benefits of both sharing information and developing a more holistic understanding of the issues. Information should be provided in an accessible way, without using complex concepts and jargon.

Communication and decision-making are not purely rational processes – people's feelings, attitudes and the ways in which they process information must be taken into account. It may be necessary to present information in different ways, e.g., as values or moral opinions, scientific facts or personal experience. Explain the objectives and goals of the process in advance, as well as what participants will be required to do.

*Support and capacity development*

Some groups may need training or other support to educate them to the level of other stakeholders. Examples include information that enables them to contribute to the discussions and data on likely impacts for their area or sector.

*Transparency*

Stakeholder groups should be identified in an open and transparent manner. From these groups, participants should also be invited in an open manner.

*Trust-building*

Stakeholder processes may bring together groups with opposing views – and with them, possibly a lack of trust. If the leaders can assure all participants that, in the engagement process, every participant's view is valued and respected, the people should feel reassured that their opinions will be heard, and they will be more likely to listen to others.

*Time for the process*

Lack of time is given as one of the most common constraints of many engagement processes. Since considerable time is required to develop the process, build partnerships and strengthen networks among stakeholders; raise awareness and build trust, and effective stakeholder engagement will take more time than conventional processes.

*Feedback and flexibility*

Participatory processes can be very flexible. If one technique is not working, another can be used or the questions changed to obtain the required information. This flexibility must be planned, and time must be allowed to get feedback on the effectiveness of the process. Are the right questions being asked? Is everyone contributing fully? If not, what are the obstacles and what could be improved? The analysis and synthesis of the outputs should be presented to stakeholders before general dissemination. Any conflicts of interest should be stated explicitly. This demonstrates a respect for differences.

---

**Box 2-4: Identifying stakeholders to involve in each Adaptation Policy Framework Component**

Ultimately, the question of who participates at any stage in an adaptation process is determined by the methods used to identify stakeholders. A simple but effective method is to ask the initial group of stakeholders (identified by the project team in Component 1) to suggest other stakeholders who are, in turn, asked the same question until no more individuals can be identified. This iterative method can be applied in each of the five APF Components. However, limited time and other resources will ultimately limit the number of stakeholders involved.

In addition to having the power to influence the adaptation process or being part of a group that would be directly affected by a predicted climatic impact, identified stakeholders must also be willing to participate in the process. In many cases, the stakeholders involved are the "usual suspects", i.e., government and NGO representatives, local dignitaries, businessmen and academics – people who are both familiar with the existing institutions and comfortable voicing their opinions. Other groups, particularly highly vulnerable individuals, may likely require more support to engage as they may not be able to attend meetings at certain times, they may feel uncomfortable in voicing their opinions or embarrassed about their lack of knowledge or education. Their involvement in the process is fundamental, as these individuals will play a key role in adapting to the impacts of critical climatic, environmental or socio-economic events. Also, they have rich experience and knowledge about the practical aspects of adaptation.

---

familiar with the project from the beginning may mean that they are more likely to take note of the project outputs and include them in their decision-making processes and policy design.

Stakeholders bring a range of interests to the APF process. Some examples are given in Table 2-2.

### 2.6.2.   *Component 2: Assessing current vulnerability*

*Who is involved?*

Component 2 would likely involve the people and groups who would be increasingly affected by the foreseen impacts, either positively or negatively, as well as those who have a role in influencing adaptation. Ideally, it would engage the most vulnerable, as identified in the first stage of the project. Regional climate, history and socio-economic experts could give advice on current conditions in the study region.

*Tasks in Component 2*

It is important to develop a common understanding among the stakeholders of what is meant by the words used. For example, the meaning of the words "vulnerability", "adaptation", "coping range" and "climatic hazard" should be discussed and agreed. Having this shared understanding is the first step to finding realistic solutions and building capacity. The project team and the regional experts may want to prepare a brief initial description of current climate and its variability in the region, as well as a description of the current socio-economic conditions and trends, which can be disseminated and discussed with key stakeholders.

Successful examples of coping strategies used in the past, or examples with a useful learning point can also be presented to the stakeholder group. Such discussions can provoke conflicts

between stakeholders. The project team must be aware that is not the objective of the APF to solve such conflicts, but to reach consensus on the issues where there is convergence or common ground (Box 2-4). At this point, the priority areas of concern, as well as the coping strategies adopted in the past, should be identified. An agreed assessment can then be elaborated, including the strategies currently accepted as successful. This information can be acquired through meetings, focus groups or workshops, where a number of different techniques (e.g., diagrams, tables, flow charts) are used to obtain information. Information about "conceptual models", which can be used at this stage, is given in TP4. Examples of how to engage stakeholders at a community level to obtain this information can be found in several case studies (Box 2-5). The team will want to identify those techniques that are appropriate to their region.

Access to and presentation of information is an important part of levelling out power differences between the stakeholders and with the project team. This can be difficult, as some may be reluctant to present their work or ideas in a manner they perceive to be an oversimplification of reality, while other stakeholders may feel alienated and disengage from the process if information is presented in a manner that is at too complex a level or relies on the use of jargon. A local-level process may need to be preceded by an awareness-raising campaign in order to engage people and give them a clearer understanding of what may happen and how it might affect them or the group that they represent.

As outlined in the Nigeria case study (Box 2-5), historical climate data also needs to be obtained for this Component of the APF (e.g., climatic variables, frequency or intensity of extreme events and documentation on the immediate impacts). Stakeholders can document the measures or strategies they use or have used in the past to cope with those events. This provides a collective understanding of how the various social, economic and environmental systems might behave under different climatic conditions (see TP4, Figure 4-2 for a schematic overview).

*Table 2-2:* Potential Adaptation Policy Framework stakeholders (adapted from Aguilar, Y., 2001).

| Stakeholders | Interests and Roles |
|---|---|
| Global Environmental Facility (GEF) | • Support capacity development for adaptation where this is a national priority<br>• Support adaptation projects agreed under the UNFCCC, such as Second National Communications and National Adaptation Programmes of Action |
| National government and ministries (e.g., agriculture, health, environment, education); early warning systems and disaster prevention institutions | • Honour international agreements and participate in international negotiations on regional programmes<br>• Implement sectoral policies, programmes and plans<br>• Improve local human development<br>• Build capacity and develop effective mechanisms to solve local problems<br>• Reduce the risk of local, climate-related damage |
| Local governments | • Solve local problems<br>• Develop local capacity<br>• Finance local plans and programmes<br>• Strengthen local institutions<br>• Prevent local climate damage and disasters |
| National/regional research centres and universities | • Contribute to solving national and regional climate problems affecting vulnerable human systems and ecosystems<br>• Build permanent national and regional capacity for addressing climate change<br>• Develop national and regional approaches to address climate change with a developing country perspective |
| Local environmental/ development NGOs | • Facilitate the organisation of local people and identify action to fulfil local needs<br>• Finance local development programmes and projects<br>• Develop capacity (e.g., technical, financial, human, institutional)<br>• Strengthen local institutions |
| Local communities/people affected by climate risks and damages | • Improve or preserve health, education and housing<br>• Improve or preserve land and aquatic productivity<br>• Decrease local vulnerability to climatic risks<br>• Improve or preserve adaptive capacity for coping with climatic risks |

Once the basic information has been collected and summarised, the links may be identified between climate and the chosen regions and/or sectors in relation to the socio-economic situation and the current state of vulnerability. A report containing a summary of the stakeholder discussions and this initial analysis can be presented back to all the stakeholders who have been involved in the process up to this stage to enable them to check that it is a fair account. Indicators and models that relate climate events, the socio-economic context, and the impacts of climatic hazards can then be identified, tested and agreed either using data in the report or with the stakeholders themselves. These can then be used to evaluate future vulnerability.

### 2.6.3. Component 3: Assessing future climate risks

*Who is involved?*

Essentially, the same stakeholders engaged in Component 2 will be involved in Component 3 – stakeholders involved in the pol-icy-making process and in decision-making in the relevant sector, and stakeholders that have been involved in developing scenarios of the possible climatic and socio-economic futures.

*Tasks in Component 3*

Adaptation projects that undertake Component 3 should, at this stage, have a brief but clear description of climate change projections, the socio-economic future scenarios related to these projections and a brief review of previous impact studies (e.g., done by the project team in Component 2). Stakeholders involved in the policy-making process and in decision making in the relevant sector (Table 2-1) will decide what planning horizons to work toward for the chosen region/sectors (TP5).

Much adaptation in the developing world relies on people's previous experience in dealing with climate-related risks. Their perceptions of the risks they encounter currently, and how they

**Box 2-5: Using rapid rural appraisal techniques to elicit information from stakeholders**
*Jos Plateau, Nigeria, Environmental Resources Development Programme*

The objective of this study was to identify viable projects to address resource problems faced by people in the tin-mining region of Nigeria's Jos Plateau. Researchers focused on two communities – Marit and Wereng. Identifying priority projects required reliable, yet quick and cost-effective, appraisals to be performed by researchers in collaboration with community residents, members of the relevant departments of Jos University and representatives of local government and non-governmental offices.

In the past, rapid appraisals had been criticised for only studying areas that were easily accessible, for focusing exclusively on the elite or affluent community members, and for scheduling according to needs of researchers rather than the needs of the local communities. Researchers had also failed to recognise the value of indigenous knowledge and did not report back to the communities on what they had learned, or how the information would be used.

To avoid these biases, the study team used the Rapid Rural Appraisal (RRA) approach, which incorporates the following concepts:

**Appropriate precision** – gathering information at a sufficient level of accuracy. If you need monthly rainfall information, do not collect daily data.

**Optimal ignorance** – understand what you don't need to know and don't waste time getting it.

**Value of indigenous knowledge** – local people can have important information to share, and should also be informed of the findings of studies.

**Triangulation/Iteration** – ensure that you are getting a realistic picture by comparing the information from one source with that from other sources.

**Flexibility** – this turned out to be a key concept for this study, as logistical problems shifted the timeframe considerably.

**Interactive teamwork** – a small team with mixed skills, each member assigned a specific role.

The study areas were identified using a Rapid Rural Reconnaissance process (Chambers, 1983). In this process, the local people identified the most vulnerable areas. This is important when secondary data sources (maps, reports, etc.) are of poor quality or out-of-date.

**Data collection** – The team used a number of techniques to create a history of the communities: past events, how they had affected the community, and effective responses. **Qualitative methods**: in-depth interviews; informal, spontaneous conversational interviews; semi-structured interviews (topics were pre-selected, but not the actual questions) and standardised, open-ended interviews (structured questions). **Diagram techniques**: participatory mapping of the community; transect walks through agricultural zones; Venn/Chapatti diagrams of organisational structures. **Trend analysis**: daily activity charts (chart people's locations throughout the day); seasonal and annual calendars.

Having synthesised the RRA data, the team – together with the community – identified the key issues, grouped and prioritised them. The Marit team decided to take a multi-purpose approach and identify projects that could involve more than one key issue at the same time. They came up with 22 possible projects, and reduced these to nine "best bet" projects. The Wereng team undertook a similar project identification process. To assess project viability, the Wereng team used the following criteria: productivity, sustainability, stability, equity, cost, time to benefit, social, technical and institutional feasibility.

*Conclusions*

Considerable, but perhaps not unusual, logistical problems were encountered during the study (e.g., vehicular failure, inadequate catering facilities, lack of timekeeping). Many of the lessons learned related to how to involve external agencies in rural development. Overall, team members felt that the objectives of the study were satisfactorily achieved. One issue that became apparent during the process was the absolute necessity for follow-up, including training, to institutionalise the lessons learned, and project identification, to ensure that there would be action on the identified projects.

*Source:* Presentation prepared by Anthony Nyong for the workshop "Climate Change, Vulnerability and Adaptation: AIACC Development Workshop", Third World Academy of Sciences, Trieste, Italy, 3-14 June, 2002; and interim workshop reports.

view these changing in the future, should thus be included in the design of strategies to cope with future climate change. Examples of how this could be done using a stakeholder-driven approach are given in TP4 and TP5 (also Jones, 2000; Hulme and Brown, 1998).

Participatory scenario building, simulation, role play, visioning and back-casting are techniques that can be used with stakeholders to construct possible futures resulting from the combination of possible "coping ranges" and possible future "climate change". (Descriptions of these techniques are given in Annex A.2.2.). This kind of analysis can be used to explore questions such as: What if the climate changes but the coping range does not? What if the predicted climatic changes are to be generally positive, but the socio-economic projections suggest that the coping ranges will decrease? Because both of these factors change with time, there are many more dynamic situations that can be investigated.

Future risks can also be evaluated using impact thresholds (TP4). This concept suggests that certain thresholds can be identified in a system – thresholds that, if crossed, will lead to marked deterioration in the resilience of the system. These thresholds can be established using models, as well as the knowledge and experience of stakeholders, and their perception of possible futures.

The analysis of how to recover from future climatic (or socio-economic) shocks that might weaken the capacity of a system to adjust involves significant uncertainties. Planning and policy horizons are crucial for this analysis (TP5). Groups responsible for planning and policy processes with long time horizons will need to be able to take potential climate change impacts into account. As such, they may represent an important group of stakeholders that should be involved in this Component of the APF. For example, stakeholders involved in dam construction, with a time horizon of more than 50 years, and in national park management with an even longer horizon will benefit greatly from the availability of information on future climate vulnerability and risk. Similarly, international negotiators for transboundary water use might need to know the long-term future scenarios for that resource. In other sectors the planning horizons may be much shorter, and it may thus be harder to persuade relevant stakeholders to make provisions for adaptation. In these cases, examples of climate variability impacts in the past may be useful.

The project team will likely choose to synthesise stakeholder input on the possible climatic and socio-economic futures. These syntheses can be disseminated, with an executive summary, to local or regional policy makers. Strategies to raise public awareness of these possible futures and to influence policy makers to include these results in their agendas should be discussed and agreed by the team.

The case study below (Box 2-6) shows how farmers in Mali used a participatory approach to plan for future changes to make the best use of scarce resources – in this case, by identifying methods for improving soil fertility.

### 2.6.4. Component 4: Formulating an adaptation strategy

*Who is involved?*

At this stage, all stakeholders have a role to play, particularly local, regional and national policy makers.

*Tasks for Component 4*

At this stage, stakeholders will have determined the scope of the issues of interest and identified the links between climate and the sector or region under consideration. They may have considered the future climate and socio-economic scenarios and discussed the implication of these for the sector or region. Stakeholders may undertake a cost-benefit analysis, or other evaluation and prioritisation processes, for the adaptation measures suggested to assess the feasibility of implementing such measures (TP8).

Together, the project team and the stakeholders can initiate a process for evaluating the viability of the proposed adaptation strategies and identifying key areas for further action. Policy makers play a key role in this step. Proceedings of workshops, technical reports and a summary for policy-makers can be disseminated and used as a guide to the next stage of the adaptation process.

### 2.6.5. Component 5: Continuing the adaptation process

*Who is involved?*

All stakeholders, including the range of policy makers.

*Tasks for Component 5*

The aim of this task is to sustain the adaptation process, including the selection of appropriate adaptation mechanisms (TP9). The national and/or regional meetings described in Activity 4 should have resulted in an in-depth evaluation of the results and the identification of a list of priority areas for action to reduce vulnerability.

In some countries the adaptation policies designed during the APF process might not influence immediately the policy-making processes, or even may not be included in the national or regional agendas. However, those goals can be achieved on the long run, if this process is sustained through the stakeholders and if they are able to replicate the process in other sectors or regions.

Activity 5 is the point at which the project team and stakeholders start implementing an action plan to address these priority areas, begin crafting realistic next steps to achieve these goals, and determine how the results can be included in existing plans and budgets. This can be done in a formalised way, as outlined in Table 2-3.

Other actions that could be considered include: increasing

---

**Box 2-6. Participatory approaches to plan future changes: A case study from Mali**

A participatory action research process was developed by the Malian Farming Systems Research team to assist farmers in southern Mali to improve their soil fertility management practices. As more land is being brought under cultivation, the traditional practice of allowing land to lie fallow to restore soil fertility is becoming increasingly rare, leading to widespread depletion of organic matter and nutrient reserves of the soil. As there are a variety of farming and soil fertility management systems in Mali, solutions for an "average" farmer and an "average" field would not be sufficient.

A collaborative learning and action approach was used, which enabled the farmers to play an active role in finding solutions. The Participatory Action Research (PAR) process had been developed by the Farming Systems Research team (Equipe Systèmes de Production et Gestion des Ressources Naturelles) of the Malian Agricultural Research Institute (IER: Institut d'Economie Rurale), with the aim of assisting farmers to improve their soil fertility management practices. The PAR process comprises four phases: (i) diagnosis/analysis, (ii) planning, (iii) implementation, and (iv) evaluation. After the diagnosis phase, the planning, implementation, and evaluation phases are repeated on a yearly basis, in a continuous active learning cycle.

The first element of the diagnosis stage is to ask the participants to list the criteria that they feel reflect the diversity of soil fertility management strategies. The participants were separated into groups of older men, women and younger men in order to show the different perspectives these groups have on the issue. The criteria were divided into two types – indicators that refer to "proper" soil fertility management, and socio-economic characteristics of the households that might influence soil fertility management. After this, all the farming households in the village were classified as "good", "average" or "poor", according to their ability to manage soil fertility. Five test farmers from each group were then asked to participate in the remaining PAR process. The villagers themselves, in consultation with the researchers, selected farmers on the basis of their interest in learning and their capacity to exchange information with their peers.

Farm level Resource Flow Models (RFMs) were used to analyse the soil fertility strategies. On large sheets of packing paper, test farmers drew the different elements of their farms, such as grain stores, fields, animal pens, and compost piles. For each field, both present and preceding crops were noted. Afterwards, farmers drew arrows to represent resource flows entering and leaving the farm, as well as flows between fields and other farm components. Quantities were given in units used locally, e.g., cart loads, baskets. The arrows were labelled with approximate quantities. By visualising these flows and how they were managed, the farmers were able to discuss the present situation and to identify any improvements they could make with scarce resources. The RFMs also became a means of communicating with other farmers. The next stage was the development of a planning map. The test farmers were asked to visualise their plans for the next year. Improvements to be made were marked onto a new map of the farm, with estimated resource uses added, and other flows marked on as before. These were then presented to other farmers at a village meeting where the technical implications were discussed. As the work was done, the actual resource flows were marked on to the planning RFMs, and discrepancies between what was planned and the final usage were discussed.

The RFMs' advantage over formal surveys is that the flows are visualised, allowing more reliable and complete data collection; omissions or mistakes are easier to spot. RFMs are context-specific and easily understood. It was shown that the RFMs used by the farmers allow for the collection of information that can be successfully transformed into management performance indicators, soil nutrient flows and partial balances. This process improves both the farmers' and the researchers' understanding and knowledge, and creates a common ground for creative interaction between researchers and farmers that can lead to finding ways to use the scarce resources more efficiently.

*Source:* Defoer, Toon (2002) "Methodology on the Move: Case studies from Mali and Kenya on methodology development for improved soil fertility management". In *Agricultural Systems Special Issue: Deepening the Basis of Rural Resource Management*. Gujit, I., J.A. Berdegué & M. Loevinsohn (Co-ordinating Editors) and Hall, F. (Supporting Editor). A collaboration of ISNAR, RIMISP, IIED, ISG, CIRAD-TERA, INTA, ECOFORÇA with the aid of grants from the European Commission and the International Development Research Centre, Canada.

---

farmer access to micro-insurance schemes, developing indigenous seed banks or providing access to agricultural machinery through co-operative structures. For each of those or other next-step actions, the questions in the top row of Table 2-3 must be thought through.

At the action planning stage, the project team may wish to scale back its facilitation and guidance role. If the process has managed to build sufficient capacity among the stakeholders, they, or a network of them, can step in to undertake the roles formerly played by the team. If this handover is successful, the responsibility for carrying out the action plan is taken on by these stakeholder groups and an "adaptation community" is essentially formed. Alternatively, the project team can continue to play a mentoring role for some time before the stakeholder groups feel confident enough to take the lead. In any

*Table 2-3:* Examples of next action steps

| Example action and actors | Who can help us? | Who may be resistant? | What resources do we need? (time, money, skills) | Where can we get support for resources not currently available? | Who will take a lead on the prioritised action? |
|---|---|---|---|---|---|
| Increasing farmer access to markets through support of rural road building schemes | Ministry of Rural Affairs, local businesses, Chamber of Commerce, farmer co-operatives | Ministry of Transport, Ministry of Finance, environmental groups | $1,000,000 in first 10 years would provide many jobs locally for low skilled workers | Local businesses, NGOs, multinational corporations with an interest in the area (cash crops) | Ministry of Rural Affairs<br><br>Regional or local governmental agencies<br><br>Farmers' organisations |

case, the project team and stakeholders will both have a role in monitoring and evaluating the performance of the adaptation measures and the next steps of the adaptation.

## 2.7. Conclusions

In synthesis, there is no "one size fits all" solution to engaging stakeholders for enhancing adaptive capacity. However, a few key points can help guide the process:

- Why engage stakeholders? Because they have knowledge and ideas that are relevant to the process, decisions made will affect them, and they are more likely to consent to such decisions if they feel they have contributed to making them.
- Decide what level of engagement is appropriate (Fig. 2-2: Ladder of participation) and which are the key stakeholders related to each APF Component.
- Be clear about the aims and objectives of the engagement, how it should operate and what is expected of participants.
- Encourage and support those who are unfamiliar with voicing ideas and information.
- Use techniques that are appropriate for the group involved and type of information required.
- Decide which techniques are appropriate and feasible to feed back useful information and results to the stakeholders involved.

Stakeholder involvement will be developed in a context where political differences, inequalities or conflicts might come up. The project team should find ways to build agreements and to resolve such issues where possible.

Every situation is different. Having decided the kind of information it requires, the team then needs to decide who should provide it, and the most appropriate technique to obtain it, cross-checking, if necessary, with another technique (triangulation). Annex A.2.1 suggests sources of information that may be

useful in designing the team's approach. A variety of techniques related to the participatory approach are described in Annex A.2.2. Some require planning and others take only a few minutes to complete. Some are quite formal and others less so. In the end, people will engage more if the process is enjoyable.

## References

**Aguilar**, Yvette (2001). Personal communication.

**Chambers**, Robert (1983). Whose reality counts? *Putting the Last First*, ITDG Publishing.

**Hemmati**, M. (2002). Multi-stakeholder Processes for Governance and Sustainability, London: Earthscan,.

**Hulme**. M., Brown, O. (1998). Portraying climate scenario uncertainties in relation to tolerable regional climate change. *Clim. Res.* **10**: 1-14.

**Jones**, R. (2000). Analysing the risk of climate change using an irrigation demand model. *Clim Res.* **14**: 89-100.

**Kaner**, S., Lind, L.,Toldi, C.,Fisk, S. and Berger, D. (1998). *Facilitator's Guide to Participatory Decision-Making*, Philadelphia: New Society Publishers.

**Nyong**, A. (2002). Presentation at AIACC meeting in Trieste, Italy, (2002). http://www.start.org/Projects/AIACC_Project/meetings/meetings.html and unpublished Interim Project Reports.

**Pretty**, J., 1994, *Typology of Community Participation*, quoted in Bass, S., Dalal-Clayton, B. and Pretty, J., 1995, *Participation in Strategies for Sustainable Development*, London, Environmental Planning Group, International Institute for Environment and Development.

**Thomas**, H., 1996, Public participation in planning in: Tewdwr-Jones, M., ed., *British Planning Policy in Transition; Planning in the 1990s*, London, UCL Press.

**Twigg**, J., 1999, *The Age of Accountability: Community Involvement in Disaster Reduction*. Paper presented at the Sixth Annual Conference of The International Emergency Management Society (TIEMS '99), on the theme of "Contingencies, Emergency, Crisis and Disaster Management: Defining the Agenda for the Third Millennium", Delft, The Netherlands, 8-11 June 1999.

## Additional Bibliography

**Conde**, C., R. Ferrer, C. Gay, V. Magaña, J.L: Perez, T. Morales, S. Orozco. (1999). El Niño y la agricultura. In: *Los impactos del Niño en México*. Magaña, V. (ed.). UNAM. CONACYT. SG. IAI. 103-135. http://ccaunam.atmosfcu.unam.mx/ cambio/nino.htm.

**Conde**, C., Eakin, H. (2003). Adaptation to climatic variability and change in

Tlaxcala, Mexico. (2003).Chapter in *Climate Change, Adaptive Capacity and Development*. J. Smith, R. Klein, S- Huq, eds., London: Imperial College Press, pp. 241-259.

**Defoer**, T, De Groote, H., Hillhorst, T., Kante, S. and Budelman, A. (1998). Participatory action research and quantitative analysis for nutrient management in southern Mali: a fruitful marriage? *Agricultural Ecosystems and Environment*, **71**, 215-228.

**Gay**, C. (ed.). (2000). *México: Una visión hacia el siglo XXI. El cambio climático en México. Resultados de los estudios de vulnerabilidad de los países coordinados por el INE con el apoyo del U.S. Country Studies Program*. SEMARNAP, UNAM, USCSP.220 pp. http://ccaunam.atmosfcu.unam.mx/cambio/

**IPCC**, WGII. (2001). Summary for Policy Makers, A Report of Working Group II of the Intergovernmental Panel of Climate Change.

**Jones**, P.J.S., Burgess, J. and Bhattachary, D. (2001). An evaluation of approaches for promoting relevant authority and stakeholder participation in European marine sites in the United Kingdom, English Nature (United Kingdom Marine SACs Project).

**Magaña**, V. (ed.). (1999). *Los impactos del Niño en México*. UNAM. CONACYT. SG. IAI. 229.
http://ccaunam.atmosfcu.unam.mx/cambio/nino.htm.

# ANNEXES

### Annex A.2.1. Sources of information about different methods of participatory approaches

#### Books

*Participatory Workshops: A Source Book of 21 Sets of Ideas and Activities*
Robert Chambers (2002), Earthscan, ISBN 1 185383 862 4 (paperback). Available from http://www.earthscan.co.uk. This text is a good sourcebook of information on how to run workshops including lots of practical advice and common mistakes.

*Participatory Learning and Action: A Trainer's Guide*
Jules N. Pretty, Irene Guijt, Ian Scoones and John Thompson (1995) International Institute for Environment and Development (IIED). ISBN 1 8998 2500 2. Available from: http://www.earthprint.com. This guide is a valuable collection of advice, tips and methods for participatory approaches. The focus is mostly on participatory rural appraisal but much would also be relevant for APF workshops.

*Enhancing Ownership and Sustainability: A Resource Book on Participation*
International Fund for Agricultural Development (IFAD), Coalition for Agrarian Reform and Rural Development (ANGOC) and International Institute of Rural Reconstruction (IIRR) (2001). ISBN 1 930261 004. Email: publications@iirr.org. This publication is a collection of short reviews of participatory approaches and experience.

*Facilitator's Guide to Participatory Decision-Making*
Sam Kaner with Lenny Lind, Catherine Toldi, Sarah Fisk and Duane Berger (1996), New Society Publishers. Available from http://www.newsociety.com/bookid/3705. This is a useful introduction to how to build consensus and make sustainable agreements with groups. Also gives advice on how to handle difficult group dynamics and individuals.

*Power, Process and Participation: Tools for Change*
Rachel Slocum, Lori Wischhart, Dianne Rocheleau, Barbara Thomas-Slater, eds. (1995), London, Intermediate Technology Publishers. This book talks about the history of participatory processes, how to apply them and some methods.

*Embracing Participation in Development: Wisdom from the Field*
Meera Kaul Shah, Sarah Dengan Kambou and Barbara Monahan (1999). Care-US. Available online from: http://www.careinternational.org.uk/resource_centre/civilsociety/embracing_participation_in_development.pdf. This is a field guide to participatory tools and techniques. It contains a lot of insight from experience mainly based on the Participatory Learning and Action (PLA) approach.

*Developing Technology with Farmers: A Trainer, Guide to Participatory Learning*
Laurens van Veldhuizen, Ann Waters-Bayer and Henk de Zeeuw (1997). London: Zed Books. Available from: http://zedbooks.co.uk. This book is focused on farmers, but much of the material is more widely relevant. It is designed to stimulate active learning.

#### Resources on the web

PRAXIS, Institute for Participatory Practices. http://www.praxisindia.org This site has a collection of guidelines, examples, tips for trainers and experience gathered at a workshop.

Participation Resource Centre, Institute of Development Studies, University of Sussex. http://www.ids.ac.uk/ids/particip/index.html. This site holds over 4000 documents. A limited document delivery service is available. Email: participation@ids.ac.uk.

#### Sources of information about running stakeholder engagement processes

*Multi-stakeholder processes for governance and sustainability*
Minu Hemmati (2002). London: Earthscan, ISBN 1 85383 870 5. http://www.earthscan.co.uk. A practical guide that explains how multi-stakeholder processes can be organised and implemented in order to solve complex issues related to sustainable development.

*The Power of Participation*
Institute of Development Studies Policy Briefing Issue No. 7 (1996). Available online at http://www.ids.ac.uk/ids/bookshop/briefs/brief7.html. This publication is a summary of Participatory Rural Appraisal: what it is, how to do it and some of the problems.

The new orthodoxy and old truths: participation, empowerment and other buzz words. Stirrat, R.L. (1996). In *Assessing Participation: A debate from South Asia*, Bastian, S., Bastian, N, eds., New Delhi: Duryog Nivaran/Konark Publishers. This publication provides a useful critique of participation.

### Annex A.2.2. Tool box of exercises for running a participatory workshop

The tools described below are examples of techniques that can be used at different stages of a participatory workshop. This is by no means an exhaustive list. (For more ideas and information about techniques, see the sources list in Annex A.2.1). Participatory processes are numerous and flexible. If one method does not appear to be working, you can try another.

Adapting existing methods or making up your own exercises will make the process more appropriate to your own set of circumstances.

### Techniques for the start

#### Paired interviews

This is useful for finding out what the participants' expectations are. It can be a useful way to raise questions and uncertainties or address misconceptions.

Participants are split into pairs and each is asked to interview their partner. Questions focus on their background, reasons for attending and what they hope to achieve by participating. After five minutes they report back to the whole group. If it is a large group, feedback can be restricted, e.g., to saying "Name two things you hope to achieve in this process". If group consent has been given, these can be recorded, and the record can then be referred to in the evaluation of the effectiveness of the process.

*Source:* Participatory Learning and Action: A Trainers Guide, Jules N. Pretty, Irene Guijt, Ian Scoones and John Thompson, International Institute for Environment and Development (IIED) (1995). ISBN 1 8998 2500 2. Available from: http://www.earthprint.com.

#### Hopes and fears

This is a good way to step back from the content of the process and allow participants to share any worries or misconceptions they might have brought with them.

Participants are divided into small groups of four to six people and each group is given a piece of paper. Each group is asked to write down any fears or concerns they may have had before coming. This should be done quickly (five minutes). Each group is then asked to report back to the larger group. The facilitator then has the opportunity to empathise and reassure the participants, and give any relevant information about the process that may previously have been unclear. The facilitator can then ask the question "What can I do to reduce your concerns?" This may lead to a discussion of ground rules.

*Source:* Newstrom, J.W. and Scannell, E.E. (1980). Games Trainers Play, United States, McGraw-Hill Inc.

#### Expectations and ground rules

This helps to determine what participants do and do not want from the process in terms of the content of the session, the format of the meeting and the practical details. It can provide insight into how much consensus there is.

Each participant is given a number of small pieces of paper. On each piece they are asked to write one thing that they do or do not want from the session in terms of content, format, etc. These are then grouped and fed back to the group. They can form the basis of ground rules. It also gives the facilitator an opportunity to address expectations that may not be met.

*Source: Participatory Learning and Action: A Trainers Guide,* Jules N. Pretty, Irene Guijt, Ian Scoones and John Thompson, International Institute for Environment and Development (IIED) (1995). ISBN 1 8998 2500 2. Available from: http://www.earthprint.com.

#### Agenda setting

If the agenda is to meet the needs of the participants, there has to be a certain amount of flexibility in the planning process. At the workshop, participants could be asked to write on a piece of card one item they would like to be addressed. The cards could then be sorted and an agenda drawn up to cover these items. The group could prioritise the items: each participant is given a number (three to five) of sticky dots (or crosses made with a pen) and is asked to mark those items they perceive to be the most important.

### Techniques to promote discussion, scope issues and identify gaps

#### Buzz groups

This is a method for putting aside time to think. It allows participants to work through their emerging thoughts before presenting them to the whole group. Buzz groups can be used in many situations – e.g., after a presentation of new material and before questions are asked from the audience. A buzz group can enable participants to think through any parts they were unclear about in the presentation or would like further information on. Having had this opportunity, they will then be more ready to contribute questions.

Participants are divided into pairs and the facilitator proposes a topic for discussion. One starts as the listener and the other is the thinker. At half time the roles reverse. During the thinking turn each person is encouraged to think out loud. They do not have to make sense; this is an opportunity to collect and develop thoughts at one's own pace and in one's own way. The listener says nothing but listens attentively. The roles then swap.

*Source:* Langford, A. (1998). *Designing Productive Meetings and Events: How to Increase Participation and Enjoyment,* South Oxfordshire District Council, Permaculture Academy and South Oxford District Council.

*Brainstorming*

A brainstorm is a quick way to get a group to produce a list of ideas, questions, issues or topics for later discussion. An appointed person records the suggestions. The meaning can be clarified, but the recorder should not comment on, judge or praise the suggestions as they come in. The recorder does not participate in providing suggestions. The participants should be encouraged to think as creatively as possible and not be too concerned about practical realities at this stage. The list can later be sorted and prioritised (see Delphi technique, next).

*Card sorting, Delphi technique*

This is a similar process to brainstorming except that suggestions are recorded on small pieces of card, one suggestion per card. The participants or the facilitator then clusters the cards into themes on the wall or on the floor. Duplicated ideas can be removed. The list can be prioritised if necessary.

*Spider diagrams*

This can be used to both generate ideas and link ideas together into themes. Write the issue of interest – e.g., institutional barriers to adaptation in Peru – in the centre of a large piece of paper. Then write down any interconnected ideas, thoughts, and/or questions, and draw lines between the ones that are linked. Continue until no more can be found. This can either be done in one large group, or by smaller groups that can later compare and contrast their different diagrams.

*Source: Participatory Learning and Action: A Trainers Guide*, Jules N. Pretty, Irene Guijt, Ian Scoones and John Thompson, International Institute for Environment and Development (IIED) (1995). ISBN 1 8998 2500 2. Available from: http://www.earthprint.com.

*Nominal group technique*

This gives participants the opportunity to generate solutions to problems as individuals, and then come to a collective view on priorities. Each participant is asked to write down solutions to a question, e.g., how to encourage the business community to consider climate change impacts. This is done in silence. Participants are then given the opportunity to feed back to the group and the ideas are recorded. Any misunderstandings are clarified and a final list prepared. Participants are asked to prioritise the solutions by marking the five items they consider to be most important with a pen or sticky dot. The result is a set of independent views rather than a group view. Independent thinking is generally more creative, as there is less pressure to conform.

*Source: Oomkes and Thomas (1992). quoted in Participatory Learning and Action: A Trainers Guide*, Jules N. Pretty, Irene

Guijt, Ian Scoones and John Thompson, International Institute for Environment and Development (IIED) (1995). ISBN 1 8998 2500 2. Available from: http://www.earthprint.com

*Carousel*

This is a semi-active technique to get people addressing different problems in a single issue or different aspects of the same problem, e.g., what are the barriers to effective participation for different groups (children, elderly, women, disabled people)? A series of questions or topics (two to five) are posed at different stations in a room or in different rooms. The group is divided into smaller subgroups (the same number as there are stations). Each station has a recorder who notes down responses. After a set time (5-10 minutes) the group moves on to the next station and repeats the process until all the questions have been covered.

*Johari's Window*

This technique explores the difference between professional and local people's knowledge, and helps to highlight inherent prejudices and preconceptions about the value of each.

Participants are asked to fill in the following matrix with examples from their own experience. This can be done on a general level – for professionals and locals – or on a more specific level, for administrators, small businesses versus landless people, small farmers, etc.

|  | They know | They don't know |
|---|---|---|
| We know |  |  |
| We don't know |  |  |

*Sources:* Luft, J, (1970). Introduction to group dynamics, quoted *in Participatory Learning and Action: A Trainers Guide*, Jules N Pretty, Irene Guijt, Ian Scoones and John Thompson. International Institute for Environment and Development (IIED), (1995). ISBN 1 8998 2500 2. Available from: www.earthprint.com *and Chambers*, R. (2002). *Participatory Workshops: A Sourcebook of 21 Sets of Ideas and Activities*, London: Earthscan.

**Techniques for participatory analysis**

*Sources:* Various, see Annex A.2.1

*Maps*

Maps provide a holistic picture of an area; they are useful in discussions of location, distribution, access to resources and vulnerability. Maps can illustrate social, economic or environmental features (or combinations of these) and can be provided for discussion or developed by the participants using paper or other materials such as sand or clay. The discussions that result from developing or using maps indicate the relative importance of the various features on the map for the participants. For example, maps drawn by women of their local community generally differ quite considerably from those drawn by men in the importance placed on the different buildings and facilities.

*Listing and combining*

Similar to the brainstorming and Delphi techniques described above.

*Calendars and timelines*

Calendars organise information in chronological or seasonal order. This helps in recognising patterns that are related to time. This is useful in working out community work patterns.

Timelines show a sequence of activities or changes over time. Their impact on the community can then be investigated by overlaying other trends such as migration from the area, changes in farming practices, etc.

*Ranking and scoring*

Ranking is used for comparison of items based on criteria set by the group. For example, households could be ranked in terms of their wealth or well-being. Scoring can be used to identify strengths and weaknesses of different items so that they may be compared. This could be done by individuals or the group. Scores can be compared with past scores or scores for items from different areas to observe trends. This technique can be used to prioritise adaptation measures (TP8).

*Diagrams*

This tool helps participants to visualise information and how it relates in a system. Diagrams show how different elements interact, and how strong these links are. Venn diagrams show organisational linkages. Flow charts can be used to illustrate flows of information.

**Techniques for evaluation**

*Sources*: *Participatory Learning and Action: A Trainers Guide.* Jules N. Pretty, Irene Guijt, Ian Scoones and John Thompson (1995). International Institute for Environment and Development (IIED). ISBN 1 8998 2500 2. Available from: http://www.earthprint.com; and *Participatory Workshops: A Sourcebook of 21 Sets of Ideas and Activities*. Chambers, R.

(2002), London: Earthscan.

*Smiley sheets*

A simple sheet is given to each participant. One side has a smiley face on it. On this side, participants are asked to write something they like about the process or activity. On the other side, there is a sad face. On this side, participants write something they found difficult about the process or activity, and how they would have done it differently.

*Evaluation wheel*

The group should first decide the criteria to be used for evaluation. These could be based on the expectations discussed at the beginning of the process. There should not be too many criteria (fewer than ten). Each participant is then asked to draw a wheel with the same number of spokes, as there are criteria. The spokes should then be labelled with one criterion each. The spokes represent scales from low or zero in the centre, to high or ten at the edge. Participants are then asked to indicate on the spoke their assessment of the course with respect to each criterion. The dots can then be joined. If done on overhead transparencies, the different evaluations can be compared to give the degree of consensus between individuals.

*Hopes and fears scoring*

Take the hopes and fears given by the participants at the beginning of the process (see *Techniques for the start* section). Turn any negative comments into positive or neutral ones, e.g., "I am worried that I won't have a chance to give my opinions" could become opportunities to speak. A matrix is then drawn up with the hopes and fears listed down the side and five columns to the right of this with a face at the top of each. The expressions on the faces vary from very sad the far left, to very happy at the far right, with a neutral face in the middle. Participants are then asked to indicate with a pen mark or a sticky dot how they feel the different hopes and fears have been dealt with overall.

*Feedback boards*

These boards provide an opportunity for participants to write anonymous comments about the process and ideas for improvements. They can be present throughout the process. In addition to voicing their problem, participants should be encouraged to suggest practical solutions to the difficulties they encounter. Comments can be read back to the group, with ideas for how they might be tackled.

*Representatives*

Ask the participants to suggest one or two representatives. Participants could tell these people any concerns they have and the representatives would then report back to the facilitators. Any changes suggested would then be fed back to the whole group.

*Paired interviews*

See above: *Techniques for the start*

## Other techniques

*Source:* Van Asselt, M.B.A., Mellors, J., Rijkens-Klomp, N., Greeuw, S.C.H., Molendijk, K.G.P., Beers, P.J. and van Notten, P. (2001) *Building Blocks for Participation in Integrated Assessment: A Review of Participatory Methods.* International Centre for Integrative Studies (ICIC) Working Paper: I01 – E003. Langford, A. (1998). *Designing Productive Meetings and Events: How to Increase Participation and Enjoyment,* South Oxfordshire District Council, Permaculture Academy and South Oxford District Council.

*Consensus conferences*

A consensus conference is a public enquiry centred around a group of citizens who are asked to assess a socially controversial topic. These lay people put questions to a panel of experts, discuss the experts' answers, and then negotiate amongst themselves. This results in a consensus statement in the form of a written report for policy-makers and the general public. The report expresses their expectations, concerns and recommendations at the end of the conference.

The lay panel should have no vested interests in the issues but should be chosen to represent different attitudes towards the issue. The group is balanced according to relevant factors such as age, gender, education, occupation and area of residence.

*Focus groups*

A focus group is a planned discussion in a small (four to 12 members) group of stakeholders facilitated by a skilled moderator. It is designed to obtain information about preferences and opinions in a relaxed, non-threatening environment. The topic is introduced and, in the ensuing discussion, group members influence each other by responding to ideas and comments. In focus groups, scientists are not usually involved as full participants and play the role of facilitator or observer.

In one-to-one interviews, it is assumed that individuals know what they feel and that they form ideas in isolation. When a new idea is being tested or the issue is controversial, social scientists have noted that people often need to listen to other opinions before they form their own viewpoint. The opinion of an individual may also shift during the course of a discussion. The focus group thus enables viewpoints that might not have

come forth in individual interviews and allows analysis of what might influence shifts in opinion.

Group members are generally strangers to each other, but all have something in common; this has been shown to make them more likely to communicate freely. Being strangers, they know that they are unlikely to see each other again and are thus less inhibited about sharing their thoughts and opinions.

*Citizen's jury*

Citizen's juries are based on the rationale that, given adequate information and opportunity to discuss an issue, a group of stakeholders can be trusted to make a decision on behalf of their community, even though others may be considered more technically competent. Citizen's juries are most suited to issues where a selection needs to be made from a limited number of choices. The process works better on value questions than on technical issues.

The jury is made up of a number (12-24) of stakeholders (with no special training) who listen to a panel of experts (witnesses) who are called to provide information related to the issue. The stakeholders are chosen at random from a population appropriate to the scale and nature of the problem. Selection of the members of the jury is based upon several characteristics, largely gender, education, age, race, education, geographic location, and attitude toward the question in hand. The jury is supposed to represent a microcosm of the community, including its diverse interests and subgroups. There are some doubts as to whether such a small group can really be representative of the diversity of opinion in the larger community. Does a middle-aged woman represent all middle-aged women? Some think it can only represent the community in a symbolic sense.

A panel chooses experts with no interest (or stake) in the outcome. They represent several points of view, and additional experts can be called by the jurors to clarify points or provide extra information.

*Scenario building*

In scenario analysis, stakeholders create and explore scenarios of the future in order to learn about the external environment and to understand the decision-making behaviour of the organisations involved. This approach enables the exchange and synthesis of ideas and encourages creative thinking. This method is particularly useful for addressing complex issues and uncertain futures, where decision-making is generally based on non-quantifiable factors, and where it is important to establish a dialogue between the key actors in order to plan for the future.

All stakeholders, including decision makers and scientists, will be actively involved in the process. Key issues or questions relevant to the subject are identified. From this, key trends and driving forces can be determined. These may then be prioritised to determine which are the most important or uncertain. These strands can then be fleshed out to create the "story line", from a beginning to an end. Following the initial workshop, there may be a period of reflection where the trends and indicators developed for the different scenarios may be tested for robustness and plausibility.

*Visioning*

Visioning gives people the opportunity and the space to say how they would like things to be in the future, without having to sort out the problems of today. A vision is a statement of how one would like the world to be. Goals are the practical components of visions. For example, one person's vision may be for a car-free society. Their goal might then be reducing their family's car use by 50% by the end of the year. Visioning may sound like dreaming, but holding a well-developed vision of the future helps to give a realistic appraisal of the current situation. Having developed a vision, a process of "back-casting" may then be used to bring the vision back to the present day and, thereby, identify steps that may be taken today to reach the ideal future.

# 3

# Assessing Vulnerability for Climate Adaptation

THOMAS E. DOWNING[1] AND ANAND PATWARDHAN[2]

Contributing Authors

*Richard J. T. Klein[3], Elija Mukhala[4], Linda Stephen[1], Manuel Winograd[5],*
*and Gina Ziervogel[1]*

Reviewers

*Mozaharul Alam[6], Suruchi Bhawal[7], Henk Bosch[8], T. Hyera[9], Roger Jones[10], Ulka*
*Kelkar[7], Maynard Lugenja[9], H. E. Meena[9], Mohan Munasinghe[11], Anthony Nyong[12],*
*Atiq Rahman[6], Samir Safi[13], Juan-Pedro Searle Solar[14], and Barry Smit[15]*

[1] Stockholm Environment Institute Oxford Office, Oxford, United Kingdom

[2] S J Mehta School of Management, Indian Institute of Technology, Powai, Mumbai, India

[3] Potsdam Institute for Climate Impacts Research, Potsdam, Germany

[4] Food and Agriculture Organisation/Southern African Development Community Regional Remote Sensing Project, Harare, Zimbabwe

[5] International Center for Tropical Agriculture, Cali, Colombia

[6] Bangladesh Centre for Advanced Studies, Dhaka, Bangladesh

[7] The Energy and Resources Institute, New Delhi, India

[8] Government Support Group for Energy and Environment, The Hague, The Netherlands

[9] The Centre for Energy, Environment, Science & Technology, Dar Es Salaam, Tanzania

[10] Commonwealth Scientific & Industrial Research Organisation, Atmospheric Research, Aspendale, Australia

[11] Munasinghe Institute for Development, Colombo, Sri Lanka

[12] University of Jos, Jos, Nigeria

[13] Lebanese University, Faculty of Sciences II, Beirut, Lebanon

[14] Comisión Nacional Del Medio Ambiente, Santiago, Chile

[15] University of Guelph, Guelph, Canada

# CONTENTS

## 3.1. Introduction

Adaptation involves the management of risks posed by climate change, including variability. The identification and characterisation of the manner in which human and natural systems are sensitive to climate become key inputs for targeting, formulating and evaluating adaptation policies. With the guidance presented here, users should be equipped to carry out a vulnerability assessment at the appropriate level of detail and rigour. Not every Adaptation Policy Framework (APF) user will need to undertake a vulnerability assessment; those who do will likely be motivated by a specific need to raise awareness of vulnerability, to target adaptation strategies toward key vulnerabilities and to monitor exposure to climatic stresses. These users can tap the guidance outlined here to hone in on key groups, sectors, geographic areas, etc., assess current and future vulnerability, and integrate observations into adaptation planning and policy making.

If we take the example of human health, climate change is likely to affect the distribution and prevalence of infectious disease vectors, which might lead to increased mortality and morbidity from diseases such as malaria and cholera. However, this outcome is dependent on non-climate factors, including environmental controls, public health systems, and the availability and use of drugs and vaccines. A first step in designing effective adaptation strategies would be to clearly establish the importance of climate change, including variability, in terms of the final health outcomes. In this instance, a vulnerability assessment would target those regions most affected by the health impacts of climatic variability, focus adaptation options on effective interventions for the most vulnerable populations, and produce baseline data and indices for monitoring responses.

While a vulnerability assessment (VA) is important for responding to future climate risks (TP5), the assessment process may also help improve the management of current climate risks (TP4). For example, the vulnerability assessment can be used to address the following questions of immediate relevance to policy-makers and development planners: To what extent are the anticipated benefits from existing development projects sensitive to the risk of climate change, including variability? In what way can considerations of future climate risk be incorporated into the design of development projects?

These questions are particularly germane in developing countries that are witnessing the rapid build-up of long-lived civil infrastructure (such as irrigation systems, transportation systems and urban settlements) and in conditions where natural resources are rapidly degrading (such as desertification, water quality and scarcity, and the loss of other environmental services).

Methods of vulnerability assessment have been developed over the past several decades in the fields of natural hazards, food security, poverty analysis, sustainable livelihoods and related

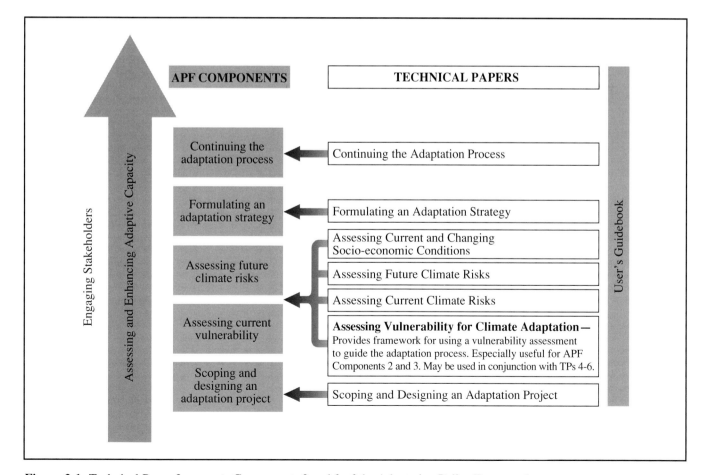

**Figure 3-1:** Technical Paper 3 supports Components 2 and 3 of the Adaptation Policy Framework

areas. These approaches – each with its own nuances – provide best practices for use in studies of climate change vulnerability and adaptation.

This Technical Paper (TP) presents a structured approach to climate change vulnerability assessment; the emphasis is on the activities and techniques that a technical team could readily implement. The paper recommends five activities and suggests methods that are suitable for different levels of analysis. The five activities link a conceptual framework of vulnerability to the identification of vulnerable conditions, analytical tools and stakeholders. The annexes give further examples and background.

## 3.2.    Relationship with the Adaptation Policy Framework as a whole

An APF vulnerability study can include analyses of current and future climate risks, and socio-economic conditions and prospects, to varying and appropriate levels of detail. Depending upon the choices made in project design (Component 1) regarding adaptation priorities and assessment methods, the guidance in this paper may be used in conjunction with the guidance in TPs 4, 5 and 6. Specifically, elements of socio-economic conditions and prospects (TP6) can be incor-

porated in the vulnerability assessment; the vulnerability assessment can in turn be used to characterise present (TP4) and future risks (TP5). Completion of the APF Components 2 and 3 provides the basis for targeting and formulating robust and coherent adaptation strategies, policies and measures (TP8), that can be implemented and continued (TP9). In this TP, readers will find an overview of the vulnerability-based approach to an adaptation project, and ways in which this approach can be integrated with others (see TP1, Sections 1.3 and 1.4.4 for an overview of the four major approaches).

The vulnerability assessment is broken down into five activities with close links to the APF Components (Figure 3-1) and the tasks suggested in the User's Guidebook (Figure 3-2). The first activity matches the overall scoping of the project (TP1). The questions described below should be considered in Component 1 of the APF (TP1), where the project team scopes and designs an adaptation project, including reviewing existing projects and analyses, planning the approach to be taken, and planning and using stakeholder input. The vulnerability assessment has implications for each of these tasks. The remainder of the activities focus on APF Components 2 and 3.

This structured approach[1] begins with a qualitative understanding of the conditions of vulnerability, (see Annex A.3.3 for the sequence of activities) and progresses towards the

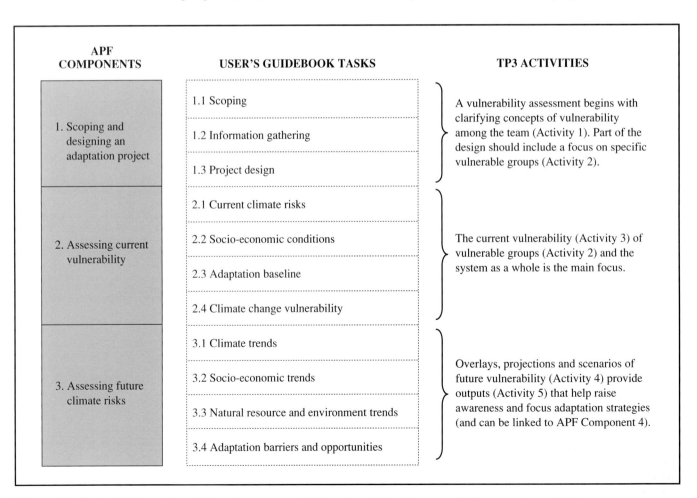

**Figure 3-2:** Technical Paper 3 activities relate to several Adaptation Policy Framework Components and tasks

development of quantitative indicators. (See Annexes A.3.5 and 3.6 for an illustration of different quantitative approaches). Links to formal models (such as environmental impact models) can be readily integrated into a vulnerability assessment, depending on the user's needs and capabilities.

## 3.3. Key concepts: About vulnerability

Vulnerability varies widely across communities, sectors and regions. This diversity of the "real world" is the starting place for a vulnerability assessment. International comparisons of vulnerability tend to focus on national indicators, e.g., to group less developed countries or to compare progress in human development among countries with similar economic conditions. At a national level, vulnerability assessments contribute to setting development priorities and monitoring progress. Sectoral assessments provide more detail and targets for strategic development plans. At a local or community level, vulnerable groups can be identified and coping strategies implemented, often employing participatory methods (TP2).

Although vulnerability assessments are often carried out at a particular scale, there are significant cross-scale interactions, due to the interconnectedness of economic and climate systems. For example, drought might affect a farmer's agricultural yield due to lack of rainfall and pests, reduced water in a major river basin allocated for irrigation, or changes in world prices driven by impacts in one of the "bread baskets". At the same time, the selected priority system for an adaptation project will be affected by linkages to other sectors.

The literature on vulnerability has grown enormously over the past few years.[2] Key articles from a development and sectoral perspective include Bohle and Watts (1993) and Chambers (1989). Extensions related to natural hazards are Blaikie et al. (1994), Clark et al. (1998), and Stephen and Downing (2001). Climate change explorations include Adger and Kelly (1999), Bohle et al. (1994), Downing et al. (2001), Handmer et al. (1999), Kasperson et al. (2002), and Leichenko and O'Brien (2002).

Vulnerability has no universally accepted definition (see Annex A.3.1 and the Glossary). The literature on risk, hazards, poverty and development is concerned with underdevelopment and exposure to climatic variability – among other perturbations and threats. In this view, vulnerability is systemic, and a consequence of the state of development. It is often manifested in some aspect of the human condition, such as under-nourishment, poverty or lack of shelter. Final outcomes are determined by a combination of climate hazards and system vulnerability. In this approach, the focus is on coping or adaptive capacity as the means for vulnerability reduction.

**Hazards literature:**
Risk = Hazard (climate) x Vulnerability (exposure)

The Intergovernmental Panel on Climate Change (IPCC) tuned its definition of vulnerability specifically to climate change.[3] Using this lens, vulnerability is seen as the residual impacts of climate change after adaptation measures have been implemented. The uncertainty surrounding climate change, impacts scenarios and adaptive processes is such that very little can be said with confidence about vulnerability to long-term climate change.

**Climate change (IPCC):**
Vulnerability = Risk (predicted adverse climate impacts) – Adaptation

Regardless of which framing is adopted, it is important to ensure that the choice is made explicit, and that the analysts and stakeholders are clear about the interpretation of the different terms. The formal methods proposed below require a tractable analytical definition.

Vulnerability by default corresponds to the hazards tradition, focusing on exposure and sensitivity to adverse consequences. In this TP, vulnerability corresponds to the present conditions (i.e., the vulnerability baseline defined by socio-economic conditions). However, it can be extended to the future as a reference scenario of socio-economic vulnerability. Where the authors refer to future vulnerability related to climate change, the term climate change vulnerability is used, corresponding to the IPCC definition. This requires explicit additions to the default term relating to the future (with climate change):

• Climate change is explicitly forecast
• Socio-economic exposure is forecast: who is vulnerable, why, etc.
• Adaptation to prospective impacts of climate change is included (although there is little agreement as to what sort of adaptation should be considered – whether autonomous, most likely, potential, maladaptive, etc.)

The result can be a plausibly integrated scenario of future vulnerability. Users should be clear that such scenarios cannot be validated or considered forecasts; they are contingent upon too many scientific and socio-economic uncertainties, as well as the iterative nature of human decision making.

## 3.4. Guidance for assessing current and future vulnerability

The five activities outlined below enable the user to prepare a vulnerability assessment that can serve as a stand-alone indication of

---

[1] The suggested approach must be considered with some flexibility. Depending on the current status of climate change studies in each country and the specific needs (target group, sector, etc.), the sequence of the different tasks can be interchanged or carried out simultaneously.

[2] Bibliographies, key publications, briefing notes and discussion forums are part of the Vulnerability Network, led by the SEI, IIED, PIK, START and others. The network promotes research and policy on vulnerability/adaptation science: See www.vulnerabilitynet.org

[3] From the glossary of the Third Assessment Report of the IPCC, see www.ipcc.ch/pub/shrgloss.pdf

current vulnerability, or can be integrated with climate change forecasts for an assessment of future climate vulnerability.

### 3.4.1.  *Activity 1: Structuring the vulnerability assessment: Definitions, frameworks and objectives*

The first activity of the vulnerability assessment team is to clarify the conceptual framework being used, and the analytical definitions of vulnerability. A shared language will facilitate new insights and help communicate to key stakeholders.[4] (See TP2 for an in-depth discussion of stakeholder engagement.)

In the overall scoping, the team likely reviewed existing regional or national assessments that relate to vulnerability, for instance, national development plans, Poverty Reduction Strategy Papers, environmental sustainability plans and natural hazards assessments. If there is a common approach already in use – for instance, in development planning or mapping hazards – then it makes sense to begin with that framework. It may need to be extended to incorporate climatic risks and climate change.

If existing reviews and plans are not available or suitable, then the team will need to develop its own conceptual and analytical framework (see Annex A.3.2 for a team exercise). Stakeholder-led exercises are valuable at this point. The process of developing a conceptual and analytical framework should clarify differences between disciplines, sectors and stakeholders, and focus

on creating a working approach and practical steps to be taken, rather than a "final" conceptual model. The output of this activity is a core framework for the vulnerability assessment.

The context of the APF study and its objectives are important for determining the set of questions that the assessment is intended to address. This, in turn, has bearing on the operational definition of vulnerability used in the analysis. For example, a vulnerability assessment could be used at two different points in the APF structure. An initial assessment of vulnerability may be used to identify more vulnerable regions and sectors, or hotspots. These might be treated to more intensive assessment, as suggested in TP4. Another use of the vulnerability assessment might be to feed into the design and evaluation of adaptation policies (TP8), including indicators of vulnerability as criteria (TP7).

Table 3-1 illustrates the linkages between the objectives, the context and the set of assessment questions, using the example of adaptation to sea level rise. Identifying a core set of questions for the vulnerability assessment will also help in carrying out the design of the project, as discussed in Component 1 (TP1).

### 3.4.2.  *Activity 2: Identifying vulnerable groups: Exposure and assessment boundaries*

Having identified a working definition of vulnerability and a core set of questions for the assessment, the team needs to iden-

*Table 3-1: Objectives, context and analysis questions in vulnerability assessments*

| Objective | Context | Analysis questions |
|---|---|---|
| Gathering and organising data, identifying data and information needs | Preliminary assessment, often part of related environmental strategy documents | • What are the trends in relative sea level?<br>• What are the geomorphological characteristics of the coastline? |
| Providing estimates of abatement costs and climate damages | Input of local data to inform international estimates of the benefits of greenhouse gas stabilisation | • What are the physical impacts of sea level rise?<br>• What are the market and non-market losses associated with sea level rise? |
| Formulating and evaluating adaptation options | Input to development planning and adaptation policy | • What will be the reduction in losses due to a specific adaptation option (such as creating coastal barriers)?<br>• In what way and to what extent should the design of coastal infrastructure accommodate the possibility of sea level rise? |
| Determining the value of reducing uncertainty through research | Input to research prioritisation | • Which research and observation strategies will have the greatest benefit in reducing uncertainty?<br>• How should observation and monitoring programmes be designed? |
| Allocating resources efficiently for adaptation | Input to policy prioritisation | • Which coastal region is most vulnerable?<br>• Which region or sector can benefit the most from adaptation actions? |

---

[4] To facilitate an international language of vulnerability, a formal notation may be helpful—see Annex A.3.2 for a complete set of notations.

tify who is vulnerable, to what, in what way, and where. The characteristics of the system chosen for the assessment include sectors, stakeholders and institutions, geographical regions and scales, and time periods. These characteristics are identified in APF Component 1, when assessment boundaries are established (TP1, Section 1.4 and Annex A.1.1).

A multi-dimensional baseline of vulnerability includes:
• Target vulnerable groups (TP1, Section 1.4)
• Group socio-economic characteristics and in particular those aspects that lead to their sensitivity to climate hazard (often referred to as exposure) (TP6)
• Natural resources and adaptive resource management (TP6)
• Degree of (present and/or future) climatic risks that affect each vulnerable group
• Institutional processes of planning adaptation strategies and options

The choice of the target of the vulnerability assessment should be a direct response to the objectives and decision context of the exercise. A fundamental issue is whether the target is people, resources, economic activities, or regions.[5] For example, a focus on food security might take as the core analyses the social vulnerability of livelihoods to a range of threats (from climatic, economic and resource changes). But this would need to be placed in an understanding of regional production, exchange and distribution. Or a focus on biodiversity might begin with detailed modelling of ecosystems and species, with a subsequent analysis of the value of lost ecosystem services for a range of economic activities.

One way to picture the choice is shown in Figure 3-3. The central concern of vulnerability assessment is people – those who should be protected from the adverse consequences of present climatic variations and projected climate change. These might be demographic groups (such as young children), livelihoods (urban poor in the informal economy) or populations at risk from diseases. Even when we focus on people as the target, we have to account for the fact that they are organised into groups at various scales – from individuals to households to communities and complete settlements. At each stage there are different sets of resources, institutions and relationships that determine not only their interaction with climate but also their ability to perceive problems, formulate responses and take actions. TP6 can assist in selecting and using indicators for various socio-economic characteristics in a vulnerability analysis.

Although a focus on groups is preferred, in practice, assessment is often carried out in sectoral or regional settings. Annex A.3.5 provides an example of the link between people as the target of vulnerability assessment and development policy and practice.

The exposure of groups, regions or sectors to climate risk is typically described using indicators. Indicators may reflect different socio-economic characteristics of the targets, including demo-

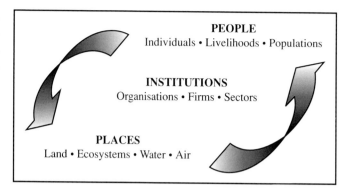

**Figure 3-3:** Units of analysis for a vulnerability assessment. The central concern of the vulnerability assessment is people, within the context of institutions and the biogeophysical resources of places. The research team and stakeholders can build up such a schema to illuminate exposure to climatic variations and to the drivers of socio-economic vulnerability. For example, "brainstorming" with boxes and arrows on a flip chart can map relationships in various ways (TP2).

graphics, composition of economic activity, infrastructure and so on. Indicators may describe stocks – e.g., stocks of human, natural and manufactured capital; or flows – e.g., flows of economic goods and services, income and trade. Developing and using indicators requires an awareness of several technical issues including their sensitivity to change, standardising indicators for comparison, the reliability of the data, mapping of indicators, collinearity among indicators, coverage of the relevant dimensions of vulnerability, etc. It is important for the assessment team to examine existing inventories and analyses, as many of these issues may have already been addressed. The literature on indicators provides examples of good practice.

The output of this activity is a set of vulnerability indicators and identification of vulnerable livelihoods (or other targets) that, together, form a vulnerability baseline of present conditions. (For additional guidance on developing socio-economic indicators, see TP6.) The collation of vulnerability indicators underpins the analyses and identification of priorities for adaptation. The process of aggregating the individual indicators into a composite view of vulnerability is covered in Activity 5.

### 3.4.3.  *Activity 3: Assessing sensitivity: Current vulnerability of the selected system and vulnerable group*

Current vulnerability can be expressed as the conjunction of the climatic hazards, socio-economic conditions, and the adaptation baseline (TP6). The first two activities in the vulnerability assessment establish the present conditions of development. Activity 3 directly links climate hazards to key socio-economic outcomes or impacts. In this activity, we develop an understanding of the process by which climate outcomes translate into risks and disasters. This may be done through a variety of approaches ranging from simple, empirical relationships to more complex,

---

[5] Using the nomenclature outlined in Annex A.3.2, these might be labeled as Vg, Vs and Vr (referring to vulnerable groups, sectors and regions).

process-based models, such as those described in TP4 and TP5. The extension of the analysis to future climate risks is covered in Activity 4.

Climate outcomes are typically described through hydrological and meteorological variables. Depending on the nature of the consequences and the nature of the impacts processes, these variables may be used directly, or secondary variables may be computed. For example, if the team is interested in the sensitivity of energy demand to climate change, a typical directly observed quantity might be daily maximum or minimum temperature, whereas heating or cooling degree-days are quantities that may be more relevant for capturing the relationship between climate and energy demand. Such quantities may need to be derived from primary climate data.

In many sectors and regions, there are already well-developed models and frameworks that describe system sensitivity. For example, there are a variety of crop models (physiology-based or empirical) that link crop yield and output to climate parameters. In many instances, detailed process models may be either unavailable, or too complex for inclusion in the assessment. In such cases, a variety of simpler techniques may be adopted, including empirical models based on analysis of historical data and events or models that look at simple climatic thresholds (e.g., the probability of drought). If it is difficult to implement even simple empirical approaches, an alternative might be to use expert opinion or examples from different, but related settings (e.g., similar countries) to develop understanding of the relationship between hazards, exposure and outcomes.

An important part of this activity is the identification of points of intervention, and options for response in the sequence leading from hazards to outcomes. Not only is this relevant for considering responses in the short-term, it is also important for the evaluation of future vulnerability (Activity 4). The evolution of vulnerability in the future depends quite critically on endogenous adaptation – planned or autonomous.

### 3.4.4.  *Activity 4: Assessing future vulnerability*

The next activity in a vulnerability assessment is to develop a more qualitative understanding of the drivers of vulnerability, in order to better understand possible future vulnerability: "What shapes future exposure to climatic risks?" "At what scales?" This analysis links the present (snapshot) with pathways of the future, pathways that may lead to sustainable development or increased vulnerability through maladaptation.

This activity requires the analyst to consider ways in which planned and autonomous adaptation may modify the manner and mechanisms by which climate is a source of risk. For example, the gradual evolution of housing stock in a coastal region might alter future outcomes following a tropical cyclone. Similarly, the availability of flood insurance might alter the perceptions of households regarding risk, leading to increased development in

flood-prone areas, and therefore to increased damage from the cyclone. In both of these cases, interventions lead to a change in the impacts associated with climate change.

Specific techniques that may be used for this purpose are likely to be qualitative in the first instance. Interactive exercises (such as cognitive mapping) among experts and stakeholders can help refine the initial vulnerability assessment framework (Activity 1) by suggesting linkages between the vulnerable groups, socio-institutional factors (e.g., social networks, regulation and governance), their resources and economic activities, and the kinds of threats (and opportunities) resulting from climatic variations. Thought experiments, case studies, in-depth semi-structured interviews, discourse analysis, and close dialogue are social science approaches that can be used in understanding the dynamics of vulnerability.

More formal techniques include cross-impact matrices, multi-attribute typologies such as the five capitals of sustainable livelihoods or the characteristics of adaptive capacity (TP7), and even quantitative approaches such as input-output models, household production functions and multi-agent social simulation. Before adopting specific quantitative analyses, a useful strategy is to start with exploratory charts and checklists, which can help identify priorities and gaps.

Extending the drivers of present socio-economic vulnerability to the future is typically based on a range of socio-economic scenarios (see TP6 for an in-depth discussion of socio-economic scenarios). Existing development scenarios are the best place to start. Are there projections for development targets? Or, are there sectoral scenarios that may be relevant, as in the visions created by the World Water Council[6]? Otherwise, stakeholder-led exercises in creating visions of the future (including worst-case fears) are worth pursuing (TP2).

Two technical issues need to be clarified in the vulnerability assessment at this stage:

*   Most indicators are snapshots of present status, e.g., GDP per capita. However, vulnerability is dynamic and indicators that foreshadow future vulnerability may be useful. For example, future wealth may be correlated with literacy and governance and only weakly correlated with present rates of growth in GDP per capita.
*   The common drivers of development need to be related to the target vulnerable groups. National and international trends, e.g., in population and income, may not map directly onto the nuances of marginalization, local land tenure, markets and poverty that characterise vulnerability. Shocks and surprises have disproportionate effects for the vulnerable – as in the macroeconomic failure in Argentina or the prolonged desiccation of the Sahel.

While we suggest that scenarios of future vulnerability are best developed at the local to national level, there are cogent reasons to place future socio-economic conditions of vulnerability in a regional to global context. The climate change policy community has its own points of reference (e.g., currently the emissions scenarios completed in Nakicenovic et al., 2000). The vulnerability assessment may benefit from coherence with such international scenarios, although it is methodologically incorrect to suggest that global socio-economic scenarios can be downscaled to local vulnerability – on theoretical, practical and empirical grounds.

Outputs of this activity are qualitative descriptions of the present structure of socio-economic vulnerability, future vulnerabilities and a revised set of vulnerability indicators that include future scenarios. Climate change overlays are included in this activity (TP5). The final activity brings together the indicators into a meaningful vulnerability assessment.

### 3.4.5. Activity 5: Linking vulnerability assessment outputs with adaptation policy

The outputs of a vulnerability assessment include:

- A description and analysis of present vulnerability, including representative vulnerable groups (for instance, specific livelihoods at risk of climatic hazards)

- Descriptions of potential vulnerabilities in the future, including an analysis of pathways that relate the present to the future;
- Comparison of vulnerability under different socio-economic conditions, climatic changes and adaptive responses;
- Identification of points and options for intervention, which can lead to formulation of adaptation responses.

The final activity is to relate the range of outputs to stakeholder decision-making, public awareness and further assessments. These topics are framed in the overall APF design and stakeholder strategy (TP1, Section 1.4.1 and TP2). Here we review technical issues regarding the representation of vulnerability. The guiding concern is to present useful information that is analytically sound and robust across the inherent uncertainties.

The first consideration is whether stakeholders and decision makers already have decision criteria that they apply to strategic and project analyses. For instance, the Millennium Development Goals (MDGs) may have been adopted in a development plan. If so, can the set of vulnerability indicators be related to the MDGs? Is there an existing map of development status that can be related to the indicators of climate vulnerability? It is always better to relate the climate change vulnerability assessment to

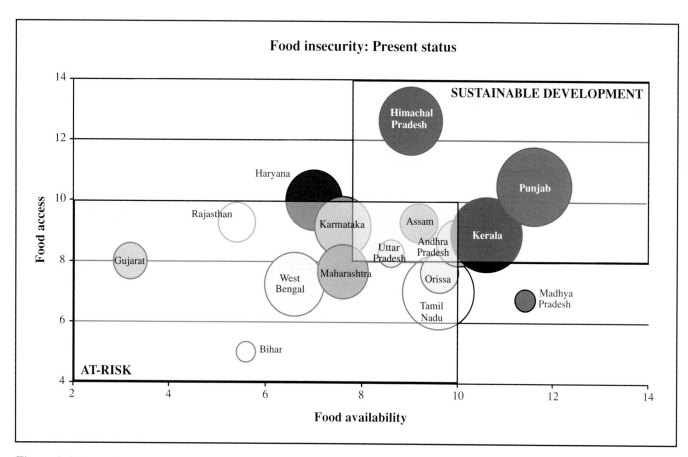

**Figure 3-4:** Rural food insecurity in India. Three dimensions of vulnerability are shown. Food availability (x-axis) is based on production indicators for each state. Food access (y-axis) aggregates indicators of market exchanges. The size of each bubble corresponds to indicators of nutritional status.
*Source:* MSSRF (2001).

existing frameworks, terminology and targets than to attempt to construct a new language solely for climate change issues.

Historically, a common approach has been to aggregate the individual indicators into an overall score, referred to as an index. For example, the Human Development Index (HDI) is a composite of five indicators, transformed into standard scores and differentially weighted (UNDP, 1999).

Do stakeholders have a formal multi-criteria framework that illuminates the choice of aggregation procedures and weights (TP8)? If so, an analogous aggregation of the vulnerability indicators data into an index may be informative for them. However, formal multi-criteria approaches are rarely generic and often contentious; the same is true for composite vulnerability indices. As a result, the use of such indices has to be done only with great caution.

A preferable device for communicating the vulnerability assessment is to use multi-attribute profiles. For example, Figure 3-4 plots the food security of states in India according to relative capacities for food production, food access and nutritional status. Many of the states would be considered food insecure. However, the structure of their vulnerability differs, and different adaptive measures are required.

Another aggregation technique is to cluster vulnerable groups (or regions) according to key indicators. For example, climatic risks might be related to different classes of vulnerability. Figure 3-5 suggests an approach that prioritises risks to sustainable livelihoods. More formal methods for clustering, such as principal Components analysis, are becoming more common as well (see Annex A.3.5 for an approach used by the World Food Programme).

The indicators in the vulnerability assessment can be used to evaluate adaptive strategies and measures (TP8). Vulnerability indicators have also been used as the baseline for monitoring development status (TP9). The technical team should consider how its

outputs could be used over a longer term. A key recommendation is likely to be improved monitoring and collection of specific data on socio-economic vulnerability.

The output should link to further steps in the APF. The focus on representative livelihoods and multiple scales of vulnerability can form the basis of an analysis of coping strategies. For instance, a multi-level assessment might include an inventory of household coping strategies and their effectiveness in different economic and climatic conditions, how local food markets might be affected by drought, and national contingency planning for drought (including food imports). A consistent analysis across these scales would inform a climate adaptation strategy with specific responsibilities for individual stakeholders (see TP8 for an in-depth discussion of adaptation strategy development).

Ultimately, the qualitative understanding of vulnerability can be developed as storylines that can be used in scenarios that describe future representative conditions (TP6, Section 6.4.6). These may be effective ways of communicating potential futures of concern. Communication methods are diverse; articles from future newspapers, radio documentaries and interviews can all be effective.

A final output might be to revisit the conceptual model (Activity 1). Are there new insights that need to be included? Does the monitoring plan capture the range of vulnerabilities and their drivers? Would the framework need to be altered to apply to different regions or vulnerable groups? Have the priorities for vulnerability assessment changed?

## 3.5.    Conclusions

Performing the five activities outlined in this TP would lead to a substantial vulnerability assessment that could meet the objectives of APF Components 2, *Assessing current vulnerability* and 3, *Assessing future climate risks*, and provide key input to Component 4, *Formulating an adaptation strategy*. The primary output is a set of priorities for adaptation and a panel of indicators for evaluating adaptation options. Further details are available from related TPs on climatic risk (TPs 4 and 5), socio-economic conditions (TP6) and future scenarios (TPs 5 and 6). We emphasise that a vulnerability assessment is a learning experience – the activities identified here are guideposts rather than a sequence of steps to be followed mechanically.

This TP closes with a set of open questions and issues in vulnerability assessment which, we hope, will be informed and refined through studies that implement the APF, as well as the next generation of vulnerability and climate impact assessment studies.

How may vulnerability be quantified? As we have seen in this TP, vulnerability can be regarded as a property or characteristic of target groups, societies and systems, but also as the outcome of a climate or other hazard process. In one case, quantification may involve the use of indicators to describe the con-

|              | Adaptive capacity |              |
|--------------|-------------------|--------------|
| **Impacts**  | *Low*             | *High*       |
| *High*       | Vulnerable Communities | Development Opportunities |
| *Low*        | Residual Risks    | Sustainability |

**Figure 3-5:** Clustering climatic risks and present development. In Figure 3-5, the quadrants are clusters of our knowledge of anticipated impacts of climate change, and the capacity of livelihoods or regions to adapt to those impacts. The high-risk cluster is labelled vulnerable communities. If impacts are high but so is adaptive capacity, there should be development opportunities to reduce the climate change burden. However, if impacts are low but uncertain, there may well be residual risks if adaptive capacity is also low. (See Downing, T.E. (2003) for a global demonstration of the approach.)

dition of the system (e.g., development, infrastructure or poverty indicators), in the other, quantification may be done through the formulation and estimation of hazard-loss relationships (e.g., the dose-response relationships used in health assessments, or the damage functions in climate impact models). Both approaches have similarities – in either case – the user gains a deep understanding of the process through which hazards translate into negative outcomes or into a disaster. It is this understanding which is critical for creating effective adaptation interventions.

Isn't socio-economic vulnerability a product of many drivers and actors? We take the view that vulnerability – as a broad condition of resource use or development – is socially constructed (or negotiated). That is, vulnerability is not just the tail of a probability distribution; it is an essential aspect of social and economic systems. Thus, multi-actor perspectives that analyse stakeholder behaviour are essential. Such methodologies focus on understanding adaptive capacity and the means to implement climate adaptation strategies.

How does vulnerability relate to ecosystems? We prefer to use the word *sensitivity* to describe the effects of driving forces and perturbations on ecosystems and natural resources. It implies a distinction between the biophysical processes and effects, and the values that people place on those changes. Clearly, ecosystem services affect vulnerable livelihoods, so there is a direct link to vulnerability assessment.

Can we predict future vulnerability? Future vulnerability is determined by the co-evolution of a number of coupled processes – the underlying climate hazards, the exposure of target groups, sectors and societies to the hazard, and planned and autonomous adaptation. In many situations, prediction of this co-evolution may be difficult, if not impossible to do. A sobering example of the difficulties in predicting the full impacts of Hurricane Mitch, despite good vulnerability assessments, is described in Ziervogel et al. (2003). In such cases, scenarios could be used as a tool to illustrate changes in vulnerability and for reviewing policy responses. Modelling approaches need to address uncertainties, as well as the difficulties of representing the processes of perception, evaluation, response, implementation and path dependency.

# References

**Adger**, N. and Kelly, M. (1999). Social vulnerability to climate change and the architecture of entitlement. *Mitigation and Adaptation Strategies for Global Change*, **4**, 253-266.

**Blaikie**, P., Cannon, T., Davies, I. and Wisner, B. (1994). *At Risk – Natural Hazards, People's Vulnerability, and Disasters*, London: Routledge.

**Bohle**, H. and Watts, M. (1993). The space of vulnerability: the causal structure of hunger and famine. *Progress in Human Geography*, **13** (1), 43-67.

**Bohle**, H., Downing, T.E. and Watts, M. (1994). Climate change and social vulnerability: the sociology and geography of food insecurity. *Global Environmental Change*, **4**(1), 37-48.

**Chambers**, R. (1989). Vulnerability, coping and policy. *IDS Bulletin*, **20**, 1-7.

**Clark**, G.E., Moser, S.C., Ratick, S.J., Dow, K., Meyer, W.B., Emani, S., Jin, W., Kasperson, J.X., Kasperson, R.E. and Schwarz, H.E. (1998). Assessing the vulnerability of coastal communities to extreme storms: the case of Revere, MA, United States. *Mitigation and Adaptation Strategies for Global Change*, **3**, 59-82.

**Downing**, T.E. (2003). Linking sustainable livelihoods and global climate change in vulnerable food systems. *Die Erde*, (**133**), 363-378.

**Downing**, T.E. et al. (2001). Vulnerability indices: Climate change impacts and adaptation. *Policy Series, 3*, Nairobi: UNEP,.

**Handmer**, J.W., Dovers, D. and Downing, T.E. (1999). Social vulnerability to climate change and variability. *Mitigation and Adaptation Strategies for Global Change*, **4**, 267-281.

**Kasperson**, J.X., Kasperson, R.E., Turner, II, B.L., Hsieh, W. and Schiller, A. (2002). Vulnerability to global environmental change. In *The Human Dimensions of Global Environmental Change*, ed. Andreas Diekmann, Thomas Dietz, Carlo C. Jaeger, and Eugene A. Rosa. Cambridge, MA : MIT Press (forthcoming).

**Kasperson**, J.X. and Kasperson, R.E. (2001). International workshop on vulnerability and global environmental change, 17-19 May 2001. Stockholm Environmental Institute Stockholm, Sweden. A workshop summary.

**Leichenko**, R. and Karen O'Brien. (2002). The dynamics of rural vulnerability to global change. *Mitigation and Adaptation Strategies for Global Change*, **7**(1), 1-18.

**Nakicenovic**, N. et al. (2000). *Special Report on Emissions Scenarios*. Cambridge: Cambridge University Press.

**Stephen**, L. and Downing, T.E. (2001). Getting the scale right: a comparison of analytical methods for vulnerability assessment and household level targeting. *Disasters*, **25**(2), 113-135.

**United Nations Development Programme** (UNDP). (1999). *Human Development Index*. New York: UNDP.

**Ziervogel**, G., Cabot, C., Winograd, M., Segnestam, L., Downing, T. and Wilson, K. (2003). Risk mapping and vulnerability assessments: Honduras before and after Hurricane Mitch. In Stephen, L., Downing, T.E. and Rahman, A., eds., *Approaches to vulnerability: food systems and environments in crisis*. London: Earthscan (forthcoming).

# ANNEXES

### Annex A.3.1. Vulnerability definitions and common usage

#### *Definitions in use*

The word vulnerability has many meanings. The User's Guide-book provides a definition developed by Kasperson et al. 2002. However, it is not the intention of the APF to impose its definitions on the wider research and policy communities concerned with climatic risks and climate change. This note summarises the main traditions in defining vulnerability and proposes a practical nomenclature. That is, it proposes a consistent terminology rather than force all authors and users to agree with a single definition.

It is essential for users to define vulnerability in their own context. The APF is meant to be useful to a wide set of users, and each will have their own views of what vulnerability is. Nevertheless, in their assessments, users need to make their definitions clear – at least to communicate among their project team and stakeholders. In many cases, those stakeholders have already formed a working definition of vulnerability. Use of those definitions may be preferable to the more arcane language sometimes adopted by the climate change community. Mainstreaming climate change means making our analyses relevant to existing decision frameworks.

Three traditions in defining vulnerability are hazards, poverty and climate change.

The longer tradition in defining vulnerability comes from ***natural hazards and epidemiology***. From this tradition, a common definition of vulnerability is:

> *The degree to which an exposure unit is susceptible to harm due to exposure, to a perturbation or stress, in conjunction with its ability (or lack thereof) to cope, recover, or fundamentally adapt (become a new system or become extinct). (Kasperson et al. 2000)*

The technical literature on disasters uses the term to mean:

> *Degree of loss (from 0% to 100%) resulting from a potential damaging phenomenon. (UNDHA Glossary of terms)*

The key aspect of these definitions is that vulnerability is distinguished from hazard – it is the underlying exposure to damaging shocks, perturbations or stresses, rather than the probability or projected incidence of those shocks themselves.

The ***poverty and development*** literature focus on present social, economic and political conditions. From this tradition, a common definition of vulnerability is:

> *An aggregate measure of human welfare that integrates environmental, social, economic and political exposure to a range of harmful perturbations. (Bohle et al., 1994)*

The important distinctions are: (1) vulnerability relates to social units (people) or systems rather than biophysical systems – which should be described as sensitive to stresses; (2) vulnerability integrates across a range of stresses (not just biophysical) and across the range of human capacities – not just food security, income or health.

In the field of ***climate change***, the IPCC promoted an alternative definition of vulnerability:

> *The degree to which a system is susceptible to, or unable to cope with, adverse effects of climate change, including climate variability and extremes. Vulnerability is a function of the character, magnitude, and rate of climate variation to which a system is exposed, its sensitivity, and its adaptive capacity.* www.ipcc.ch/pub/syrgloss.pdf.

The important distinction of the IPCC view is that it integrates hazard, exposure, consequences (impacts) and adaptive capacity. This definition corresponds more closely to the notion of risk in the natural hazards (and other) literature. The difference is that risk assessments are largely based on a probabilistic understanding of the triggering event, a risk tree of contingent impacts, quantification of outcomes and multiple criteria analysis of responses. To date, the IPCC is far from this sort of methodology, preferring to begin with scenarios of climate change and primarily first-order impact analyses.

It should be noted that within the IPCC texts, vulnerability is used in all of the above ways – the official definition has not been established as a consensus among the contributing authors.

#### *Suggested nomenclature for vulnerability definitions*

If we accept that there are always going to be many and conflicting definitions of the word vulnerability, perhaps what is needed is a nomenclature – a way of systematically referring to vulnerability in typologies and analytical exercises. For example:

$$^{T}V^{c}_{s,g}$$

*Where:*
*T = threat*
*s = sector*
*g = group*
*c = consequence*
*E.g.: climate change vulnerability in agriculture for farmers' economic welfare*

This nomenclature would results in examples such as:

- climate change vulnerability (T = climate change, no other terms specified)
- drought (T) vulnerability for food systems (s)
- drought (T) vulnerability for smallholder (g) agriculturalists (s)
- drought (T) vulnerability for smallholder (g) agriculturalists (s) at risk of starvation (c = health effects of reduced food intake)

The process of conducting a vulnerability assessment can be labelled vulnerability assessment.

If the indicators are mapped, this is extended to a vulnerability assessment map (VAM).

The database of indicators used in a vulnerability assessment (or VAM) can be labelled VI. Individual indicators ($VI_x$) might carry their own nomenclature, to specify:

t = time period (historical, present or specific projection)
g = group of people, if specific to a vulnerable population
r = region (or geographic pixel)
* = transformed indicators, as in standard scores

### Annex A.3.2. Vulnerability concepts and frameworks[7]

The following material was developed as part of a training course on climate change vulnerability and adaptation for the Assessments of Impacts and Adaptations to Climate Change in Multiple Regions and Sectors (AIACC) project (see www.start.org for further details). The objectives of the small group exercise on vulnerability concepts were to:

- introduce the range of definitions of vulnerability
- look at range of methods in vulnerability assessment
- consider ways to apply vulnerability assessment in AIACC projects

The following "vulnerability diagrams", drawn from several studies, were used to brainstorm issues regarding the framing vulnerability in the context of climate change and using vulnerability frameworks in research projects. Other sessions covered vulnerability mapping, livelihood approaches, socio-economic scenarios and the use of indicators.

In the small group exercise, the strengths and weaknesses were left blank – to be filled in by the participants. Technical teams undertaking APF projects may find the exercise useful in providing some background to conceptualising vulnerability. No one framework is "best" – all have strengths as well as weaknesses.

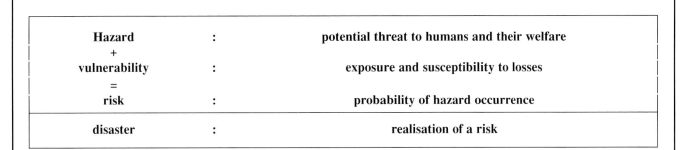

| Hazard | : | potential threat to humans and their welfare |
| + | | |
| vulnerability | : | exposure and susceptibility to losses |
| = | | |
| risk | : | probability of hazard occurrence |
| disaster | : | realisation of a risk |

▲ *Strengths:* Simple, widely used, clear definitions of key terms

■ *Weaknesses:* Not very dynamic, doesn't show what causes vulnerability, vulnerability is limited to a hazard-loss equation

● *Techniques:* Indicators, loss equations

**Figure A-3-2-1:** Definitions of hazard, vulnerability, risk and disasters

---

[7] See the TP for the references.

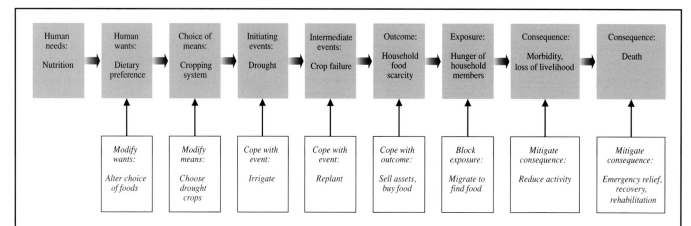

*Source:* after Downing (1991); see also Millman and Kates (1990)

▲ *Strengths:* Sequence of the drivers of vulnerability, emphasis on upstream causes, explicit ways to reduce vulnerability, multiple consequences

■ *Weaknesses:* Too linear, no feedbacks between outcomes and earlier vulnerabilities, no sense of who chooses options to modify the vulnerabilities, limited environmental forcing to only one place in the sequence

● *Techniques:* Linked models, e.g., food systems and crop model, indicators

**Figure A-3-2-2:** Causal chain of hazard development

| RESOURCES | VULNERABILITY | CAPABILITY |
|---|---|---|
| **Physical/material** | | |
| **Social/organisational** | | |
| **Motivational/attitudinal** | | |

*Source:* Anderson and Woodrow (1989)

▲ *Strengths:* Simple, flexible, brings in local knowledge, shows capability and opportunities, not just physical, includes social capital, intended for rapid use during disasters

■ *Weaknesses:* Nothing filled in, no sense of what the major issues are, not clear it would help identify vulnerable groups on its own, no drivers or assessment of future risks

● *Techniques:* Surveys, expert judgement and key informants

**Figure A-3-2-3:** Vulnerability and capability

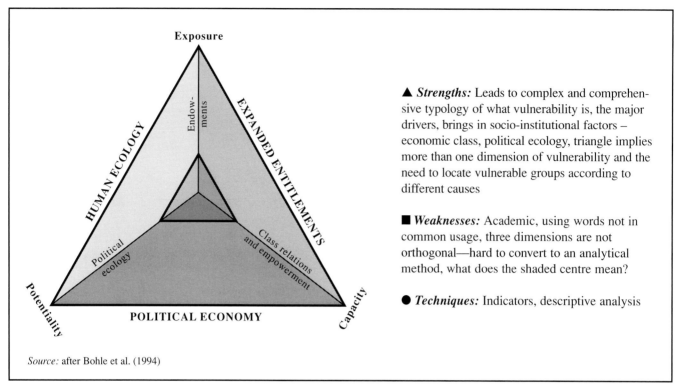

Source: after Bohle et al. (1994)

**Figure A-3-2-4:** Three dimensions of vulnerability

**PROGRESSION OF VULNERABILITY**

| ROOT CAUSES ➡ | DYNAMIC PRESSURES ➡ | UNSAFE CONDITIONS | ➡ DISASTERS ◀ HAZARDS | |
|---|---|---|---|---|
| **Limited access to** | **Lack of** | **Fragile physical environment** | | **Earthquake** |
| Resources | Institutions | Dangerous locations | **RISK** | |
| Structures | Training | Unprotected structures | | **Wind storm** |
| Power | Skills | | = | |
| | Investment | **Fragile local economy** | | **Flooding** |
| **Ideologies** | Markets | Livelihoods at risk | **HAZARD** | |
| Political systems | Press freedom | Low income | | **Volcano** |
| Economic systems | Civil society | | + | |
| | | **Vulnerable society** | | **Landslide** |
| | **Macro-forces** | Groups at risk | **VULNER-** | |
| | | Little capacity to cope | **ABILITY** | **Drought** |
| | Population growth | | | |
| | Urbanisation | | | **Virus and pest** |
| | Arms expenditure | **Public actions** | | |
| | Debt repayment | Lack of preparedness | | **Heat-wave** |
| | Deforestation | Endemic disease | | |
| | Soil degradation | | | |

Source: Blaikie et al. (1994)

▲ *Strengths:* Detail on causes, comprehensive, understandable

■ *Weaknesses:* More descriptive than analytical

● *Techniques:* Inventories, indicators

**Figure A-3-2-5:** Structure of vulnerability and disasters

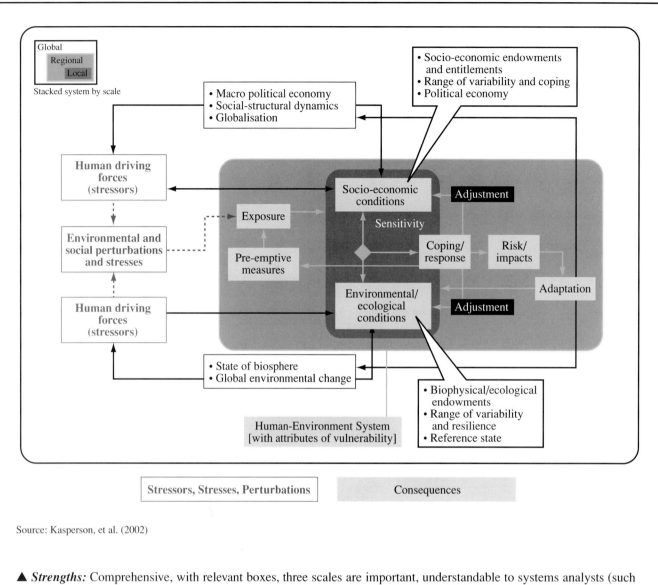

Source: Kasperson, et al. (2002)

▲ *Strengths:* Comprehensive, with relevant boxes, three scales are important, understandable to systems analysts (such as ecologists)

■ *Weaknesses:* Not clear how the dynamics at the local scale (sensitivity, adjustment, coping/response) are linked to the larger scales, would need additional material to implement

● *Techniques:* Dynamic simulation, choice of indicators

**Figure A-3-2-6:** Environmental vulnerability

### Annex A.3.3. Illustrative planning steps in vulnerability assessment for climate adaptation

The following charts illustrate the process of planning and implementing a vulnerability assessment for climate adaptation. This illustration is not a protocol – it does not include all of the possible choices and methods. Rather, it illustrates the five tasks outlined in the technical paper with specific choices and pathways through planning a project.

In the diagrams, a solid arrow indicates a positive result (Yes). A dotted arrow indicates alternative approaches in the absence of previous information (No). The outputs on the right side of the diagrams link from top to bottom. In fact, not all of the potential linkages are shown. Most importantly, the process is almost certain to be iterative. Tasks feed back to the scoping and data activities with further refinement of the information available and required.

Panes I and II show the first two activities. Scoping the technical details of the vulnerability assessment begins with a review of existing frameworks in use by national planners. If the existing development plans, poverty assessments, strategic environmental plans, etc., are not adequate for framing the climate vul-

nerability assessment, then a stakeholder-led exercise in conceptual mapping is helpful.

Panes I and II also show choices in compiling a database of indicators, initially of development conditions. This activity also identifies the vulnerable groups that are to be the target of the assessment. Thus, a two-level approach is recommended.

Panes III and IV show choices in characterising present climate risks, resulting in a climate vulnerability assessment. With the addition of scenarios of future socio-economic conditions, the set of vulnerability indicators (VI), the descriptions of their drivers and relationships to specific socio-economic groups (or vulnerable livelihoods) become the data engine for the vulnerability assessment.

Panes V and VI add in characterisations of future climate risks. This is not treated in detail in the diagram. Essentially the same choices as for activity 3 are appropriate.

The output of the vulnerability assessment requires some attention. It should be part of the scoping process – linking the vulnerability assessment data with stakeholder decision-making, identification and evaluation of adaptation strategies and the requirements for implementing adaptation policy.

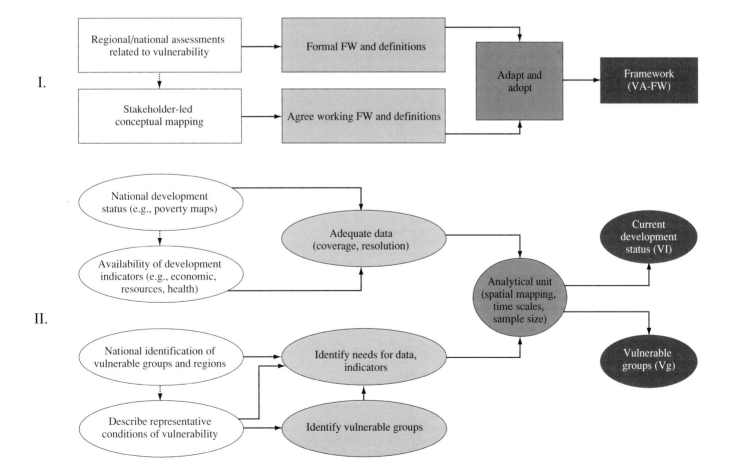

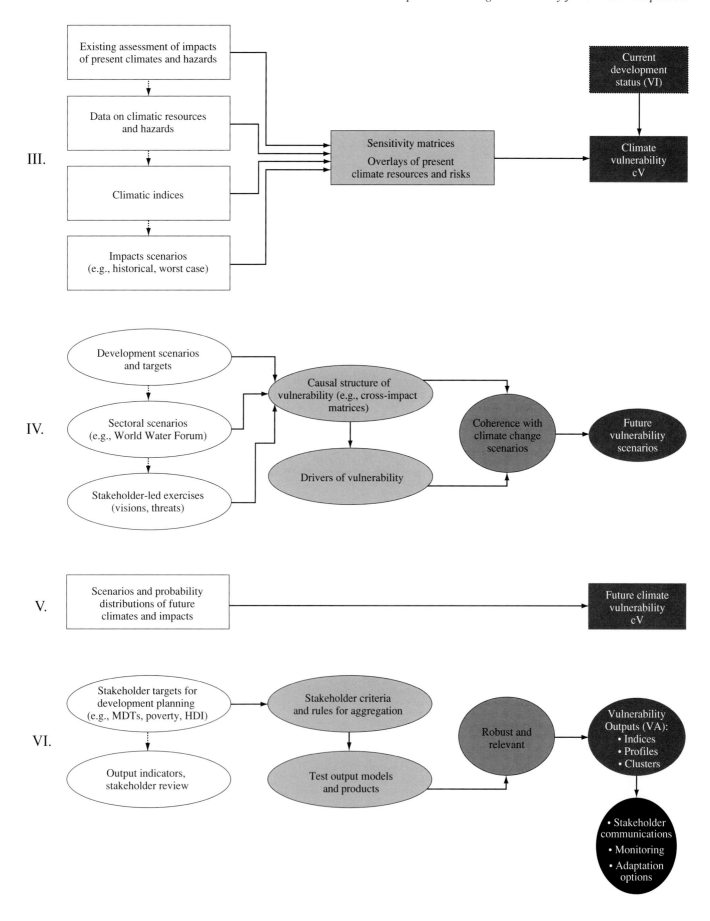

**Annex A.3.4. Vulnerability methodologies and toolkit**

*Introduction*

To gain an understanding of climate vulnerability and adaptation, four kinds of studies are appropriate:

- What if (WIf) studies are often the starting place for raising awareness among a wide variety of audiences about potential sensitivity to climate change.
- Vulnerability assessments and sustainable livelihood (VASL) approaches begin with present risks, and overlay climate change through a guided process of risk assessment.
- A focus on stakeholders and their decision-making regarding threats and opportunities (STO) leads to strategies for adapting to climate change over a range of planning periods.
- Where specific decisions need to be made, processes for evaluating additional climatic risks have been formulated in climate impacts management (CIM) studies.

For each approach, a different set of techniques is appropriate. The VASL approach is the most common. Below we describe this approach, and then we list a range of techniques for vulnerability and adaptation assessment. An expanded version of this toolkit is available, including a checklist for matching different project design criteria to the choice of methods, flow charts of common vulnerability approaches, and a set of icons for users to build their own flow charts.[8]

*Vulnerability assessment and sustainable livelihoods*

Vulnerability mapping begins with a snapshot of the present situation – whether applied to a specific hazard (e.g., hurricanes), generic disaster risks or poverty. In this approach, climate risks – both present and future – are placed in context of present vulnerability. Further elaboration provides indications of relative risks and strategies to support sustainable livelihoods.

The approach includes:

- Vulnerability mapping: ideally starting with the concepts and assessments conducted in the course of hazard management or development planning. An increasing number of such exercises have been conducted, providing a good starting place for climate change studies.
- Relating livelihoods to their exposure to risks. Often vulnerability maps do not explicitly recognise livelihoods – the exposure of specific populations to threats and opportunities. Once identified, a matrix of their exposure to development and climate risks helps to focus on the most sensitive livelihoods and those threats that can be managed.
- Description of coping strategies for the identified

livelihoods. A qualitative assessment, through interviews, secondary literature, focus groups, workshops, etc., will provide a rich context for considering the relative risks of climatic variations and potential response strategies.

- For selected livelihoods and risks, quantitative models can be constructed – following the approach that Jones terms "coping ranges" (TP4) or more dynamic decision models (as in agent-based systems).
- The qualitative and quantitative assessments can be tested against a range of scenarios of the future (including socio-institutional changes as well as climatic risks).
- It may be desirable to relate the scenario exercises to the initial vulnerability assessment. This might be simply looking at overlays of the present vulnerability and future risks. However, developing innovative techniques to deal with spatial data and relatively long time frames would be worth pursuing.

The main output of this approach should be a relatively robust presentation of present vulnerability and scenarios of future risk, accompanied by a rich understanding of coping strategies for different livelihoods. The integration of climate risk in development planning is a main goal; adopting existing development frameworks and concepts is a key strength.

*The toolkit*

The key analytical tools are vulnerability mapping and dynamic simulation of sustainable livelihoods. However, the broader techniques of stakeholder participation and risk assessment are essential.

The following table suggests further tools that may be important, with an indication of their suitability according to the following criteria:

1. **Present vulnerability** – including development policy
2. **Problem definition** – scoping of issues and options to be included in analysis and design of projects
3. **Development futures** – pathways of future development
4. **Evaluation of adaptation** – to aid decision-making between specific measures and the selection of options
5. **Strategic planning** – consideration of alternative futures, including cross-sectoral and regional issues
6. **Multi-stakeholder analysis** – analysis of individual stakeholders within an institutional context
7. **Stakeholder participation** – whether stakeholders can readily participate in the application of the tool

---

[8] The spreadsheet, ClimateScoping.xls, can be found on www.vulnerabilitynet.org in the document hotel.

*Table A-3-4-1:* Toolkit for vulnerability/adaptation assessments [9]

| Applications / Tools | Present vulnerability | Problem definition | Development futures | Evaluation of adaptation | Strategic planning | Multi-stakeholder analysis | Stakeholder participation |
|---|---|---|---|---|---|---|---|
| 1. Agent-based simulation modelling | | | X | | ? | X | ? |
| 2. Bayesian analysis | | | | X | | | |
| 3. Brainstorming | X | X | X | X | X | X | X |
| 4. Checklists/multiple attributes | X | | | X | | X | X |
| 5. Cost-effectiveness | | | X | X | | | |
| 6. Cross-impact analysis | | | X | X | | | |
| 7. Decision conferencing | | | X | X | | | |
| 8. Decision/probability trees | | | | X | | | |
| 9. Delphi technique | X | | X | X | | ? | ? |
| 10. (Strategic) environmental impact assessment | | | X | X | X | | ? |
| 11. Expert judgment | X | X | X | X | X | X | |
| 12. Focus groups | X | ? | X | ? | | ? | X |
| 13. Indicators/mapping | X | | ? | | | ? | ? |
| 14. Influence diagrams/mapping tools | X | | X | | X | | X |
| 15. Monte Carlo analysis | | | | X | | | |
| 16. Multi-criterion analysis | | | | X | | | |
| 17. Ranking/dominance analysis/ pairwise comparisons | X | | X | X | | | X |
| 18. Risk analysis | | | ? | X | | | |
| 19. Scenario analysis | ? | ? | X | ? | X | X | X |
| 20. Sensitivity/robustness analysis | | | X | X | | | |
| 21. Stakeholder consultation | X | X | X | X | | X | X |
| 22. Stakeholder Thematic Networks | X | ? | X | | ? | X | |
| 23. Uncertainty radial charts | | | | X | | | |
| 24. Vulnerability profiles | X | ? | ? | | | X | X |

## Tool Annotations

1. **Agent-based simulation modelling** – formalism of agents and their interactions at multiple levels
2. **Bayesian analysis** – used to reassess probabilistic data in light of new data; statistical analysis
3. **Brainstorming** – free flowing lists/diagrams of all ideas and options
4. **Checklists** – matrix
5. **Cost-effectiveness/ cost-benefit/ expected value** – econo-metric techniques
6. **Cross-impact analysis** – used to test robustness of risk assessment and dependencies between events
7. **Decision conferencing** – quantitative analysis of options incorporating the uncertainties in interactive modes
8. **Decision/probability trees** – charts of relationships between decision modes; helpful for generating expected value
9. **Delphi technique** – range of views of experts through itera-tive written correspondence
10. **(Strategic) environmental impact assessments** – environ-

[9] In the table above, "**X**" indicates that a tool is appropriate for the application in question, whereas, "**?**" indicates that it may be appropriate.

mental impacts taken into account before deciding on development

11. **Expert judgment** – the assessment of experts in the field on specific propositions
12. **Focus groups** – groups of stakeholders that discuss their opinions on certain topics
13. **Indicators/mapping** – compilation of indicators into aggregate indices, often mapped
14. **Influence diagrams/mapping tools** – graphic identification of options when there are a number of decisions
15. **Monte Carlo analysis** – computer based analysis that explicitly assesses uncertainty
16. **Multi-criterion analysis** – scoring and weighting of options using indicators and more than one decision criteria
17. **Ranking/dominance analysis/pairwise comparisons** – preference of options
18. **Risk analysis** – approaches to decision uncertainty including hedging and flexing, regret, minimax and maximin
19. **Scenario analysis** – fuller picture of implications of uncertainty gained through simultaneous variation of key uncertainties
20. **Sensitivity analysis/robustness analysis** – identification of variables contributing most to uncertainty
21. **Stakeholder consultation** – consultation with individuals and/or groups affected by future processes
22. **Stakeholder Thematic Networks** (STN) – mapping of the key actors and their interactions
23. **Uncertainty radial charts** – assessment of the potential uncertainty of options
24. **Vulnerability profiles** – mapping of the different indicators of vulnerability for different groups

### Annex A.3.5. Vulnerability to food insecurity in Kenya

*Source:* Haan, N., Farmer, G. and Wheeler, R. (2001). *Chronic Vulnerability to Food Insecurity in Kenya.* A WFP Pilot Study for Improving Vulnerability Analysis.

The World Food Programme (WFP) has developed the Standard Analytical Framework (SAF), based on a clear conceptual framework of food insecurity. National assessments begin with a literature review to understand contextual issues, enable the study to build from previous research and identify relevant indicators and data needs.

In Kenya, the secondary data analysis sought to identify relative differences in vulnerability to food insecurity between districts and to characterise contributing factors to vulnerability at the district level and prioritise districts for subsequent community-based analysis. (Figure A-3-5-1) A variety of data sets and

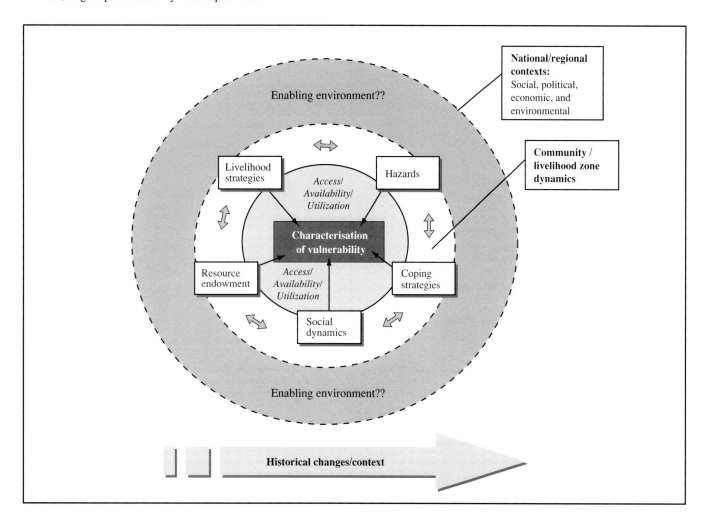

**Figure A-3-5-1:** Conceptual framework for characterising vulnerability to food insecurity

techniques were employed, allowing for verification of results and a mixture of interpretations. The Geographic Information System mapped 18 variables at the district level: life expectancy, adult literacy, stunting, wasting, livelihood diversification, access to safe water, livelihood fishing, high potential land, mean vegetation condition variation and persistence (using the NDVI), education, gender development, non-agricultural income, proximity to markets, HIV/AIDS incidence, and civil insecurity.

Two techniques were utilised to aggregate the indicators. A deductive approach used Z-Scores (not shown here). The inductive approach used Principal Components Analysis (PCA) and clustering where the raw data for each district were statistically grouped into clusters of districts with similar characteristics, and then interpreted for relative vulnerability.

The PCA (Figure A-3-5-2) indicates highest levels of vulnerability in the arid and semi-arid districts of northern Kenya. The clustering technique shows groups of similar districts (in terms of food security). This PCA and clustering (Figure A-3-5-3) is helpful to understand some of the dynamics of food insecurity. For example, Cluster 1 is strongly and negatively associated with food insecurity characterised by: low adult literacy rates, high wasting, low non-farm income, low market access, low NDVI mean, high annual variance of NDVI, high civil insecurity, and low HIV/AIDS.

The community-based analysis, called Participatory Vulnerability Profiles (PVP), covered 79 villages stratified by livelihood zones in 12 districts selected based on the SDA results and key informant discussions. The goals of the PVP were to: describe relatively homogenous livelihood zones, verify and further disaggregate results of the SDA, characterise community vulnerabilities to food insecurity, characterise and identify proportions of more vulnerable populations, identify both community-level and macro, or structural causes of food insecurity, and identify intervention opportunities.

An important emphasis of the PVP methodology was the direct links between the conceptual framework and the field techniques, enabling the field researchers to better understand the reasons for asking questions in the field. Districts were selected to represent each of the clusters from the national analysis. The field teams, in consultation with district officials, created livelihood zones (LZs) within each district (Figure A-3-5-4). The definition of LZs as used in this study is: *a relatively homogenous area with regard to four variables including main food sources, main income sources, hazards, and socio-cultural dynamics.* The creation of LZs allows the research to sample only a few villages within a large area and make a statement about the whole area. The third layer of sampling was within each community, and involved focus group interviews with various social groups, including the "typical group", the "most vulnerable", women, community leaders, and a mixed representative group.

The analysis revealed broad similarities between the district analysis and the detailed understanding by livelihood zones. Implications of hazards, coping strategies, social dynamics and health on food insecurity led to specific recommendations.

For example, one of the main hazards throughout the most vulnerable districts is drought, which is reportedly occurring more frequently. The relative drought risk by livelihood zone shows variation even within the more vulnerable districts (Figure A-3-5-5).

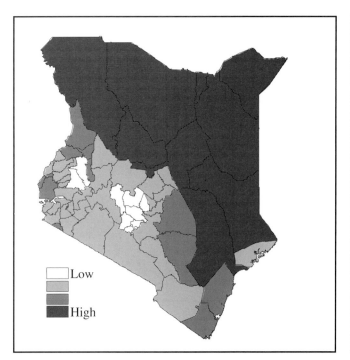

**Figure A-3-5-2:** Inductive approach: PCA and clustering relative vulnerability to chronic food insecurity

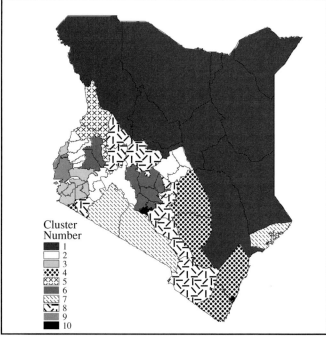

**Figure A-3-5-3:** Clusters of similar districts from PCA analysis of 18 variables

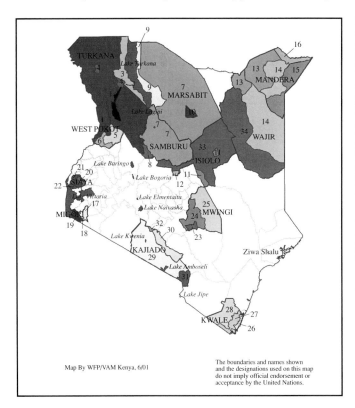

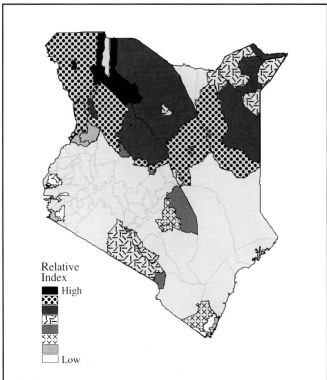

**Figure A-3-5-4:** Livelihood zones

**Figure A-3-5-5:** Relative drought risk by livelihood zone

# 4

# Assessing Current Climate Risks

ROGER JONES[1] AND RIZALDI BOER[2]

Contributing Authors
*Stephen Magezi[3] and Linda Mearns[4]*

Reviewers
*Mozaharul Alam[5], Suruchi Bhawal[6], Henk Bosch[7], Mohamed El Raey[8], Mike Hulme[9], T. Hyera[10], Ulka Kelkar[6], Mohan Munasinghe[11], Atiq Rahman[5], Samir Safi[12], Barry Smit[13], Joel B. Smith[14], and Henry David Venema[15]*

[1] Commonwealth Scientific & Industrial Research Organisation, Atmospheric Research, Aspendale, Australia
[2] Bogor Agricultural University, Bogor, Indonesia
[3] Department of Meteorology, Kampala, Uganda
[4] National Center for Atmospheric Research, Boulder, United States
[5] Bangladesh Centre for Advanced Studies, Dhaka, Bangladesh
[6] The Energy and Resources Institute, New Delhi, India
[7] Government Support Group for Energy and Environment, The Hague, The Netherlands
[8] University of Alexandria, Alexandria, Egypt
[9] Tyndall Centre for Climate Change Research, Norwich, United Kingdom
[10] The Centre for Energy, Environment, Science & Technology, Dar Es Salaam, Tanzania
[11] Munasinghe Institute for Development, Colombo, Sri Lanka
[12] Lebanese University, Faculty of Sciences II, Beirut, Lebanon
[13] University of Guelph, Guelph, Canada
[14] Stratus Consulting, Boulder, United States
[15] International Institute for Sustainable Development, Winnipeg, Canada

# CONTENTS

## 4.1.   Introduction

As part of Component 2 of the Adaptation Policy Framework (APF), *Assessing Current Vulnerability*, this Technical Paper (TP) focuses on how to assess the historical interactions between society and climate hazards. Key concepts related to current climate risks are outlined, and conceptual models that can be used to assess climate risks over short- and long-term planning horizons are introduced and described. Two major approaches to assessing those risks – a natural hazards-based approach and a vulnerability-based approach – are outlined. These two methods are complementary and can be used separately or together, as outlined in this TP and in TP3.

Understanding the historical interactions between society and climate hazards, including adaptations that have evolved to cope with these hazards, is a critical first step in developing adaptations to manage future climate risks. The characterisation of current climate hazards is also a key step towards building scenarios of future climate. In TP5, the methods described here are combined with climate scenario-building techniques to assess future risks.

This paper asserts that understanding current climate risks is a more appropriate basis for developing adaptation strategies to manage future climate risks than simply collecting baseline climate data and perturbing that data using scenarios of climate change. The relationships between current climate risks, vulnerability to those risks and the adaptations developed to manage those risks are often neglected in assessment methodologies – but not always in assessments themselves. Adaptation will be more successful if it accounts for both current and future climate risks. Even if future adaptation strategies are very different from those currently in use, today's adaptation will inform those strategies.

The main outputs that adaptation project teams can produce using this TP are:

1. Assessment of adaptive responses to past and present climate risks;
2. Knowledge of the climate drivers influencing current climate risks that will provide a basis for constructing scenarios of future climate (TP5); and
3. Understanding the relationship between current climate risks and adaptive responses that provides a basis for developing adaptive responses to possible future climate risks.

## 4.2.   Relationship with the Adaptation Policy Framework as a whole

This paper is linked directly to the APF Component 2, *Assessing Current Vulnerability*. Dealing specifically with current climate impacts and risks, TP4 takes into account natural resource drivers, socio-economic drivers, adaptation experi-

ence and the policy environment, and is thus connected to other TPs in the following way:

TP2:   *Engaging Stakeholders in the Adaptation Process* – Stakeholders are vital in identifying various aspects of the coping range, including the key climatic variables and criteria for risk assessment, including thresholds.

TP3:   *Assessing Vulnerability for Climate Adaptation* – This TP explores methods of assessing current and future vulnerability to climate change including variability. Methods of assessing vulnerability in TP3 can be combined with methods of hazard identification – outlined in this TP – to assess risk.

TP5:   *Assessing Future Climate Risks* – This TP describes how climate–society relationships may change under climate change and discusses how climatic information can be applied within a variety of risk assessments.

TP6:   *Assessing Current and Changing Socio-Economic Conditions* – This TP can be used to analyse the changing social responses to past and present climate. These techniques can be used to construct a dynamic view of changes in the ability to cope with climate over time.

TP7:   *Assessing and Enhancing Adaptive Capacity* – This TP describes the potential to respond to an anticipated or experienced climate stress. Analysis of the historical ability to cope with climate risks can indicate the adaptive capacity of a particular system.

TP8:   *Formulating an Adaptation Strategy* – This TP looks at specific choices to adapt to risks recognised in this TP and TP5.

## 4.3.   Key concepts

### *4.3.1.   Risk*

Risk is a term in everyday use, but is difficult to define in practice due to the complex relationships between its Components. Risk is the combination of the likelihood (probability of occurrence) and the consequences of an adverse event (e.g., climate hazard)[1]. In this TP, we describe the major elements of risk such as hazard, probability and vulnerability, though other terminology (e.g., exposure) can be used (TP3). These elements of risk can be applied in various ways depending on factors such as the level of uncertainty, whether the focus of an assessment is broad or specific and on the direction and emphasis of the approach used. Here, we describe two major approaches to assessing climate risk, a natural hazards-based approach and a vulnerability-based approach. These approaches rely most on whether the starting emphasis is on the biophysical or the socio-economic aspect of climate-related risk. In other words, is the emphasis on the climate hazard or on socio-economic outcomes? These two approaches are complementary and can be developed separately or together.

A *hazard* is an event with the potential to cause harm.

---

[1] Beer and Ziolkwoski, 1995; USPCC RARM, 1997.

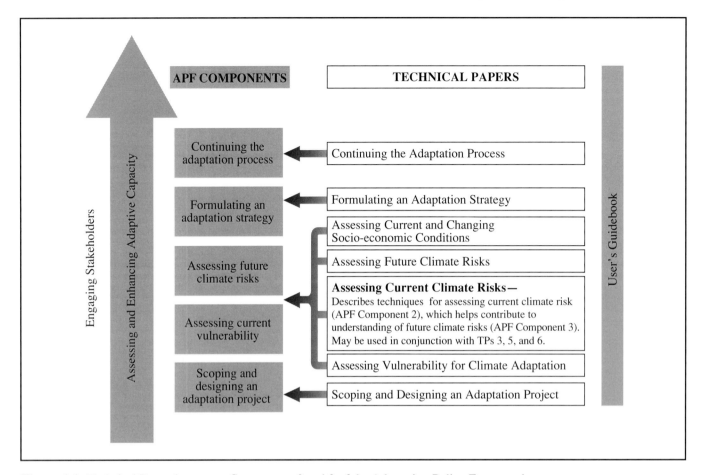

**Figure 4-1:** Technical Paper 4 supports Components 2 and 3 of the Adaptation Policy Framework

Examples of climate hazards are tropical cyclones, droughts, floods, or conditions leading to an outbreak of disease-causing organisms (plant, animal or human). Probabilities can be associated with the frequency and magnitude of a given hazard, or with the frequency of exceedance of a given socio-economic criterion (e.g., a threshold). Probability can range from being qualitative (using descriptions such as "likely" or "highly confident") to quantified ranges of possible outcomes, to single number probabilities. Vulnerability is broadly defined in TP3. Here, we limit our use of the term vulnerability to refer to climate vulnerability – specifically, the outcomes of climate hazards in terms of cost or any other value-based measure. Specific vulnerabilities (e.g., to drought, flood or storm surge) can also be assessed within the investigation of more broadly based social vulnerability, as described in TP3.

### 4.3.2.  *Natural hazards-based approach*

The natural hazards-based approach to assessing climate risk begins by characterising the climate hazard(s) and can be written as:

Risk = Probability of climate hazard x Vulnerability

Hazard is generally fixed at a given level and used to estimate changing vulnerability over space and/or time. For example, a flood of a given height or a storm with a given wind speed may increase in frequency of occurrence over time, increasing the risk faced (assuming that vulnerability remains constant).

### 4.3.3.  *Vulnerability-based approach*

The vulnerability-based approach begins by characterising vulnerability to produce criteria by which risk is assessed, e.g., by assessing the likelihood of exceeding a critical threshold.

Risk = Probability of exceeding one or more criteria of vulnerability[2]

Fixing the level of vulnerability allows the magnitude and frequency of climate-related hazards contributing to that vulnerability to be diagnosed. This is the "inverse method" as described in Carter et al. (1994). While commonly used in other disciplines, this technique has not been widely used for assessing climate change risks. If adaptation occurs, then successively larger and/or more frequent climate hazards can be coped with (e.g., a farming system adapting to drought should be able to manage

---

[2] Other formulations of risk are possible, but most will fall into the above two groups. Here, we have tried to provide a broad framework for assessing risk that will encompass more specific approaches.

more severe droughts before that system becomes vulnerable).

Two other methods mentioned in TP1 are the policy-based approach and the adaptive-bcapacity approach:

- Risk assessment techniques can be used in the policy-based approach where:
  - a new policy being framed is tested to see whether it is robust under climate change;
  - an existing policy is tested to see whether it manages anticipated risk under climate change.

- The adaptive-capacity approach investigates a system to determine whether it can increase the ability to cope with climate change, including variability. This approach will also be informed by a better knowledge of climate risks.

### 4.3.4. *Adaptation, vulnerability and the coping range*

Over time, societies have developed an understanding of climate variability in order to manage climate risk. People have learned to modify their behaviour and their environment to reduce the harmful impacts of climate hazards and to take advantage of their local climatic conditions. They have observed biophysical and socio-economic systems responding automatically to climate, and have tried to understand and manage these responses. This social learning is the basis of planned adaptation. *Planned adaptation* is undertaken by all societies, but the degree of application and the methods used vary from place to place. In mod-

ern societies, public sector adaptation may rely largely on science and government policy, and private sector adaptation on market forces, business models and regulation. Traditional societies may rely on narrative traditions, bartering of trade goods and local decision-making. All of these methods can be expressed using a common template.

This template has three climate ranges, depending on whether the outcomes are beneficial, negative but tolerable, or harmful. Beneficial and tolerable outcomes form the *coping range* (Hewitt and Burton, 1971). Beyond the coping range, the damages or losses are no longer tolerable and an identifiable group is said to be vulnerable. This structure is shown in Figure 4-2. A coping range is usually specific to an activity, group and/or sector, though society-wide coping ranges have been proposed (Yohe and Tol, 2002). The coping range provides a template that is particularly suitable for understanding the relationship between climate hazards and society. It can be utilised in risk assessments to provide a means for communication and, in some cases, may provide the basis for analysis.

The climatic stimuli and their responses for a particular locale, activity or social grouping can be used to construct a coping range if sufficient information is available. For example, in an agricultural system, this may include aspects of rainfall variability, temperature and other important prerequisites for understanding crop growth, information about crop yield and prices and knowledge of what constitutes a sustainable level of yield. Analyses can then be used to show which levels of yield are good, marginal, poor and which pose a serious threat. For a water sys-

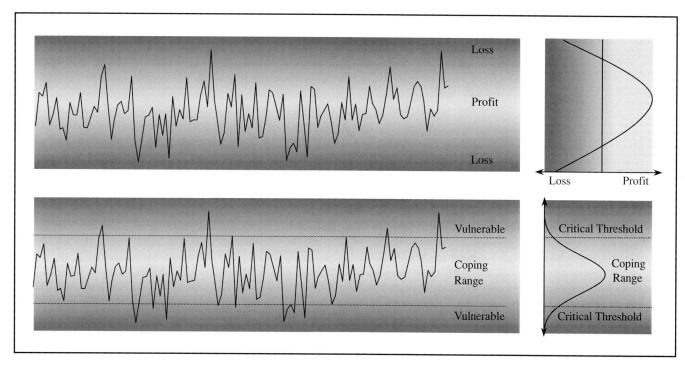

**Figure 4-2:** Simple schematic of a coping range under a stationary climate representing rainfall or temperature and crop yield. Vulnerability is assumed not to change over time. The upper time series and chart shows a relationship between climate and profit and loss. The lower time series and chart shows the same time series divided into a coping range using critical thresholds to separate the coping range from a state of vulnerability.

tem, climate drivers may include accumulated rainfall and evaporation, if supply is being addressed, or rainfall intensity and duration, if flooding is being addressed. On a coastline, climate variables contributing to storm surge, tidal regimes and sea level anomalies may be linked to thresholds related to the degree of coastal flooding or property damage. Coping range Components can range from simple "rule of thumb" estimates to accurate representations of a system based on detailed modelling.

Figure 4-2, upper left, shows a time series of a single variable, e.g., temperature or rainfall, under a stationary climate. If conditions get too hot (wet) or cold (dry), then the outcomes become negative. The response curve on the upper right represents the relationship between climate and levels of profit and loss for some measure, e.g., crop yield. Under normal circumstances, outcomes are positive but become negative in response to extremes of climate variability.

Using a response relationship between climate and other drivers and specific outcomes, we can select criteria or indicators representing different levels of performance for the purposes of assessing risk (Figure 4-2, lower left). For example, a yield relationship can be divided into good, poor or disastrous segments or coping capacity can be delimited by a critical threshold. More complex criteria, perhaps based on vulnerability analysis (TP3, Activities 2 and 3), may represent factors such as the ability to grow next season's seed supply, grow next year's food supply, break even economically, or produce sufficient surplus to pay for supplementary food and children's school fees. Note that in Figure 4-2, the critical threshold representing the ability to cope is held constant, but in the real world is dynamic, responding to internal process in addition to external climatic and non-climatic drivers (Annex A.4.3).

By adapting the knowledge of climate–society relationships held within a community, as well as within public and private institutions, the project team may be able to develop a relationship linking climate to criteria that represent a given level of vulnerability. For example, a narrative history of past droughts and the responses to those droughts can be matched with rainfall records to construct a fuller picture of climate–society relationships that can then be assessed under conditions where both climate and society may change (TP2, Activity 2; Tarhule and Woo, 1997).

Therefore, risk can be assessed by calculating how often the coping range is exceeded under given conditions (Figure 4-2, lower right). The method of assessing risk can range from qualitative to quantitative. Qualitative methods can be carried out by building or using an existing conceptual model of a specific coping range and assessing risk in terms of qualifiers such as low, medium and high. Quantitative methods will begin to assess the likelihood of exceeding given criteria, such as critical thresholds. Quantitative modelling will allow these relationships to be assessed under changing conditions. When undertaking mathematical modelling using the coping range, it is advisable to modify the mathematical models to suit the conceptual models rather than let the structure of the models dominate the assessment.

The coping range is a very useful concept because it fits the mental models that most people have concerning risk. People have an intuitive understanding of the situations they face regarding commonly encountered climatic risks – which risks can be coped with, which cannot and what the consequences may be. This understanding can be extended to other less commonly encountered risks and to never before experienced situations that may occur under climate change. Stakeholders will also have different coping ranges. An assessment may wish to explore those differences in order to gather a common activity-wide coping range for the purposes of assessment, or to explore the differences between coping ranges, e.g., why do certain groups cope better with a situation, and how do we share that capacity with others?

## 4.4.  Guidance on assessing current climate risks

The goal of this section is to guide the user through the process of assessing current climate risks, as outlined in Figure 4-3, rather than provide a tight prescription for how to proceed. There are two major paths one can use, depending on whether the starting point focuses on climate or on vulnerability to climate. For example, a project focusing on the identification of regional climate hazards and how they may alter vulnerability will probably be more suited to a natural hazards-based approach. Approaches focused on the nature of vulnerability or critical thresholds may well start at that point then work backwards to determine the magnitude and frequency of hazards contributing to that vulnerability. Natural hazards-based approaches are favoured where the probabilities of the climate hazards can be constrained, where the main drivers of impacts are known and where the chain of consequences between hazard and outcome is well understood. The vulnerability-based approach will be favoured where: the probability of the hazard is unconstrained, there are many drivers and there are multiple pathways and feedbacks leading to vulnerability. Steps can be carried out in any order to suit the needs of an assessment and can be skipped if they are not considered necessary. Previous information on risks and hazards can also be introduced. The most basic elements needed are a conceptual model of the system and a basic knowledge of the hazards and vulnerabilities in order to prioritise risk. Both qualitative and quantitative methods can be used to assess risk depending on the quality of information needed by stakeholders and the data and knowledge available to provide that information.

### 4.4.1.  *Building conceptual models*

Component 2 of the APF requires an understanding of the important climate–society relationships within the system being investigated. Those relationships are dominated by the climate impacts within the system and the sensitivity of the system response. *Climate sensitivity* is defined as the degree to which a system is affected, either beneficially or adversely, by climate-related stimuli (IPCC, 2001). Sensitivity affects the magnitude and/or rate of a climate-related perturbation or

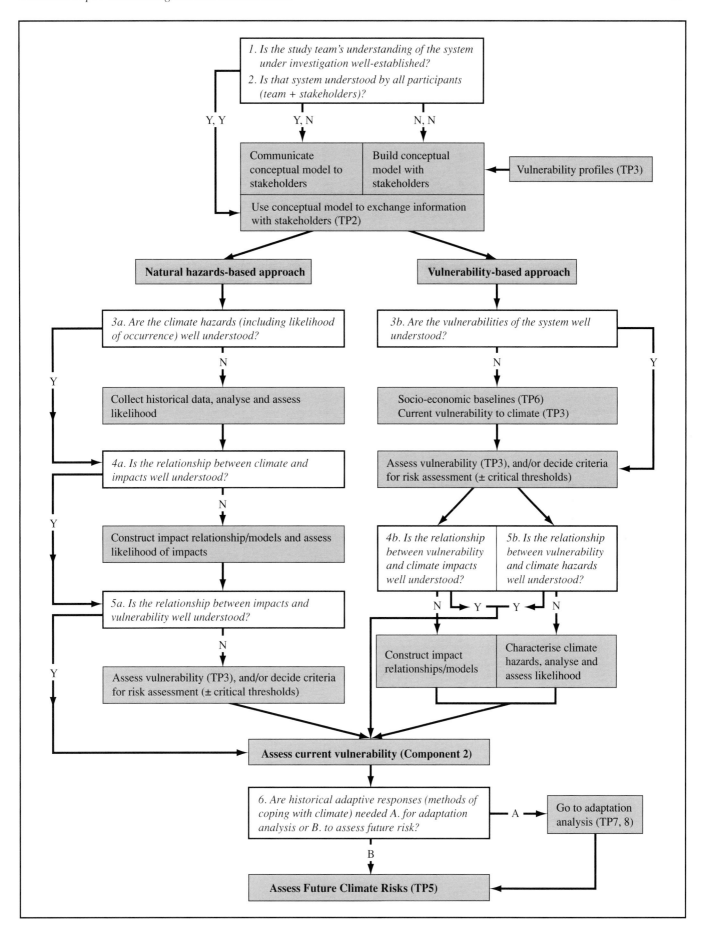

**Figure 4-3:** Flow chart for assessing current climate risk

stress, while vulnerability is the degree to which a system is susceptible to harm from that perturbation or stress (TP3 presents the development of conceptual models for assessing vulnerability).

Climate–society relationships can be identified through stakeholder workshops, or may be well known from previous work. The creation of lists, diagrams, tables, flow charts, pictograms and word pictures will create a body of information that can be further analysed. TP2 describes a number of ways this can be carried out with stakeholders. Establishing conceptual models in the early stages of an assessment can help the different participants develop a common understanding of the main relationships and can also serve as the basis for scientific modelling. In this chapter, we utilise the coping range extensively because of its utility as a template for understanding and analysing climate risks, but it is not the only such model that can be used. Other models include decision support systems, causal chains of hazard development, and mapping analysis (e.g., using geographic information systems). A comprehensive list of methods is provided in TP3.

### 4.4.2.  *Characterising climate variability, extremes and hazards*

The characterisation of climate variability begins with understanding the aspects of climate that cause harm, i.e., the climate hazards. With reference to the coping range, climate hazards are the aspects of climate variability and extremes that have the potential to exceed the ability to cope.

A starting question could be: "Are the climate hazards (affecting the system) known and understood?" There are two steps to this: the identification of the relevant climate hazards and their analysis. If the hazards for a system need to be identified, or their impact on the system investigated, the following questions can be addressed:

- Which climate variables and criteria do stakeholders use in managing climate-affected activities?
- Which climate variables most influence the ability to cope (i.e., those linked to climate hazards)?
- Which variables should be used in modelling and scenario construction?

These questions can be investigated by ways such as:

- Moving through a comprehensive checklist of climate variables in stakeholder workshops.
- Literature search, expert assessment and information from past projects.
- Exploring climate sensitivity with stakeholders, through interview, survey or focus groups.
- Building conceptual models of a system in a group environment.

Different aspects of climate variability will need to be examined. For example, rainfall can be separated into single events, daily variability and extremes, seasonal and annual totals and variability, and changes on longer (multi-annual and decadal) timescales. Daily extremes are important in urban systems for flash-flooding, inter-annual variability for disease vectors, and seasonal rains for dry-land agriculture. Temperature can be divided into mean, maximum and minimum daily averages, variability and extremes. In each system, people will have a different set of variables that they use to manage that system. Even though this management may not be scientific, it may be very sophisticated. Each of these variables involves a different level of skill in terms of climate modelling and has different degrees of predictability under climate change – information that is critical for building climate scenarios.

Hazards are not the same as extreme events, though they are related. Hazards are events and combinations of events with a propensity to cause harm, whereas extreme events are defined through rarity, impact, or a combination of both. Some extreme

*Table 4-1: Typology of climate extremes (based on Schneider and Sarukhan, 2001)*

| Type | Description | Examples of events | Typical method of characterisation |
|---|---|---|---|
| Simple | Individual local weather variables exceeding critical levels on a continuous scale | Heavy rainfall, high/low temperature, wind speed | Frequency/return period, sequence and/or duration of variable exceeding a critical level |
| Complex | Severe weather associated with particular climatic phenomena, often requiring a critical combination of variables | Tropical cyclones, droughts, ice storms, ENSO-related events | Frequency/return period, magnitude, duration of variable(s) exceeding a critical level, severity of impacts |
| Unique or singular | A plausible future climatic state with potentially extreme large-scale or global outcomes | Collapse of major ice sheets, cessation of thermohaline circulation, major circulation changes | Probability of occurrence and magnitude of impact |

events are defined as such because they occur rarely, such as a one in 100-year flood. Some more common events have extreme impacts, as in hurricanes or tropical cyclones, referred to as extreme events because of the damage they cause, rather than through rarity. Table 4-1 shows a typology of extreme climate events from the Intergovernmental Panel on Climate Change (IPCC) Third Assessment Report (TAR). A number of changes in extremes expected under climate change, and their impacts, are also associated with current extremes (Annex A.4.2).

Stress may occur in response to a shock associated with an extreme weather event, or accumulate through a series of events or a prolonged event such as drought. Risk assessment requires us to move from characterising extremes to defining hazards.

A climatic hazard is an event, or combination of climatic events, which has potentially harmful outcomes. Depending on the approach taken, hazards can be characterised in two ways: the *natural hazards-based approach*, where the focus is on the climate itself, and the *vulnerability-based approach* that stresses on the level of harm caused by an impact.

- The natural hazards-based approach is to fix a level of hazard, such as a peak wind speed of 10ms$^{-1}$, hurricane severity, or extreme temperature threshold of 35°C, then to see how that particular hazard affects vulnerability across space or time. Different social groupings will show varying degrees of vulnerability depending on their physical setting and socio-economic capacity.

- The vulnerability-based approach sets criteria based on the level of harm in the system being assessed then links that to a specific frequency, magnitude and/or combination of climate events. For example, if drought is known to harm a social group, we may choose to look at a given level of stress due to crop failure, and then determine the climatic characteristics that cause those shortages. Or if loss of property due to flooding is the level of vulnerability, then the rainfall and flood peak contributing to that level of flooding may constitute the hazard (and may be due to both climate and catchment conditions caused by land-use change). The level of vulnerability that provides this trigger can be decided jointly by researchers and stakeholders, chosen based on past experience or defined according to policy.

Figure 4-3 provides pathways for both of these approaches.

### 4.4.3. Impact assessment

Impact assessment under current climate can be used to establish a framework for how a climate hazard acts on society, or can look at vulnerability, then determine which climate hazards are involved. Qualitative methods can stand alone, or can establish the relationships prior to a modelling study.

*Qualitative methods*

Relationships between climate variables and impacts can be analysed by a number of methods such as ranking in order of importance, identifying critical control points within relationships, and quantifying interactions through sensitivity analysis (e.g., through workshops, focus groups and questionnaires). Often, this knowledge exists in institutions (e.g., agricultural extension networks) where important relationships are well known. In such cases, stakeholder workshops may allow the information to be gathered relatively easily. In other situations, several stakeholder workshops may be needed, the first to familiarise stakeholders with the issue of climate change (TP5, Figure 5-2) and to establish areas of shared knowledge and gaps, before investigating the specifics of a particular activity (TP2). Cross-impacts analysis, detailed in Annex A.4.1, can be used to manage the information gathered at such workshops.

The exploration of climate sensitivity with stakeholders is part of "learning by doing". By listing and discussing the climate variables that are important to them, stakeholders can consider the adaptations they currently use, the important thresholds or criteria they use in management and how those variables might change under climate change (TP2, Activity 3). Scenario builders and impact researchers have the opportunity to ask stakeholders which types of climatic events are important to them, and how they have responded to extreme events in the past (e.g., the relationship between climate events and changes in adaptive capacity, see TP7). This process is very useful if introduced with an overview of climate change and expected impacts. It is also an opportunity to discuss the policy and institutional environment, how non-climatic factors interact with climate in specific activities and issues of sustainable development (Activity 4, TP3). For example, in Bangladesh, damage from cyclones of the same intensity was US$1,780 million in 1991 and US$125 million in 1994. Reduction in damage was mainly due to setting up institutions after the 1991 cyclone and effective cyclone preparedness in 1994.

*Quantitative methods*

Quantitative impact assessment involves the formal assessment of climate, impacts and outcomes within a modelling framework. There is extensive literature on how to carry out impact assessment that includes IPCC assessment reports, impacts and adaptation assessment guidelines, and works within the individual disciplines (e.g., Carter and Parry, 1998; Carter et al., 1994; IPCC-TGCIA, 1999; UNEP, 1998).

In assessing current risk, impact modelling will largely concentrate on assessing the impacts of extreme events and variability, perhaps undertaking modelling to extend the results based on relatively short records of historical data (e.g., through statistical analysis). Sensitivity modelling in testing changes to variability and investigating extreme event probabilities can be of benefit later when climate scenarios are being constructed. Furthermore, given the difficulty in combining various types of climate uncertainty (discussed in TP5), sensitivity modelling of impacts under climate variability will help identify which uncertainties need to be represented in scenarios.

### 4.4.4.   *Risk assessment criteria*

As mentioned earlier, risk is a function of the likelihood of a harmful event and its consequences. Likelihood can be attached to the frequency of a hazard and/or to the frequency of given criteria being exceeded. All risk assessments need to be mindful of which criteria are important: what is to be measured and how are values to be attached to various outcomes?

Each assessment needs to develop its own criteria for the measurement of risk. Assessment criteria can be measured as a continuous function or in terms of limits or thresholds. For example, in farming, crop yields can be divided into good, moderate, poor and devastating yields depending on yield per hectare, per family or in terms of gross economic yield. There may be a minimum level of yield below which hardship becomes intolerable. This level can become a criterion by which risk is measured. It marks a reference point with known consequences to which probabilities can be attached. More sophisticated assessment may utilise different frequencies and combinations of good and bad years.

Levels of criteria that associate climate and impacts are known as *impact thresholds*, where the threshold marks a change in state. Impact thresholds can be grouped into two main categories: *biophysical* and *socio-economic*.

- Biophysical thresholds mark a physical discontinuity on a spatial or temporal scale. They represent a distinct change in conditions, such as the drying of a wetland, floods, breeding events. Climatic thresholds include frost, snow and monsoon onset. Ecological thresholds include breeding events, local to global extinction or the removal of specific conditions for survival.
- Socio-economic thresholds are set by benchmarking a level of performance. Exceeding a socio-economic threshold results in a change of the legal, managerial or regulatory state, and the economic or cultural behaviour. Examples of agricultural thresholds include the yield per unit area of a crop in weight, volume or gross income (Jones and Pittock, 1997).

*Critical thresholds* are defined as any degree of change that can link the onset of a critical biophysical or socio-economic impact to a particular climatic state (Pittock and Jones, 2000). Critical thresholds can be assessed using vulnerability assessment and mark the limit of tolerable harm (Pittock and Jones, 2000; Smit et al., 1999). For any system, a critical threshold is the combination of biophysical and socio-economic factors that marks a transition into vulnerability. The construction of a critical threshold can be used to limit the coping range. If this threshold can be linked with a level of climate hazard, then the likelihood of that threshold being exceeded can be estimated subjectively if the relationship is known qualitatively, or calculated if the relationship is quantifiable.

Table 4-2 lists a number of criteria, including thresholds, which have been used in climate risk assessments. They range from the biophysical to the socio-economic, from being universal to context-specific, and from the subjective to the objective. For example, economic write-off for infrastructure is socio-eco-

*Table 4-2: Examples of criteria used in impact and climate risk assessments (based on Jones, 2001)*

| SECTORS | CRITERIA | EXAMPLES |
|---|---|---|
| **Agriculture** | | |
| Animal health | • Temperature stress (also production) | Ahmed and El Amin (1997) |
| Animal production | • Parasites and disease | Estrada-Peña (2001); Sutherst (2001) |
| Crop production | • Carrying capacity | Hall et al. (1998) |
| | • Accumulated degree days to fruit and/or harvest | Kenny et al. (2000) |
| | • Yield | Chang (2002); Onyewotu et al. (1998); Mati (2000); Ferreyra et al. (2001) |
| Agro-meteorology | • Monsoon arrival | Smit and Cai (1996) |
| | • Multiple indices | Salinger et al. (2000); Sivakumar, (2000); Hammer et al. (2001) |
| Economic | • Net/Gross income per ha/farm/region/nation | Kumar and Parikh (2001) |

| SECTORS | CRITERIA | EXAMPLES |
|---|---|---|
| **Biodiversity** | | |
| Species or community abundance | • Vulnerable<br>• Endangered<br>• Sustainable population levels | Country/species specific |
| Species distribution | • Climate envelope shifts beyond current distribution<br>• Quantified change in core climatic distribution | Villers-Ruiz and Trejo-Vásquez, 1998) |
| Ecological processes | • Climatic thresholds affecting distribution<br>• Critical levels of mean browsing intensity<br>• Climatic threshold between eco-geomorphic systems | Kienast et al. (1999)<br>Lavee et al. (1998) |
| Phenology | • Mass bleaching events on coral reefs<br>• Winter chill – e.g., frequency of occurrence below daily min. temp. threshold<br>• Cumulative degree days for various biological thresholds<br>• Day length/temperature threshold for breeding<br>• Temperature threshold for coral bleaching | Hoegh-Guldberg (1999)<br>Hennessy and Clayton-Greene (1995), Kenny et al. (2000)<br>Spano et al. (1999)<br>Reading (1998)<br>Huppert and Stone (1998) |
| **Coastal zone** | | |
| General | • Salinity<br>• Flooding and wetlands<br>• Mangroves<br>• Planning for disasters/hazards<br>• Coastal dynamics<br>• Critical thresholds for atolls<br>• Regional assessment/multiple factors<br>• Infrastructure/economics | Nicholls et al. (1999)<br>Ewel et al. (1998)<br>Arthurton (1998)<br>Pethick (2001)<br>Dickinson (1999)<br>Perez et al. (1996); Yim (1996)<br>El Raey (1997) |
| **Forestry** | | |
| | • Distribution | Somaratne and Dhanapala (1996); Eeley et al. (1999) |
| **Hydrology** | | |
| Water quality | • Regulated water quality standards for factors such as salinity, dO, nutrients, turbidity. | Widespread and locally specific. |
| Water supply | • Regulated and/or legislated annual supply at system, district at farm level<br>• Water storage stress<br>• Renewable supply/water stress<br>• Institutional frameworks | Jones (2000); Bronstert et al. (2000)<br>Lane et al. (1999)<br>Jaber et al. (1997)<br>Arnell (1999); Savenije (2000) |
| Streamflow | • Maintenance or low-flow event frequency and duration<br>• Change in runoff and streamflow | El-Fadel et al. (2001)<br>Panagoulia and Dimou (1997)<br>Mkankam Kamga (2001) |
| Flooding | • Flood events | Panagoulia and Dimou (1997); Mirza (2002) |
| Drought | • Palmer drought severity index<br>• Drought exceptional circumstances | Palmer (1965)<br>White and Karssies (1999) |
| Hydroelectric power | • Current mean and minimum energy supply | Mimikou and Baltas (1997) |

| SECTORS | CRITERIA | EXAMPLES |
|---|---|---|
| **Human Health**<br><br>Vector-borne diseases<br><br><br><br><br>Thermal stress<br>Multiple Indices | • Aggregate epidemic potential<br>• Climatic envelope/indices of disease vector<br>• Critical density of vector to maintain virus transmission<br>• Heat and cold temperature levels and duration<br>• Disease and disaster | Patz et al. (1998)<br>McMichael (1996); Hales et al. (2002)<br>Jetten and Focks (1997); Martens et al. (1999); Lindblade et al. (2000a & b)<br>McMichael (1996)<br>Patz and Lindsay (1999); Epstein (2001); Watson and McMichael (2001) |
| **Infrastructure** | • Economic "write off", e.g., replacement less costly than repair<br>• Infrastructure condition falling below given standard | See TP8 for cost-benefit analysis |
| **Land degradation**<br>Erosion | • Threshold for overland flow erosion | Tucker and Slingerland (1997) |
| **Montane systems**<br>Glacial lakes<br>Montane cloud forests | • Catastrophic collapse and flooding<br>• Loss of ecosystem | Richardson and Reynolds (2000)<br>Foster (2001) |

nomic, context-specific and subjective, based on assumptions used in cost-benefit analysis. Degree-days to harvest for a crop is biophysical, universal and objective, but a threshold based on economic output from that crop will be socio-economic, context-specific and probably subjective.

Criteria for risk assessment can be developed using vulnerability analysis (TP3). Where criteria are context-specific, stakeholders and investigators can jointly formulate criteria that become a common and agreed metric for an assessment (Jones, 2001). These may link a series of criteria ranked according to outcomes (e.g., low to high), or be in the form of thresholds. Critical thresholds can be defined simply, as in the amount of rainfall required to distinguish a severe drought, e.g., <100 mm rainfall over a dry season, or can be complex, such as the accumulated deficit in irrigation allocations over a number of seasons (Jones and Page, 2001; TP5 Annex A.5.1). Widely applicable thresholds can be obtained from the literature. Other thresholds may be legal or regulatory (e.g., building safety standards, water quality standards).

There are no hard and fast rules for constructing thresholds – they are flexible tools that mark a change in state that is considered important. For example, stakeholders may link a given deficit of rainfall with drought hardship that leads to regional out migration, or loss of fresh water supply. Although annual and seasonal total rainfall is on a continuous scale, a change in behaviour associated with given amounts may constitute a threshold. Thresholds can vary widely over time and space, so

each assessment has to identify the adequate criteria. This will depend on a trade-off between the level of information available and what criteria are considered important.

### 4.4.5. *Assessing current climate risks*

This section demonstrates different methods of assessing risk under current climate. Within the broad framework of assessing risk, it is possible to conduct assessments that range from being qualitative to those that apply numerical techniques. As uncertainty decreases, the use of analytic and numerical methods increase, and the capacity to understand the system over changing circumstances increases. The following list outlines this development:

1. Understanding the relationships contributing to risk
2. Relating given states with a level of harm (e.g., low, medium and high risk)
3. Using statistical analysis, regression relationships
4. Using dynamic simulation
5. Using integrated assessment (multiple models or methods)

These methods can be used to undertake the following investigations:

• Understanding the relationship between climate and society at a given point in time
• Establishing current climate and society relationships

prior to investigating how climate change may affect these relationships (e.g., setting an adaptation baseline)
- Developing an understanding of how past adaptations have affected climate risks
- Assessing how technology, social change and climate are influencing a system, in order to be able to separate changes due to climate variability from changes due to ongoing adaptation (e.g., Viglizzo et al., 1997)
- Assessing how known adaptation strategies can further reduce current climate risks

*Choice of method*

The following examples show that there are a number of ways to assess climate risk. The method applied in Box 4-1 is hazard-driven, starting with the frequency and magnitude of extremes and their relationship to property damage and insurance claims. The assessment in Box 4-2 deals with famine, and in Box 4-3 with malarial outbreaks. In both cases, they have begun with the impacts causing vulnerability, and then identified the climate hazard driving those impacts. Adaptation in the form of early warning systems has been applied in the first case and recommended in the second. In both cases, socio-economic factors also affect the level of vulnerability. In Box 4-2, high prices and conflict make populations more vulnerable to drought. In Box 4-3, land-use change is exacerbating the climate hazard, specifically high minimum temperatures, increasing the survival of malaria vectors. Box 4-4 begins with an impact factor, crop yield, then identifies how deviations in yields are increasing over time; although average yields are increasing, so is vulnerability to bad years.

These differences help to explain why this TP does not offer tight prescriptions for constructing risk relationships in Section 4.4. Likewise, Figure 4-3 is not meant to provide similarly tight prescriptions. Either the right- or left-hand path, or both, can be taken. Questions can be missed. Perhaps this information already exists or is not needed for a particular assessment. It is also possible to start with impacts in the middle of the diagram and work forward to vulnerability and backwards towards hazards. In that case, techniques from TP3, this paper and TP6 could be utilised.

The natural hazards-based approach has been the traditional approach for assessing climate risks but, where the link between hazards and vulnerability are unclear, or where there are complex relationships between climate and non-climatic drivers, a vulnerability-based approach could be considered. This may involve setting desirable or undesirable criteria in the form of thresholds, then determining how hazards contributed to meeting or avoiding those criteria. For example, how achievable are given levels of water yield and quality, and food security, if the criteria for those are set first, then levels of exposure to climate hazards are determined? If the type and magnitude of hazard that may breach a given level of vulnerability is known, adaptation can then ensure that even larger hazards are managed.

*Examples*

Box 4-1 describes the vulnerability of property to wind damage in the south-eastern United States. This assessment takes a natural hazards-based approach (the left-hand path in Figure 4-3), where relationships between effective mean wind speed and property damage have been created and expressed in annual insurance claim and damage ratios. Having created these relationships, it would be possible to set thresholds for exceedance, e.g., the level where an insurance company may decide to charge higher premiums or to withdraw protection altogether. Alternatively, such criteria could be used to increase building-strength regulations in high-risk zones.

Box 4-2 describes a natural hazards-based approach to disaster prevention, where an early warning system is used to reduce the risk of famine accompanying drought and to increase the ability of people to cope with drought. The development of a Famine Early Warning System (FEWS) has increased the coping range of local populations, but incomplete uptake of the system, and the short-term nature of adaptation strategies means that significant risks still exist. This suggests that although the FEWS has increased the coping range to current climate variability, the delivery of its outputs needs to be fine-tuned and more widely disseminated. Continuing shocks are continuing to reduce the coping capacity of populations, requiring short-term risk management before considering longer-term adaptation options under climate change. This example is one where the current risks are so high, detailed risk assessment of possible future conditions are not required to prioritise adaptation options. In addition to short-term food aid, productive assets and viable livelihoods can only be restored by promoting longer-term development strategies and investments aimed at addressing the root causes of vulnerability to drought and food insecurity (FEWS NET, March 19, 2003).

Box 4-3 is an example of a risk assessment that follows the right-hand path of Figure 4-3. The investigation begins with an impact – malarial outbreak in highland East Africa – aiming to identify the hazards leading to those impacts. The major reason for the increase in malarial outbreaks was an increase in warmer micro-climates in villages near cleared swamps. This indicated that land use change is a factor in increasing malaria risk through increasing minimum temperatures. However, the basic climatic hazard was associated with the warmer temperatures of the El Niño event of 1997/98, which caused a malaria epidemic in the region. Lindblade et al. (2000a and b) also identified critical thresholds for *Anopheles* mosquito density that is associated with minimum temperatures. These densities could be used to develop sampling strategies to contribute to early warning systems. The identified hazards were of climatic (El Niño) and socio-economic (land-use change) origin. Further risk assessment under climate change would need to include both climatic and socio-economic drivers of change.

Box 4-4 shows an assessment of current climate risks within a system that is also changing due to non-climatic influences. Changing technology and cropping area have influenced rice production in Indonesia, creating a trend that masks the impacts

**Box 4-1: Assessing property damage from extreme winds**

The following example from Huang et al. (2001) assesses property damage from a model of extremes winds. Figures 4-4 and 4-5 show two damage relationships between effective mean wind speed and weighted claim and damage ratios from the southeastern United States. These ratios are the proportion of claims and damages made observed from Hurricanes Andrew and Hugo. One hundred percent of weighted claims or damages indicates that the maximum damage has been reached. Using Monte Carlo modelling of wind fields based on historical hurricane data and the data in Figures 4-4 and 4-5, Huang et al. (2001) estimated the spatial vulnerability to damage in Florida as expected annual claim and damage ratios for Florida (Figures 4-6 and 4-7).

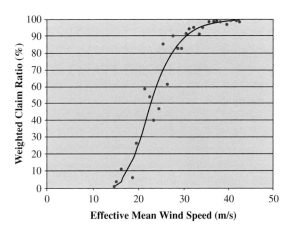

**Figure 4-4:** Claim ratio vs. effective mean surface wind speed

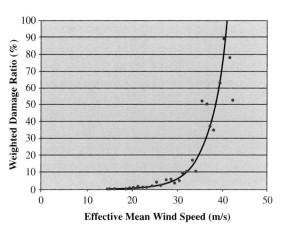

**Figure 4-5:** Damage ratio vs. effective mean surface wind speed

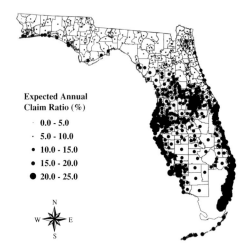

**Figure 4-6:** Expected annual claim ratio for each zip code in Florida

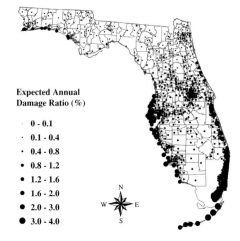

**Figure 4-7:** Expected annual damage ratio for each zip code in Florida

What critical thresholds or any other criteria measuring vulnerability could be used for the above information? Based on mean wind speed, weighted claims data increase markedly at >20 ms[-1]; damage ratios increase markedly at >30 ms[-1] and are a maximum at 41.4 ms[-1]. Huang et al. (2001) also include information about 50-year return interval wind gusts. Based on levels of property damage, a 2% expected annual damage ratio would see damage occurring to the total value of a building at least once in its 50-year design life. Thresholds could also be set by the insurance industry at levels where damage rates exceed returns. Under climate change, such thresholds may change spatially, or may change in likelihood of exceedance in a single location.

## Box 4-2: The use of climate forecasts in adapting to climate extremes in Ethiopia

### *Introduction*

In Ethiopia, famine has long been associated with fluctuations in rainfall. For example, a serious humanitarian disaster occurred during the 1984–5 Ethiopian drought when close to one million people perished. During 2000–1, a more serious drought affected most of Ethiopia. The failure of the 2000 *Belg* (secondary) rains was more critical compared to the case of 1984. This followed consecutive years of drought in 1998 and 1999, which had killed livestock and over-stretched the coping capacities of local populations. During the year 2000 however, a humanitarian crisis was averted due to a functioning Famine Early Warning System (FEWS) which had been put in place. However, another drought in 2002 has continued to decrease the ability of populations to cope.

### *Hazard assessment*

The mean rainfall in Ethiopia ranges from about 2000 mm in the southeast to <150 mm in the northeast. There are three seasons: *Bega*, a dry season (October – January); *Belg*, a short rain season (February – May) and *Kiremt*, a long rain season (June – September). Trend analysis showed declining rainfall over the northern half and south-western areas of Ethiopia. A vulnerability assessment showed that a decrease in rainfall over the northern parts of Ethiopia was expected. An investigation with three global climate models also indicated a risk of more frequent droughts under climate change.

### *Impacts*

The major negative impact is on food supply, since Ethiopia is dependent on rain-fed agriculture. Droughts affect the Greater Horn of Africa regularly and the resulting food crisis can easily affect up to twenty million people in Ethiopia alone. Apart from widespread famine, livestock perish and there is potential for armed conflict among communities. Increases in both climate variability and the intensity of drought in Ethiopia are anticipated under climate change.

### *Adaptation measures*

Following the human disaster in 1984, Ethiopia developed a comprehensive Famine Early Warning System, which integrated climate forecasts for Ethiopia with other information such as harvest assessments, vegetation indices and field reports. By 1999, early warning signals showed that a major famine was likely by 2000, due to drought and the border conflict between Ethiopia and Eritrea. As a result, the United States Agency for International Development (USAID) and European Union significantly increased their food aid commitments. Although there was a significant loss of livestock and livelihoods, a humanitarian disaster was averted. The FEWS played a significant role in sensitising the government and the famine early warning community. This also encouraged small anticipatory actions by affected populations, which improved their coping capacity.

### *Constraints*

In spite of a reasonable FEWS by the year 2000, government and donor decisions were not entirely driven by the FEWS. This meant that the potential maximum coping range could not be achieved in Ethiopia. Often the early warning bulletins did not target the appropriate audience. Secondly, the application of seasonal climate forecasts emphasised short-term responses, increasing the risk of reinforcing short-term strategies at the expense of longer-term adaptations and limiting resilience to increased climate change including variability. By early 2003, yet another drought and high prices had reduced the coping capacity of populations even further, and the FEWS had issued a pre-famine alert for 11.3 million people.

### *Conclusion*

Despite the probabilistic nature of climate forecasts and early warning systems, a well-designed FEWS can improve the resilience and coping capacity of communities to the impacts of climate variability and change. Early warning systems combined with good seasonal climate forecasts are cost-effective. Early warning information must be disseminated in a timely way to all stakeholders in formats they can understand or appreciate. However, as the events of 2002–3 show, repeated shocks can reduce coping capacities, requiring even greater intervention by outside agencies.

*This text is based on Kenneth Broad and Shardul Agrawala's report in Science Vol. 289, 8 September 2000; the Initial National Communication of Ethiopia to the UNFCCC and on-line at: http://www.fews.net.*

of climate. Despite this upward trend, drought still poses a risk to the majority of farmers in Indonesia. By developing a regression relationship to remove the production-based trend, it is possible to independently analyse the impacts of poor years on production and therefore, to assess the role of climate on drought risk. It shows that although adaptation is improving crop yields, individual poor years still constitute a risk.

This example has investigated question 4a in Figure 4-3: "Is the relationship between current climate and impacts well understood?" A vulnerability analysis of which populations were affected by low yields in bad years and how they were affected would help link climate hazards in terms of the El Niño–Southern Oscillation (ENSO) to vulnerabilities related to crop failure.

### 4.4.6.    Defining the climate risk baseline

An assessment of current climate risks (baseline) is needed for assessing future risks. Planned adaptation to future climate will be based on current individual, community and institutional behaviours that, in part, have been developed as a response to current climate. Existing adaptation is a response to the net effects of current climate (change, including variability) as expressed by the coping range. Adaptation analogues show that adapting to a future climate is influenced by past behaviour (Glantz, 1996; Parry, 1986; Warrick et al., 1986). This includes both autonomous and planned responses. Adaptation measures need to be consistent with current behaviour and future expectations if they are to be accepted by stakeholders. The analysis of behavioural responses to current climate variability also aids in the construction of climate scenarios.

Because the interactions between climate and society are dynamic (see Annex A.4.3 for a detailed explanation, also TP6), a climate-risk baseline needs to be created. This is an initial risk assessment at time = $t_0$, or even $t$-10, which provides the refer-

---

**Box 4-3: Investigating Malaria risks in highland East Africa**

*Impacts and vulnerability*

As highland regions of Africa historically have been considered free of malaria, recent epidemics in these areas have raised concerns that high elevation malaria transmission may be increasing. Hypotheses about the reasons for this include changes in climate, land use and demographic patterns. The effect of land use change on malaria transmission in the southwestern highlands of Uganda was investigated. Two related studies investigated the role of climate and malaria in highland Uganda and devised critical thresholds of vector density to provide early warnings of new outbreaks (Lindblade et al., 2000a and b).

*Hazard assessment*

From December 1997 to July 1998, during an epidemic associated with the 1997-8 El Niño, mosquito density, biting rates, sporozoite rates and entomological inoculation rates were compared between eight villages located along natural papyrus swamps and eight villages located along swamps that have been drained and cultivated. Since vegetation changes affect evapotranspiration patterns and thus, local climate, differences in temperature, humidity and saturation deficit between natural and cultivated swamps were also investigated. On average, all malaria indices were higher near cultivated swamps, although differences between cultivated and natural swamps were not statistically significant. However, maximum and minimum temperatures were significantly higher in communities bordering cultivated swamps. In multivariate analysis using a generalized estimating equation approach to Poisson regression, the average minimum temperature of a village was significantly associated with the number of *Anopheles gambiae s.l.* per house after adjustment for potential confounding variables. It appears that replacement of natural swamp vegetation with agricultural crops has led to increased temperatures, which may be responsible for elevated malaria transmission risk in cultivated areas.

*Critical thresholds linking vector density with malarial outbreaks*

Because malaria transmission is unstable and the population has little or no immunity, these highlands are prone to explosive outbreaks when densities of *Anopheles* exceed critical levels and conditions favour transmission. If an incipient epidemic can be detected early enough, control efforts may reduce morbidity, mortality and transmission. Three methods (direct, minimum sample size and sequential sampling approaches) were used to determine whether the household indoor resting density of *Anopheles gambiae* s.l, exceeded critical levels associated with epidemic transmission. A density of 0.25 Anopheles mosquitoes per house was associated with epidemic transmission, whereas 0.05 mosquitoes per house was chosen as a normal level expected during non-epidemic months. It is feasible, and probably expedient, to include monitoring of *Anopheles* density in highland malaria epidemic early warning systems. Although the local severity of the malaria epidemic was associated with changing microclimates associated with land use, the positive correlation between average minimum temperature and household densities of *Anopheles* mosquitoes shows that warmer seasons associated with El Niño and global warming pose a continuing threat.

**Box 4-4: Calculating climate-driven anomalies in the rice production system of Indonesia**

This assessment analysed 20 years of national rice production in Indonesia (BPS, 2000) to determine the impact of annual climate anomalies in a cropping system with an upward trend in yields. In the period 1980–1989, national rice production in Indonesia increased consistently from year to year, the increase slowing after 1989 (Figure 4-8). This increasing trend was due to improvements in crop management technology, variety and expansion of the rice planting area. In order to obtain anomaly data, this trend was removed by applying a regression equation. The steps of analysis are as follows:

1. Develop a regression equation to fit the rice production data
2. Calculate the deviation of observed data from the regression line as anomaly data
3. Separate the production anomalies between normal years and extreme years (Figure 4-8)
4. Evaluate trend of the anomalies between good years and bad years. Good years represent normal climate, while bad years represent extreme dry years due to the ENSO phenomenon.

Figure 4-9 shows that the anomalies for the bad years (squares) became more negative with time while those for good years (diamonds) became more positive over time. This indicates that the production loss due to extreme climate events tends to increase, or that the rice production system is becoming more vulnerable.

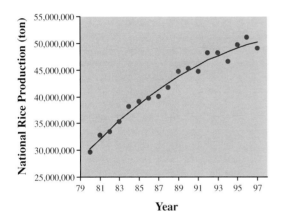

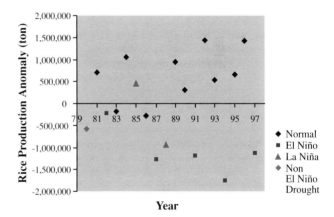

**Figure 4-8:** Rice production data and regression line     **Figure 4-9:** Rice production anomalies

ence on which future risks are measured. It is not the same as a climate baseline, which may be 1961–90, or longer. The climate-risk baseline can be tied to a period when both socio-economic and climate data are available, or to a period when particular infrastructure or policy was put in place. For example, when undertaking a risk assessment of water resources, Jones and Page (2001) used a climate baseline of 1890–1996, but the catchment and water resource management model they used was adjusted at flow rules set in 1996, so the risk became a measure of how the 1996 catchment would have behaved under historical climate. This allows a climate-risk baseline to be established using the full range of historical climate with modern catchment management rules.

### 4.5. Conclusions

By applying the methods outlined in this TP, the team can assess adaptive responses to past and present climate risks, and gain an understanding of the relationship between current climate risks and adaptive responses. This understanding will provide a basis

for developing adaptive responses to possible future climate risks. The assessment of climate hazards causing present climate vulnerability will also help decide which climate hazards need to be incorporated into scenario development.

Although an understanding of current climate–society interactions is an important starting point for adaptation to future climate, it would be dangerous to assume that new hazards will not arise and that new adaptations may not be needed. In most cases both current and future risk will need to be investigated. If knowledge of current climate risks is already established, then the team may move straight to TPs 5 and 6 to develop an understanding of how climate and socio-economic change may affect future climate risks. However, where current climate vulnerability is high, then adaptation to those risks will be required to develop sufficient capacity to cope with future risks (e.g., Box 4-3). In this case, basic information about how climate may affect those risks in the future could be sufficient.

The assessment of future climate risks is described in TP5.

# References

**Ahmed**, M.M.M. and El Amin, A.I. (1997). Effect of hot dry summer tropical climate on forage intake and milk yield in Holstein-Friesian and indigenous Zebu cows in Sudan, *Journal of Arid Environments*, **35**, 737–745.

**Arnell**, N.W. (1999). Climate change and global water resources. *Global Environmental Change*, **9**, S31–S49.

**Arthurton**, R.S. (1998). Marine-related physical natural hazards affecting coastal mega-cities of the Asia-Pacific region – awareness and mitigation. *Ocean and Coastal Management*, **40**, 65–85.

**Broad**, K. and Agrawala, S. (2000). Policy forum. Climate – The Ethiopia food crises – Uses and limits of climate forecasts. *Science*, 289, 1693-1694.

**Bronstert**, A., Jaeger, A., Güntner, A., Hauschild, M., Döll, P. and Krol, M. (2000). Integrated modelling of water availability and water use in the semi-arid northeast of Brazil. *Physics and Chemistry of the Earth, Part B: Hydrology, Oceans and Atmosphere*, **25**, 227–232.

**Carter**, T.R., Parry, M.L., Harasawa, H. and Nishioka, S. (1994). *IPCC Technical Guidelines for Assessing Climate Change Impacts and Adaptations*, London: University College, and Japan: Centre for Global Environmental Research.

**Carter**, T.R. and Parry, M. (1998). *Climate Impact and Adaptation Assessment: A Guide to the IPCC Approach*, London: Earthscan.

**Carter**, T.R. and La Rovere, E.L. (2001). Developing and applying scenarios. In: McCarthy, J.J., Canziani, O.F., Leary, N.A., Dokken, D.J. and White, K.S. eds., *Climate Change 2001: Impacts, Adaptation, and Vulnerability*, Contribution of Working Group II to the Third Assessment Report of the Intergovernmental Panel on Climate Change. Cambridge: Cambridge University Press, pp. 145–190.

**Chang**, C.C. (2002). The potential impact of climate change on Taiwan's agriculture. *Agricultural Economics*, **27**, 51–64.

**De Vries**, J. (1985). Analysis of historical climate society interaction. In Kates, K.W., Ausubel, J.H. and Berberian M. eds., *Climate Impact Assessment: Studies in the Interaction of Climate and Society*. Chichester, United Kingdom: Wiley, pp. 273-293.

**Dickinson**, W.R. (1999). Holocene sea-level record on funafuti and potential impact of global warming on central Pacific atolls. *Quaternary Research*, **51**, 124–132.

**Eeley**, H.A.C., Lawes, M.J. and Piper, S.E. (1999). The influence of climate change on the distribution of indigenous forest in KwaZulu-Natal, South Africa. *Journal of Biogeography*, **26**, 595–617.

**El-Fadel**, M., Zeinati, M. and Jamali, D. (2001). Water resources management in Lebanon: institutional capacity and policy options. *Water Policy*, **3**, 425–448.

**El-Raey**, M. (1997). Vulnerability assessment of the coastal zone of the Nile delta of Egypt to the impacts of sea level rise. *Ocean and Coastal Management*, **37**, 29–40.

**Epstein**, P.R. (2001). Climate change and emerging infectious diseases. *Microbes and Infection*, **3**, 747–754.

**Estrada-Peña**, A. (2001). Forecasting habitat suitability for ticks and prevention of tick-borne diseases. *Veterinary Parasitology*, **98**, 111–132.

**Ewel**, K., Twilley, R. and Ong, J. (1998). Different kinds of mangrove forests different kinds of goods and services. *Global Ecology and Biogeography Letters*, **7**, 83–94.

**Ferreyra**, R.A., Podestá G.P., Messina, C.D., Letson, D., Dardanelli, J., Guevara, E. and Meira, S. (2001). A linked-modeling framework to estimate maize production risk associated with ENSO-related climate variability in Argentina. *Agricultural and Forest Meteorology*, **107**, 177–192.

**Foster**, P. (2001). The potential negative impacts of global climate change on tropical montane cloud forests. *Earth-Science Reviews*, **55**, 73–106.

**Glantz**, M.H. (1996). Forecasting by analogy: local responses to global climate change. In: Smith, J., N. Bhatti, G. Menzhulin, R. Benioff, M.I. Budyko, M. Campos, B. Jallow, and F. Rijsberman (eds.), *Adapting to Climate Change: An International Perspective*. New York, NY, United States: Springer-Verlag, 407–426.

**Hales**, S., de Wet, N., Maindonald, J. and Woodward, A. (2002). Potential effect of population and climate changes on global distribution of dengue fever: an empirical model. *The Lancet*, **360**, 830–834.

**Hall**, W.B., McKeon, G.M., Carter, J.O., Day, K.A., Howden, S.M., Scanlan, J.C., Johnston, P.W. and Burrows, W.H. (1998). Climate change in Queensland's grazing lands: II. An assessment of the impact on animal production from native pastures. *Rangeland Journal*, **20**, 177–205.

**Hammer**, G.L., Hansen, J.W., Phillips, J.G., Mjelde, J.W., Hill, H., Love, A. and Potgieter, A. (2001). Advances in application of climate prediction in agriculture. *Agricultural Systems*, **70**, 515–553.

**Hennessy**, K.J. and Jones, R.N. (1999). *Climate Change Impacts in the Hunter Valley: Stakeholder Workshop Report*, CSIRO Atmospheric Research, Melbourne.

**Hennessy**, K.J. and Clayton-Greene, K. (1995). Greenhouse warming and vernalisation of high-chill fruit in southern Australia. *Climatic Change*, **30**, 327–348.

**Hoegh-Guldberg**, O. (1999). Climate change, coral bleaching and the future of the world's coral reefs. *Marine and Freshwater Research*, **50**, 839–866.

**Hewitt**, K. and Burton, I. (1971). *The Hazardousness of a Place: A Regional Ecology of Damaging Events*, Toronto: University of Toronto.

**Huang**, Z., Rosowsky, D.V. and Sparks, P.R. (2001). Long-term hurricane risk assessment and expected damage to residential structures. *Reliability Engineering and System Safety*, **74**, 239–249.

**Huppert**, A. and Stone, L. (1998). Chaos in the Pacific's coral reef bleaching cycle. *American Naturalist*, **152**, 447–459.

**IPCC-TGCIA** (1999). *Guidelines on the Use of Scenario Data for Climate Impact and Adaptation Assessment*. Version 1. Prepared by Carter, T.R., Hulme, M. and Lal, M., Intergovernmental Panel on Climate Change, Task Group on Scenarios for Climate Impact Assessment.

**IPCC** (2001) Summary for Policy-makers, in Houghton, J.T., Ding, Y., Griggs, D.J., Noguer, M., Van Der Linden, P.J. and Xioaosu, D., eds., *Climate Change 2001: The Scientific Basis*, Contribution of Working Group I to the Third Assessment Report of the Intergovernmental Panel on Climate Change, Cambridge University Press, Cambridge.

**Jaber**, J. O., Probert, S. D. and Badr, O. (1997). Water Scarcity: A Fundamental Crisis for Jordan. *Applied Energy*, **57**, 103–127.

**Jetten**, T.H. and Focks, D.A. (1997). Potential changes in the distribution of dengue transmission under climate warming. *American Journal of Tropical Medicine and Hygiene*, **57**, 285–297.

**Jones**, R.N. and Pittock, A.B. (1997). Assessing the impacts of climate change: the challenge for ecology, in Klomp N. and Lunt, I, eds, *Frontiers in Ecology: Building the Links*, Amsterdam: Elsevier Science Ltd, pp. 311–322.

**Jones**, R.N. (2000). Analysing the risk of climate change using an irrigation demand model, *Climate Research*, **14**, 89–100.

**Jones**, R.N. and Page, C.M. (2001). Assessing the risk of climate change on the water resources of the Macquarie River Catchment. Ghassemi, F., Whetton, P., Little, R. and Littleboy, M. eds., *Integrating Models for Natural Resources Management Across Disciplines, Issues and Scales (Part 2)*, Modsim 2001 International Congress on Modelling and Simulation, Modelling and Simulation Society of Australia and New Zealand, Canberra, pp. 673–678.

**Kienast**, F., Fritschi, J., Bissegger, M. and Abderhalden, W. (1999). Modeling successional patterns of high-elevation forests under changing herbivore pressure – responses at the landscape level. *Forest Ecology and Management*, **120**, 35–46.

**Kenny**, G.J., Warrick, R.A., Campbell, B.D., Sims, G.C., Camilleri, M., Jamieson, P.D., Mitchell, N.D., McPherson, H.G. and Salinger, M.J. (2000). Investigating climate change impacts and thresholds: an application of the CLIMPACTS integrated assessment model for New Zealand agriculture, *Climate Change*, **46**, 91–113.

**Kumar**, K.S.K. and Parikh, J. (2001). Indian agriculture and climate sensitivity. *Global Environmental Change*, **11**, 147–154.

**Lane**, M.E., Kirshen, P.H. and Vogel, R.M. (1999). Indicators of impacts of global climate change on US water resources. *Journal of Water Resources Planning M.-ASCE*, **125**, 194–204.

**Lavee**, H. Imeson, A.C. and Sarah, P. (1998). The impact of climate change on geomorphology and desertification along a Mediterranean-arid transect. *Land Degradation and Development*, **9**, 407–422.

**Lindblade**, K.A., Walker, E.D., Onapa, A.W., Katungu, J. and Wilson, M.L.

(2000). Land use change alters malaria transmission parameters by modifying temperature in a highland area of Uganda. *Tropical Medicine and International Health*, 5, 263–274.

Lindblade, K.A., Walker, E.D. and Wilson, M.L. (2000). Early warning of malaria epidemics in African highlands using Anopheles (Diptera: Culicidae) indoor resting density., *Journal of Medical Entomology*, 37, 664–674.

Martens, P., Kovats, R.S., Nijhof, S., de Vries, P., Livermore, M.T.J., Bradley, D.J., Cox, J. and McMichael, A.J. (1999) Climate change and future populations at risk of malaria, *Global Environmental Change*, 9, S89–S107.

Mati, B.M. (2000). The influence of climate change on maize production in the semi-humid–semi-arid areas of Kenya. *Journal of Arid Environments*, 46, 333–344.

McMichael, A.J. (1996). Human population health, in Watson, R.T., Zinyowera, M.C. and Moss, R.H. (eds), *Climate Change 1995: Impacts, Adaptations and Mitigation of Climate Change: Scientific-Technical Analyses*, Contribution of Working Group II to the Second Assessment Report of the Intergovernmental Panel on Climate Change, Cambridge: Cambridge University Press, pp. 561–584.

Mimikou, M.A. and Baltas, E.A. (1997). Climate change impacts on the reliability of hydroelectric energy production. *Hydrological Science Journal*, 42, 661–678.

Mirza, M.M.Q. (2002). Global warming and changes in the probability of occurrence of floods in Bangladesh and implications. *Global Environmental Change*, 12, 127–138.

Mkankam Kamga, F. (2001). Impact of greenhouse gas induced climate change on the runoff of the Upper Benue River (Cameroon). *Journal of Hydrology*, 252, 145–156.

Nicholls, R.J., Hoozemans, F.M.J. and Marchand, M. (1999). Increasing flood risk and wetland losses due to global sea-level rise: regional and global analyses. *Global Environmental Change*, 9, S69–S87.

Ogallo, L.A., Boulahya, M.S. and Keane, T. (2000). Applications of seasonal to interannual climate prediction in agricultural planning and operations. *Agricultural and Forest Meteorology*, 103, 159–166.

Onyewotu, L.O.Z., Stigter, C.J., Oladipo, E.O. and Owonubi J.J. (1998). Yields of millet between shelterbelts in semi-arid northern Nigeria, with a traditional and a scientific method of determining sowing date, and at two levels of organic manuring. *Netherlands Journal of Agricultural Science*, 46 53–64.

Panagoulia, D. and Dimou, G. (1997). Sensitivity of flood events to global climate change. *Journal of Hydrology*, 191, 208–222.

Parry, M.L. (1986). Some implications of climatic change for human development. In Clark, W.C. and Munn, R.E. eds., *Sustainable Development of the Biosphere*, Laxenburg, Austria: International Institute for Applied Systems Analysis, 378–406.

Patz, J.A. and Lindsay, S.W. (1999). New challenges, new tools: the impact of climate change on infectious diseases Commentary. *Current Opinion in Microbiology*, 2, 445–451.

Patz, J.A., Martens, W. J. M., Focks, D.A. and Jettson, T.H.: 1998, Dengue fever epidemic potential as projected by general circulation models of global climate change. *Environmental Health Perspectives*, 106, 147-153.

Perez, R.T., Feir, R.B. and Carandang, E.G. (1996). Potential impacts of sea level rise on the coastal resources of Manila Bay: a preliminary vulnerability assessment. *Water Air and Soil Pollution*, 42, 137–147.

Pethick, J. (2001). Coastal management and sea-level rise, *Catena*, 42, 307–322.

Pittock, A.B. and Jones R.N. (2000). Adaptation to what, and why? *Environmental Monitoring and Assessment*, 61, 9–35.

Reading, C.J. (1998). The effect on winter temperatures on the timing of breeding activity in the common toad. *Bufo bufo, Oecologia*, 117, 469–475.

Richardson, S.D., Reynolds, J.M. (2000). An overview of glacial hazards in the Himalayas. *Quaternary International*, 65–66, 31–47.

Salinger, M.J., Stigter, C.J. and Das, H.P. (2000). Agrometeorological adaptation strategies to increasing climate variability and climate change. *Agricultural and Forest Meteorology*, 103, 167–184.

Savenije, H.H.G. (2000). Water scarcity indicators; the deception of the num-

bers. *Physics and Chemistry of the Earth, Part B: Hydrology, Oceans and Atmosphere*, 25, 199–204.

Schneider, S.H. and Sarukhan, J. (2001). Overview of impacts, adaptation, and vulnerability to climate change, In McCarthy, J.J., Canziani, O.F., Leary, N.A., Dokken, D.J. and White, K.S. eds. *Climate Change 2001: Impacts, Adaptation, and Vulnerability*, Contribution of Working Group II to the Third Assessment Report of the Intergovernmental Panel on Climate Change, Cambridge: Cambridge University Press, 75–103.

Sivakumar, M.V.K., Gommes R. and Baier, W. (2000). Agrometeorology and sustainable agriculture, *Agricultural and Forest Meteorology*, 103, 11–26.

Smit, B., Burton, I., Klein, R.J.T. and Street, R. (1999). The science of adaptation: a framework for assessment, *Mitigation and Adaptation Strategies*, 4, 199–213.

Smit, B. and Cai, Y.L. (1996). Climate change and agriculture in China. *Global Environmental Change*, 6, 205–214.

Smit, B. and Pilifosova, O. (2002). From adaptation to adaptive capacity and vulnerability reduction, In Smith, J.B., Klein, R.J.T. and Hug, S. eds., *Climate Change, Adaptive Capacity and Development*, Imperial College Press, London.

Somaratne, S. and Dhanapala, A.H. (1996). Potential impact of global climate change on forest distribution in Sri Lanka, *Water, Air, and Soil Pollution*, 92, 129–135.

Spano, D., Cesaraccio, C., Duce, P. and Snyder, R.L. (1999). Phenological stages of natural species and their use as climate indicators. *International Journal of Biometrics*, 42, 124–133.

Sutherst, Robert W. (2001). The vulnerability of animal and human health to parasites under global change. *International Journal for Parasitology*, 31, 933–948.

Tarhule, A. and Woo, M.-K. (1997). Towards an interpretation of historical droughts in northern Nigeria. *Climatic Change*, 37, 601-616.

Tucker, G.E. and Slingerland, R. (1997). Drainage basin responses to climate change. *Water Resources Research*, 33, 2031–2047.

UNEP (1998). *Handbook on Methods for Climate Change Impact Assessment and Adaptation Strategies, Version 2.0*, Feenstra, J.F., Burton, I. Smith, J.B. and Tol, R.S.J. (eds.) United Nations Environment Programme, Vrije Universiteit Amsterdam, Institute for Environmental Studies, http://www.vu.nl/english/o_o/instituten/IVM/research/climate-change/Handbook.htm.

Viglizzo, E.F., Roberto, Z.E., Lértora, F., Gay, E.L. and Bernardos, J. (1997). Climate and land-use change in field-crop ecosystems of Argentina. *Agriculture. Ecosystems and Environment*, 66, 61–70.

Villers-Ruíz, Lourdes; Trejo-Vázquez, Irma, (1998). Climate change on Mexican forests and natural protected areas. *Global Environmental Change*, 8, 141–157.

Warrick, R.A, Gifford, R.M. and Parry, M.L., (1986). $CO_2$, climatic change and agriculture. In Bolin, B., B.R. Döös, J. Jager, and R.A. Warrick (eds.), *The Greenhouse Effect, Climatic Change and Ecosystems*, New York: John Wiley and Sons, pp. 393–473.

Watson, R.T. and McMichael, A.J. (2001). Global Climate Change – the Latest Assessment: Does Global Warming Warrant a Health Warning? *Global Change and Human Health*, 2, 64–75.

White, D.H. and Karssies, L. (1999). Australia's national drought policy: aims, analyses and implementation. *Water International*, 24, 2–9.

Yim, W.W.-S. (1996). Vulnerability and adaptation of Hong Kong to hazards under climatic change conditions. *Oceanographic Water Air and Soil Pollution*, 92, 181–190.

Yohe, G. and Tol, R.S.J. (2002). Indicators for social and economic coping capacity – moving toward a working definition of adaptive capacity. *Global Environmental Change*, 12, 25–40.

# ANNEXES

### Annex A.4.1. Cross-impacts analysis

The results from a sectoral or regional investigation can be collated and analysed through the use of sensitivity and cross-impacts matrices[3]. The feedback from stakeholders is usually positive when such matrices are used. The activity/variable matrix shown in Table A-4-1-1 is an example from a project carried out in the Hunter Valley, Australia. From a stakeholder workshop, key climate and climate-related variables were listed and linked to selected activities or exposure units. The questions asked were what aspects of climate currently cause impacts in your region, and what activities are affected? The climate variables were then linked to how they affected each activity using a weighting of 3, 2, or 1 to denote strong, moderate or weak influences. Activities were divided into four main groupings: agriculture, coastal and marine, catchment and the built environment. The row and column values were summed and the results shown in Table A-4-1-2.

Table A-4-1-2 shows two outcomes of the analysis. The climate variables having the greatest impact are aspects of rainfall variability, with a lesser emphasis on temperature. Moisture levels

*Table A-4-1-1: Weighted sensitivity matrix of key climate variables and climate-related variables compared with selected activities or exposure units based on Table 1 of the workshop report (Hennessy and Jones, 1999).*

| Sensitivity matrix linking climate drivers (below) with activities (across) | Poultry | Dairy | Grazing | Cropping | Wine | Horses | Marine (esp. fisheries) | Beach | Coastal water supply | Harbour | Inland water supply | River management | Dryland/irrigation salinity | Forest & biodiversity | Urban infrastructure | Air quality | Waste | Health | Industry, coal & power | Driving force |
|---|---|---|---|---|---|---|---|---|---|---|---|---|---|---|---|---|---|---|---|---|
| Rainfall - average |  | 2 | 1 | 2 | 2 |  |  |  | 1 |  | 3 | 2 | 2 | 2 | 2 | 1 |  |  | 1 | 21 |
| Rainfall - extreme |  | 1 | 2 | 2 | 1 | 1 | 1 | 2 | 2 |  | 2 | 3 | 3 | 2 | 3 |  | 2 | 2 | 1 | 30 |
| Rainfall - variability |  | 2 | 3 | 3 | 2 |  | 2 |  | 1 |  | 3 | 3 | 1 | 1 | 2 |  |  |  |  | 23 |
| Drought |  | 2 | 3 | 3 | 2 | 1 | 1 |  | 2 |  | 3 | 2 | 2 | 3 |  |  | 2 | 2 |  | 28 |
| Temperature - average | 1 | 2 | 1 | 2 | 2 |  | 2 |  |  | 1 |  | 2 | 1 | 2 |  | 2 | 1 |  | 1 | 20 |
| Temperature - max | 3 | 2 | 2 | 2 |  | 2 | 1 |  |  |  |  |  | 3 | 1 | 3 |  | 2 | 1 | 2 | 24 |
| Temperature - min |  | 2 | 2 | 2 | 2 | 2 |  |  |  |  |  |  |  | 1 | 2 |  | 1 | 1 | 1 | 16 |
| CO$_2$ |  | 2 | 2 | 2 | 2 | 1 |  |  |  |  |  |  |  | 2 |  |  |  |  |  | 11 |
| Cloud |  |  | 1 |  |  |  |  |  |  |  |  |  |  |  | 1 | 1 |  |  |  | 3 |
| Pressure |  |  |  |  |  |  |  |  |  |  |  |  |  |  |  | 1 |  |  |  | 1 |
| Humidity | 2 | 1 |  |  | 2 |  |  |  |  |  |  |  |  | 1 | 2 |  |  | 1 | 1 | 10 |
| Wind |  | 1 | 1 | 1 | 1 |  | 1 | 2 |  | 2 |  |  |  | 1 | 2 | 2 |  | 2 |  | 16 |
| Evaporation | 2 |  |  |  |  |  | 1 |  |  |  | 2 | 1 | 2 | 1 |  |  | 1 |  |  | 10 |
| Soil moisture |  | 3 | 3 | 3 | 3 | 2 |  |  |  |  | 1 |  | 2 |  | 1 |  | 2 |  |  | 20 |
| Stream flow |  |  |  |  |  |  | 2 |  | 1 |  | 3 | 3 | 2 | 1 | 1 |  |  | 3 | 1 | 17 |
| Flood |  | 2 | 1 | 1 | 1 | 1 | 1 |  |  | 3 | 2 | 3 | 2 | 1 | 3 |  | 3 | 3 | 2 | 29 |
| Water table |  |  |  |  |  |  | 2 |  | 3 |  | 1 | 1 | 1 | 1 | 1 |  | 2 |  |  | 12 |
| Water salinity |  |  | 1 | 1 |  |  | 1 |  | 3 |  | 2 | 2 | 3 | 1 |  |  |  | 2 | 1 | 17 |
| Irrigation |  | 2 | 1 | 2 | 3 |  |  |  |  |  | 2 | 2 | 1 |  |  |  |  |  |  | 13 |
| Sea level |  |  |  |  |  |  | 1 | 3 | 3 | 3 |  |  |  |  | 2 |  |  |  |  | 12 |
| Storm surge |  |  |  |  |  |  |  | 3 |  | 3 |  |  |  |  | 1 |  |  |  |  | 7 |
| Waves |  |  |  |  |  |  | 2 | 3 | 1 | 2 |  |  |  |  |  |  |  |  |  | 8 |
| Lightning |  |  |  |  |  |  |  |  |  |  |  |  |  |  | 1 | 1 |  |  | 1 | 3 |
| Hail |  |  | 2 | 3 |  |  |  |  |  |  |  |  |  |  | 2 |  |  |  |  | 7 |
| Fire |  |  | 1 |  | 1 | 1 |  |  |  |  |  |  |  | 3 | 1 | 2 |  | 1 | 2 | 12 |
| Total sensitivity | 8 | 24 | 23 | 29 | 28 | 11 | 18 | 13 | 17 | 14 | 24 | 27 | 20 | 27 | 30 | 9 | 14 | 18 | 16 |  |

[3] These matrices were illustrated in Carter et al. (1994) but have not been widely used.

on land or in the atmosphere are also important. The activities showing the largest climatic sensitivity influence are largely rural land-based activities. Coastal aspects have a moderate exposure to climate variables due to a few ocean-related variables being very important while most others have little influence. Those activities with a broad exposure to climate are difficult to assess due to the number of forcing variables and feedbacks. The criteria of low, medium and high have been chosen subjectively, and are intended to indicate the relative importance of the various results.

Cross-impacts analysis can be used to map the relationships between drivers and dependent variables in a system. Table A-4-1-3 contains all climate variables, catchment-related variables and major activities shown in Table A-4-1-1 on both axes (some variables more important to the urban, agricultural and coastal systems were removed or combined). Each variable on the vertical axis was examined to determine whether it is likely to force a change in all other variables on the horizontal axis. Where this condition was true, an entry was made in the appropriate cell. Where variables act upon each other, both cells are marked. Note that economic and social activities affecting the

catchment have been omitted. Table A-4-1-3 is a cross-impacts matrix based on the variables in Tables A-4-1-1 and A-4-1-2. A caveat with this type of analysis is that the identification of cause-and-effect is subjective, where:

i. two variables may be interdependent, but this interdependence is not well understood, or

ii. a sequence of consequences may indirectly link a variable and an activity.

Figure A-4-1-1 shows the results from Table A-4-1-3 on a forcing/dependency graph. The variables on the upper left are those that show strong external forcing but are not affected much by what is going on inside the system. Those labelled autonomous on the lower left may be important in specific cases but have a minor role overall. The upper right variables are relay variables that are highly dependent on factors within the system, but are also strong influences on other variables. These variables are likely to exhibit feedbacks. On the lower right are the dependent variables that are sensitive to many other variables above and to the left of them. These latter variables are the important

*Table A-4-1-2: Results of sensitivity matrix showing the climate and related variables with the greatest forcing and activities with the broadest sensitivity to climate*

| Forcing and sensitivity category and range of weighted values | Climate and related variables (forcing) | Activities (sensitivity) |
|---|---|---|
| High (21-30) | Rainfall – extreme<br>Flood<br>Drought<br>Temperature – max<br>Rainfall – variability<br>Rainfall – average | Urban infrastructure<br>Cropping<br>Wine<br>River management<br>Forest & biodiversity<br>Inland water supply<br>Dairy<br>Grazing |
| Moderate (11-20) | Temperature – average<br>Soil moisture stream flow<br>Water salinity<br>Temperature – min<br>Wind<br>Irrigation<br>Water table<br>Sea level<br>Fire | Dryland/irrigation salinity<br>Industry, coal & power<br>Marine (esp. fisheries)<br>Coastal water supply<br>Health<br>Harbour<br>Waste<br>Beach<br>Horses |
| Low (1-10) | Humidity<br>Evaporation<br>Waves<br>Storm surge<br>Hail<br>Cloud<br>Lightning<br>Pressure | Air quality<br>Poultry |

Table A-4-1-3: Interaction matrix for climate change – catchment interactions for the Hunter River in New South Wales under climate change. (Forcing marks where a change in the variable on the vertical axis forces a change in a variable on the horizontal axis; read across the matrix. Dependency is read down the matrix.)

| Interaction Matrix for Catchment-Climate Change Impacts | Forcing | Dependency |
|---|---|---|
| Temperature (average) | 18 | 7 |
| Temperature (maximum) | 12 | 3 |
| Temperature (minimum) | 9 | 4 |
| Precipitation (average) | 29 | 3 |
| Precipitation (intensity) | 25 | 2 |
| Precipitation (variability) | 30 | 7 |
| Drought | 33 | 15 |
| Radiation/sunshine | 15 | 3 |
| Cloudiness | 12 | 5 |
| Humidity | 11 | 10 |
| Evapotranspiration | 25 | 15 |
| Evaporation (open water) | 8 | 7 |
| Wind | 9 | 1 |
| $CO_2$ | 8 | 5 |
| Soil Moisture | 20 | 16 |
| Recharge | 4 | 15 |
| Runoff | 19 | 12 |
| Streamflow (peak) | 18 | 15 |
| Streamflow (average) | 15 | 15 |
| Streamflow (variability) | 13 | 19 |
| Water storage | 16 | 19 |
| Water supply (domestic) | 7 | 16 |
| Water supply (irrigation) | 10 | 18 |
| Water supply (power) | 9 | 11 |
| Riverine aquifer | 11 | 11 |
| Mine dewatering | 5 | 12 |
| Water quality (nutrients) | 4 | 22 |
| Water quality (turbidity) | 5 | 23 |
| Water quality (dO) | 1 | 21 |
| Water quality (salinity) | 5 | 18 |
| Stream/wetland ecology | 6 | 34 |
| Algal blooms | 6 | 21 |
| Land degradation (erosion) | 19 | 14 |
| Land degradation (salinity) | 13 | 13 |
| Fire | 17 | 17 |
| Vegetation cover | 22 | 20 |
| Biodiversity | 11 | 24 |
| Forest | 24 | 22 |
| Agriculture | 22 | 29 |
| Urban areas | 24 | 22 |
| Coastal zone | 5 | 22 |

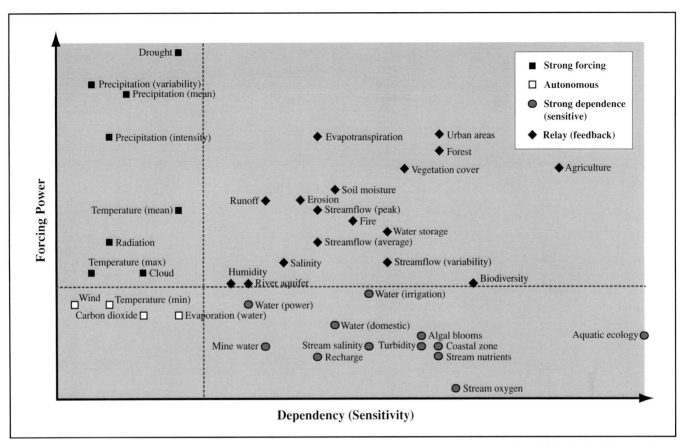

**Figure A-4-1-1:** Forcing/dependency chart for climate, catchment processes and catchment-based activities in the Hunter River Valley (based on the cross-impacts analysis presented in Table A-4-1-3.)

outputs for the system. They are used to construct measures of environmental quality and to monitor how well the system is working. They are also the most vulnerable. This type of analysis can show:

- which drivers are external to the system (in the top left quadrant),
- which variables are important drivers but are themselves modified by feedbacks within the system (top right), and
- the most important indicators of health and water quality (shown in the lower right of the system and affected by everything else).

The results may be no surprise to the research team but this type of analysis is useful for managers and other stakeholders who are dealing with complex environmental systems.

**Annex A.4.2. Examples of impacts resulting from projected changes in extreme climate events (from Carter and La Rovere, 2001)**

| Projected Changes during the 21st Century in Extreme Climate Phenomena and their Likelihood[a] | Representative Examples of Projected Impacts[b] (all high confidence of occurrence in some areas[c]) |
|---|---|
| **Simple Extremes** | |
| Higher maximum temperatures; more hot days and heat waves[d] over nearly all land areas (Very Likely[a]) | • Increased incidence of death and serious illness in older age groups and urban poor<br>• Increased heat stress in livestock and wildlife<br>• Shift in tourist destinations<br>• Increased risk of damage to a number of crops<br>• Increased electric cooling demand and reduced energy supply reliability |
| Higher (increasing) minimum temperatures; fewer cold days, frost days, and cold waves[d] over nearly all land areas (Very Likely[a]) | • Decreased cold-related human morbidity and mortality<br>• Decreased risk of damage to a number of crops, and increased risk to others<br>• Extended range and activity of some pest and disease vectors<br>• Reduced heating energy demand |
| More intense precipitation events (Very Likely[a] over many areas) | • Increased flood, landslide, avalanche, and mudslide damage<br>• Increased soil erosion<br>• Increased flood runoff could increase recharge of some floodplain aquifers<br>• Increased pressure on government and private flood insurance systems and disaster relief |
| **Complex Extremes** | |
| Increased summer drying over most mid-latitude continental interiors and associated risk of drought (Likely[a]) | • Decreased crop yields<br>• Increased damage to building foundations caused by ground shrinkage<br>• Decreased water resource quantity and quality<br>• Increased risk of forest fire |
| Increase in tropical cyclone peak wind intensities, mean and peak precipitation intensities (Likely[a] over some areas)[e] | • Increased risks to human life, risk of infectious disease epidemics, and many other risks<br>• Increased coastal erosion and damage to coastal buildings and infrastructure<br>• Increased damage to coastal ecosystems such as coral reefs and mangroves |
| Intensified droughts and floods associated with El Niño events in many different regions (Likely[a]) | • Decreased agricultural and rangeland productivity in drought- and flood-prone regions<br>• Decreased hydro-power potential in drought-prone regions |
| Increased Asian summer monsoon precipitation variability (Likely[a]) | • Increase in flood and drought magnitude and damages in temperate and tropical Asia |
| Increased intensity of mid-latitude storms (little agreement between current models)[d] | • Increased risks to human life and health<br>• Increased property and infrastructure losses<br>• Increased damage to coastal ecosystems |

[a] Likelihood refers to judgmental estimates of confidence used by TAR WGI: very likely (90-99% chance); likely (66-90% chance). Unless otherwise stated, information on climate phenomena is taken from the *Summary for Policymakers, TAR WGI.*

[b] These impacts can be lessened by appropriate response measures.

[c] Based on information from chapters in this report; high confidence refers to probabilities between 67 and 95% as described in *Footnote 6 of TAR WGII, Summary for Policymakers.*

[d] Information from *TAR WGI, Technical Summary, Section F.5.*

[e] Changes in regional distribution of tropical cyclones are possible but have not been established.

## Annex A.4.3. Coping range structure and dynamics

The coping range is a conceptual framework that provides a structure for showing how a system, or an activity, has coped historically and how it copes now, e.g., how has the system responded to past and present climate risk. If the team is to use the coping range they first need to be aware of its basic dynamics, in order to be able to adapt it to the specific circumstances of an assessment. The coping range, response relationships and thresholds can be constructed independently of climate change scenarios, and that information will continue to be relevant even if projections of climate change alter.

Climate–society relationships, and by implication coping ranges, are dynamic. The coping range has two main dynamic influences that can affect the sensitivity of the system:

- Changes in climate drivers can change the frequency and magnitude of hazards, and
- Changes in socio-economic drivers can alter the capacity of the system to cope with hazards.

If a system moves beyond its coping range, the level of harm suffered can threaten sustainability in a number of ways. People may be harmed through loss of livelihood, injury or death. An activity could cease, the coping range may narrow through reduced socio-economic capacity, system sensitivity may increase, or adaptive capacity may be reduced (i.e., the system survives the current stress but its ability to adjust to future change is reduced).

The climatic phenomena used to describe coping range may be simple (as in a single driver such as average temperature or total rainfall), a combination of factors influencing a process (e.g., temperature, rainfall, photosynthetically active radiation and $CO_2$ for crop production) or indirect variables that can be linked to climate (such as stream flow or crop yield). The coping range can be expressed in a number of ways, ranging from narrative to mathematical. Graphically, one climate or climate-related driver can be shown as a time series, two drivers can be expressed on a response surface, and three in three-dimensional charts.

Within the coping range, an activity is resistant – able to withstand stress – or resilient – able to weather stresses without undergoing significant change). Beyond this range is a zone of vulnerability. Some losses may be so large that people's livelihoods are threatened by losses to environmental security. In many systems, this may take several seasons of loss to occur, and in the most vulnerable systems, only one season. Often, when people are coping poorly, they have lost environmental security through previous events that may, or may not be, climate related.

There are several ways to show outcomes in terms of the coping range. They can be portrayed in terms of continuous output, such as the relationship between crop yield and climate. They can be segmented into good, moderate or poor outcomes; or we can choose trigger points, or thresholds, where either the system changes, or a change in management is indicated. For example, drought policy may stipulate a level of rainfall, or an aridity index, and if conditions remain below these levels for a sustained period, drought conditions are declared. These outcomes can become the criteria for a risk assessment where changes in the frequency of drought declarations over time are measured. It is also possible to use a critical threshold to measure risk. This is the point where the level of harm is too high to be tolerated and a system moves beyond the coping range into a state of vulnerability.

The width of the coping range is a function of historical adaptation. For developing countries, in cases where the capacity to adapt has been limited by factors such as access to technology and financial resources, climate variability is large and the reliance on climate is high (e.g., Ogallo et al., 2000), the coping range may be small compared to the range of climate variability. Small coping ranges are likely to be breached by numerous single events. Large coping ranges, typical of developed countries where resilience is high, may experience a sequence of extreme events such as a string of droughts before impacts become unacceptable (e.g., Smit et al., 1999)[4]. Historical adaptation influences the behaviour upon which any response to climate change will be based. Adaptation to current climate stress is influenced by past behaviour (Glantz, 1996; Parry, 1986; Warrick et al., 1986). Adaptive capacity is the ability to adjust to change through adaptation, and is thus a potential that can be brought into play by an experience of stress or information about a potential future stress. The level of adaptive capacity will influence the evolution of the coping range.

### *Relationship between coping range and adaptive capacity*

Adaptive capacity is a measure of the potential to adapt (TP7). When realised, it becomes coping capacity or the ability to cope. Adaptive capacity describes the potential of the coping range to expand or contract in response to autonomous (unplanned) or planned changes to the environment. Most systems affected by climate will also be affected by other drivers of change. For example, as well as climate, farming systems are affected by land tenure, cost structure and commodity prices and trade relationships. These can be independent of climate or can interact with climate in complex ways affecting the dynamics of adaptive and coping capacity. This is true of many other systems where natural resources are being managed, and for health, where complex social interactions can affect climate-driven exposure to disease (e.g., mosquito vectors).

Figure A-4-3-1 shows four different relationships between climate variability and coping capacity that can be called Decreasing Resilience, Increasing Resilience, Suffering Climate Shocks and Responding to Climate Shocks. Graphs 1 and 2 rep-

---

[4] This is a generalisation that is consistent with the overall findings of the IPCC Third Assessment Report. However, some coping ranges in developing countries are substantial and some coping ranges in developed countries are extremely limited. Each situated needs to be assessed individually.

resent autonomous changes occurring independently of climatic responses. For example, environmental degradation may make people more vulnerable to climate extremes, and economic diversification may make them less vulnerable. Graphs 3 and 4 show where the coping range changes directly in response to climate extremes. In graph 3, where adaptive capacity is low to non-existent, the coping range will decrease in response to climate shocks. In graph 4, where adaptive capacity is moderate to high, the coping range will increase in response to climate shocks. In most systems, all four of these influences are likely to be interacting, and

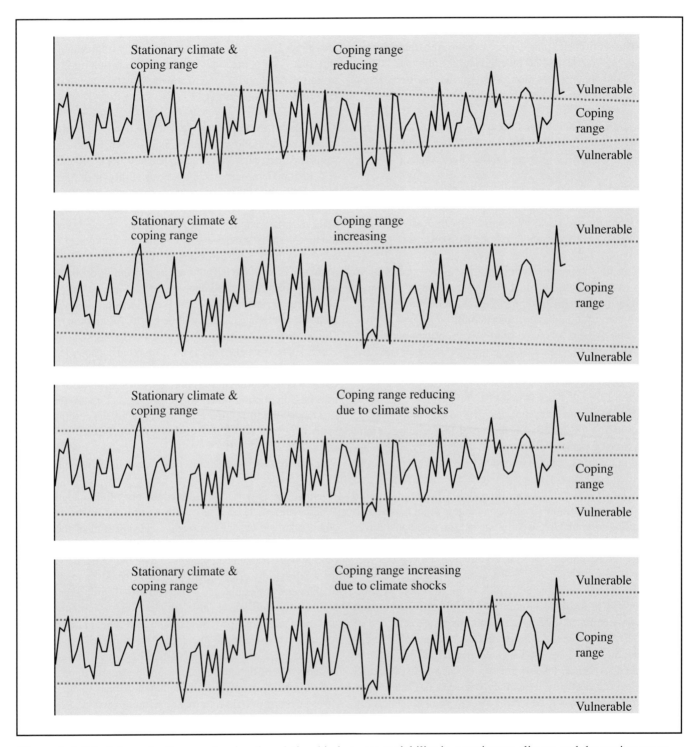

**Figure A-4-3-1:** Schematic diagram showing the relationship between variability in a stationary climate and the coping range, showing four different mechanisms that can be called (1) decreasing resilience, (2) increasing resilience, (3) suffering climate shocks and (4) responding to climate shocks. Graph 1 shows gradual decreases in coping capacity over time; Graph 2 shows gradual increases in coping capacity over time; Graph 3 shows climate shocks reducing coping capacity over time (adaptive capacity here would be low to non-existent); and Graph 4 shows climate shocks producing an increase in coping capacity (where adaptive capacity is high). See also de Vries (1985) and Smit and Pilifosova (2002).

analysis needs to identify the over-riding determinants of changing responses. This is the "bumpy road" of irregular socio-economic change mentioned in TP6. Not shown are dynamics, where following a change, conditions relax back to the original situation (e.g., where water conservation measures are gradually discarded following a period of enforced restrictions).

The coping range can be utilised in various ways. One is to assess vulnerability assuming the climate will change, while holding the ability to cope constant, to test what adaptation may be needed in response to climate change. Another way is to change climate and the coping range according to expectations of adaptive capacity being developed and generating an adaptive response to climate. This is a more dynamic situation where both climate and the coping range change over the time.

# 5

# Assessing Future Climate Risks

ROGER JONES[1] AND LINDA MEARNS[2]

Contributing Authors
*Stephen Magezi[3] and Rizaldi Boer[4]*

Reviewers
*Mozaharul Alam[5], Suruchi Bhawal[6], Henk Bosch[7], Mohamed El Raey[8], Ulka Kelkar[6], Mohan Munasinghe[9], Stephen M. Mwakifwamba[10], Atiq Rahman[5], Samir Safi[11], Barry Smit[12], Joel Smith[13], and Darren Swanson[14]*

[1] Commonwealth Scientific & Industrial Research Organisation, Atmospheric Research, Aspendale, Australia
[2] National Center for Atmospheric Research, Boulder, United States
[3] Department of Meteorology, Kampala, Uganda
[4] Bogor Agricultural University, Bogor, Indonesia
[5] Bangladesh Centre for Advanced Studies, Dhaka, Bangladesh
[6] The Energy and Resources Institute, New Delhi, India
[7] Government Support Group for Energy and Environment, The Hague, The Netherlands
[8] University of Alexandria, Alexandria, Egypt
[9] Munasinghe Institute for Development, Colombo, Sri Lanka
[10] The Centre for Energy, Environment, Science & Technology, Dar Es Salaam, Tanzania
[11] Lebanese University, Faculty of Sciences II, Beirut, Lebanon
[12] University of Guelph, Guelph, Canada
[13] Stratus Consulting, Boulder, United States
[14] International Institute for Sustainable Development, Winnipeg, Canada

# CONTENTS

## 5.1. Introduction

This technical paper (TP) describes techniques for assessing future climate risk and therefore, adaptation needs, under a changing climate. In doing so, this TP outlines a process that is consistent with Adaptation Policy Framework (APF) Component 3, *Assessing Future Climate Risks*. The techniques described here utilise information about future climate in assessments that build on an understanding of current climate risks. Two pathways to assessing risk are described, the hazards-based approach and the vulnerability-based approach. The former begins with plausible changes in future climate, then projects biophysical and socio-economic conditions from those changes. The vulnerability-based approach sets criteria based on socio-economic or biophysical outcomes, then determines how likely these criteria are to be met or exceeded (this approach was introduced in TP3). The climate risks that are described using either pathway can be managed through policy changes that reduce a population's exposure to current and future climate hazards.

The material presented here builds on the concepts addressed in TP4 for assessing current risks by adding information on climate change to assess future risks. Unless historically unprecedented hazards are indicated by climate studies, criteria for risk management of future climate can be based on an understanding of current climate risks (TPs 3 and 4). If knowledge of those current risks is established, then assessments may commence by characterising how climate risks may change due to future climate and socio-economic changes (TP6).

The paper briefly describes the latest climate information and summarises methods on scenario development, directing the researcher towards source material on how to develop climate scenarios. It then outlines how climate scenario information can be used to extend our understanding of current climate–society relationships into the future, how to analyse risk relevant to different planning horizons, and how to assess planned adaptations as a form of risk management.

## 5.2. Relationship with the Adaptation Policy Framework as a whole

With its focus on future climate risks, this paper contributes primarily to Component 3 of the APF. Yet it is closely linked to the other TPs, as outlined here.

TP2 – *Engaging stakeholders in the adaptation process:* Engaged stakeholders can be a key element of modern risk assessments, and can contribute by extrapolating their current experience to possible future climate and identifying adaptations to address changing risks.

TP3 – *Assessing vulnerability for climate adaptation:* Assessment of the consequences of climate change form a key part of climate risk assessment. TP3 describes the tools required to characterise vulnerability in preparation for assessing both current and future climate risks.

TP4 – *Assessing current climate risks:* Knowledge of current climate risks, and adaptation to those risks provide a sound basis for assessing future adaptation needs. TP4 describes how climate risk is a combination of the likelihood of a climate event (or a combination of events) and its consequences. This paper builds on the techniques described in TP4, describing methods for incorporating information about future climate into the risk assessment. TP4 is paired with the current paper within the APF.

TP6 – *Assessing current and changing socio-economic conditions:* A dynamic understanding of future risk requires knowledge of both biophysical and socio-economic change. Socio-economic analysis can be used to describe change in human systems that will affect a group's ability to cope with future climates, as outlined in TP6.

TP7 – *Assessing and enhancing adaptive capacity:* Adaptive capacity is the ability of a group to expand their coping range in response to an anticipated or experienced climate stress. Analysis of historical changes in the coping range can indicate the adaptive capacity of a particular group or activity.

TP8 – *Formulating an adaptation strategy:* The process of preparing an adaptation strategy involves making decisions on specific adaptation options – choices that respond to the risks recognised in this paper.

## 5.3. Key concepts

Climate risk arises from interactions between climate and society, and can be approached from its social aspect through vulnerability-based assessment, from its climatic aspect through natural hazards-based assessment, or through complementary approaches that combine elements of both. The coping range, described and illustrated in TP4, provides a framework that can accommodate these approaches under climate change. As such, it can be used as an analytic tool or communication device in assessments.

When carrying out a risk assessment, the team needs to be aware of what type of information is needed to apply the results to planning or policy. In some cases, qualitative information may be all that is needed. For instance, in a region under water stress, an indication that drought risks are likely to increase in the future may be sufficient to warrant adaptation (Box 5-3, Section 5.5.5). In other cases, decisions about natural resource allocations based on climate change may be open to legal challenge, requiring outcomes based on scientific assessments that can stand up in court (where scientific evidence will be assessed on the balance of probabilities). However, uncertainty also limits choice. Sometimes, although stakeholders want hard numbers, uncertainty may only allow qualitative responses. In this case, a compromise is to rely less on analytic techniques and modelling, and rely more on techniques from the social sciences, such as eliciting information from different stakeholders (TP2) on how they perceive climate risks, to provide semi-quantitative assessments.

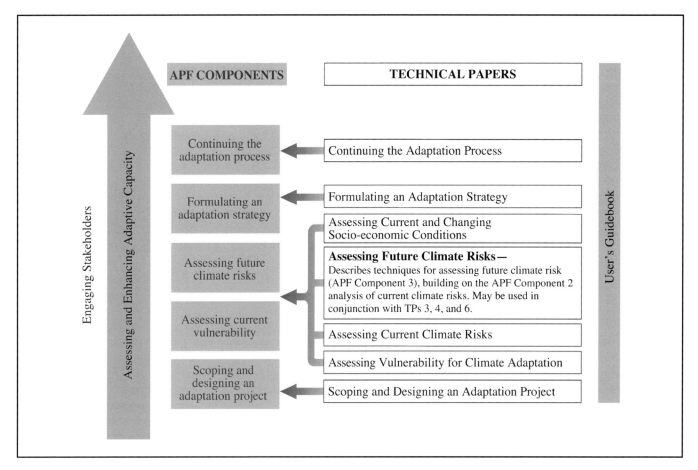

**Figure 5-1:** Technical Paper 5 supports Components 2 and 3 of the Adaptation Policy Framework

### 5.3.1.  *Uncertainty*

Climate change assessments are permeated by uncertainty, requiring the use of specialised methods such as climate scenarios. This is a principle reason to recommend that adaptation assessments be anchored with an understanding of current climate risk; it helps to provide a road map from known territory into uncertain futures. Risk assessment also utilises a formalised set of techniques for managing uncertainty that can be used to expand the methods developed and utilised in Intergovernmental Panel on Climate Change (IPCC) assessments. For example, Moss and Schneider (2000) prepared a cross-cutting paper on uncertainty for the IPCC Third Assessment Report (TAR) that provides valuable guidance on framing and communicating uncertainty. Particularly valuable is the advice on providing guidance on the confidence used in terms such as *likely*, *unlikely*, *possible* and *probable*. Further guidance on managing uncertainty within assessments (both qualitative and statistical) is provided by Morgan and Henrion (1990) and, on communicating risk, in Morgan et al. (2001).

The major tool used to assess the impacts of future climate is the climate scenario. *A scenario is a coherent, internally consistent and plausible description of a possible future state of the world.* It is one of the main tools for assessing future developments in complex systems that may be unpredictable, are insufficiently under-

stood and have high scientific uncertainties (Carter and La Rovere, 2001). Scenarios can range from the simple to the complex, and from the qualitative to quantitative, encompassing narrative descriptions of possible futures to complex mathematical descriptions combining mean climate changes with climate extremes. Climate scenarios are not restricted to Global Climate Models (GCM) output – any information about future climate utilised in an assessment will suffice. Even when scenarios are constructed in narrative form, or are based on broad projections of climate change (e.g., Section 5.5.1.2), plausibility and consistency should be maintained as much as possible. Usually, a scenario has no likelihood attached to it beyond being plausible. However, it is the basic building block of risk assessment approaches under climate change that use scenarios, ranges of uncertainty, probability distribution functions and Bayesian analysis. Section 5.5.4 contains examples of how to apply some of these techniques.

### 5.3.2.  *Coping ranges*

The coping range was introduced in TP4 (Section 4.3.4) to show how current climate can be related to socially-related outcomes in order to carry out risk assessment. It can be used to assess how the ability to cope is affected by a perturbed climate (Figure 5-2) and to assess the changing ability to cope over time (TP4, Annex A.4.3).

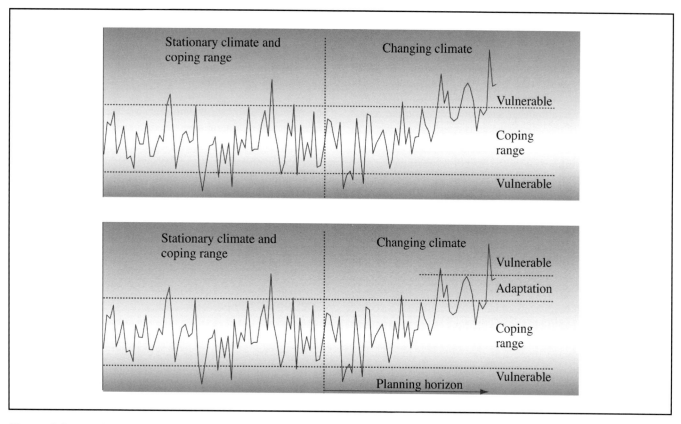

**Figure 5-2:** Relationships between climate change, coping range, vulnerability thresholds and adaptation

The upper panel shows how a coping range may be breached under climate change if the ability to cope is held constant. If that range is represented in terms of temperature (or rainfall), the upper hot (or wet) baseline or reference threshold is exceeded more frequently, while the exceedance of the lower cold (or dry) baseline threshold reduces over time. Vulnerability will increase to extreme levels for the hot (wet) threshold over time. The lower panel represents the expansion of the coping range through adaptation and the consequent reduction of vulnerability. The amount of adaptation required depends on the planning horizon under assessment and the likelihood of exceeding given criteria over that planning horizon.

The coping range can also be used to explore how both climate and the ability to cope may interact over time. For example, an agricultural assessment could account for projected growth in technology, yield and income that broadens the coping range. An assessment could then determine whether these changes are adequate to cope with projected changes in climate. These assessments should be carried out on an appropriate planning horizon.

### 5.3.3.  *Risk quantification*

Approaches for quantifying risk and the use of coping ranges under climate change are emerging areas and, as yet, there are limited assessments to draw from for guidance. Introductory

material is described in the IPCC Third Assessment Report: Mearns and Hulme (2001) for risk, and Smit and Pilifosova (2001) for coping ranges. This is developed further in Jones et al. (2003). An illustrative approach to using coping ranges is described by Yohe and Tol (2002). Methods for undertaking risk assessments utilising critical thresholds built around the conventional seven-step method of Carter et al. (1994) are described in Jones (2001). A guide for assessing risk (Willows and Connell, 2003), principally designed for decision-makers, contains participatory, qualitative and quantitative approaches.[1] Further information on setting risk criteria and thresholds can be found in TP4 (Section 4.4.4).

While the qualitative aspects of risk and coping ranges can be readily utilised in conceptual models (i.e., by stakeholders identifying the point where the level of harm exceeds tolerance levels), the more applied methods require a well-developed research capacity. The probability of exceeding a given level of vulnerability is an exceedingly useful concept to develop in methodological terms, and is discussed by Jones et al. (2003). While it would be useful, it is not always possible to have models linking the entire process from climate change to socio-economic outcomes.

For example, if only biophysical models are available, or if vulnerability cannot be adequately quantified, stakeholders may decide to identify levels of vulnerability in biophysical

[1] This guide, *Climate adaptation: risk, uncertainty and decision-making*, can be found at http://www.ukcip.org.uk/risk_uncert/risk_uncert.html

terms where there is an agreed consensus about the degree of vulnerability:

- in terms of flooding, there may be a particular water level associated with widespread damage.
- if only rainfall data is available, researchers may quantify the rainfall amounts preceding similar levels of inundations. These amounts can then be used to construct a threshold providing the bounds of the coping range for a community within a catchment.
- for agriculture, rainfall may be used as a proxy for loss of production or given levels of food security. In terms of sustainability, stakeholders may identify a level of crop production that they think is sustainable and assess how they may reach that target under climate change.

Socio-economic scenarios may need to be developed to explore how coping ranges may evolve (TP6). More applied methods of exploring vulnerability are detailed in TP3.

The "learning by doing" aspect of the APF will help in this regard. Assessments that build capacities and tools that then become available for successive assessments will consequently build the capacity to develop new techniques. Meanwhile, policy makers and stakeholders, once they have learned that firm forecasts of climate change are not forthcoming, are generally receptive to working with risk, especially if it is framed in terms of what they already know (i.e., couched in terms of their current exposure to climate risk). An example of a quantitative risk assessment for the water sector detailing the methods used and policy response is described in Annex A.5.1 (Jones and Page, 2001).

## 5.4.    Guidance on assessing future climate risks

A broad structure for assessing future climate risks is provided in Figure 5-3. Included are some initial activities to carry out with stakeholders, such as exchanging information on what is already known. At this point in the process, some level of prior knowledge of climate change is assumed to exist in most countries, including that generated by National Communications to the United Nations Convention on Climate Change (UNFCCC). This flowchart is meant to provide guidance for constructing a risk assessment – it is not meant to be followed step-by-step if the assessment, material and circumstances do not readily permit it.

There are several ways an assessment can be approached. It may build on an assessment of current climate risks as described in TP4, or may be based on pre-existing knowledge. It is also possible to integrate an assessment of current and future climate risks. One way to do this would be to take important elements from Figure 5-3 and Figure 4-2 in TP4, and order them to create a logical sequence relevant to a particular assessment. Elements from TPs 3 and 6 could be introduced in the same way. The decision of what elements need to be included

can be carried out jointly by researchers and stakeholders as part of conceptual model development.

### 5.4.1.    Selecting an approach

The two major pathways through risk assessment are the natural hazards and the vulnerability-based approaches. (TP4, Section 4.4.) The natural hazards approach is a climate scenario-driven approach. It starts off with climate scenarios, applies them to impact models and determines possible changes in vulnerability. The vulnerability-based approach starts with possible future outcomes in the form of biophysical or socio-economic criteria that represent a given state of vulnerability. It then determines how likely those criteria are to be met/exceeded under different future climates, again by applying a range of climate scenarios. Outcomes used as criteria for risk assessment can be desirable (e.g., a future sustainable state) or undesirable (e.g., an important activity losing viability).

1.  The natural hazards-based approach fixes a level of hazard (such as a peak wind speed of $10ms^{-1}$, hurricane severity, or extreme temperature threshold of 35°C), and then assesses how changing that particular hazard, according to one or more climate scenarios, changes vulnerability. Limitations in climate modelling often mean that changing hazards cannot be represented specifically but scenario-building methods are continually evolving. A broad formulation is *Risk = Probability of climate hazard x Vulnerability*.

2.  The vulnerability-based approach sets criteria based on the level of harm in the system being assessed, then links that to a specific frequency, magnitude and/or combination of climate events. For example, loss of livelihood linked to severe drought, loss of property due to flooding, critical thresholds for management, or system viability. The level of vulnerability that provides this "trigger" can be decided jointly by researchers and stakeholders, chosen based on past experience, or defined according to policy guidelines. With this approach, *Risk = Probability of exceeding one or more criteria of vulnerability*.

These methods are complementary and can be used separately or together. Table 5-1 provides a quick checklist that may help to decide which technique may be most appropriate. If the ranges of uncertainty described by climate scenarios and/or characterisation of hazard under climate change are well-calibrated and if the drivers of change and the processes by which change can be represented are understood, then the natural hazard approach may be best suited. If the climate hazards cannot easily be characterised under climate change, there are many drivers of change and many pathways along which change can take place, then a vulnerability-based approach may be best suited. Another important distinction is that the natural hazard method is largely exploratory, i.e., given the underlying assumptions and conditions, a specific outcome is predicted; and the vulnerability-based method is more normative, i.e., a future outcome is proposed (either positive or negative) and the risk of attaining or exceeding that outcome is assessed. Adaptation will aim to reduce the likelihood if that outcome is negative, or increase the likelihood if it is positive.

### 5.4.2. *Gathering information on future climate*

Information on what future climate may be like has increased substantially in the past decade. The most recent and complete information on the climate change science community's assessment of this subject is found in the IPCC TARs (Houghton et

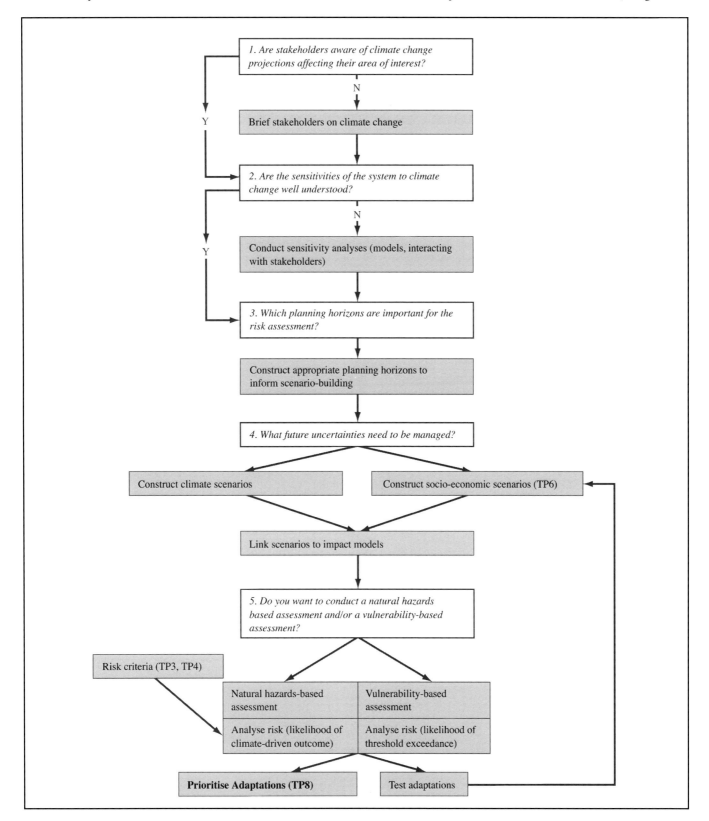

**Figure 5-3:** Flow chart for assessing future climate risks *(as described in this chapter)*

*Table 5-1: Checklist to determine the efficacy of using the natural hazard- and vulnerability-based approaches in an assessment*

| Method | Natural hazard-based approach | Vulnerability-based approach |
|---|---|---|
| Hazard characterisation | Ranges of uncertainty described by climate scenarios and/or characterisation of hazard under climate change well-calibrated | Ranges of uncertainty described by climate scenarios and/or characterisation of hazard under climate change not well-calibrated |
| Drivers of change | Main drivers known and understood | Many drivers with multiple uncertainties |
| Structure | Chain of consequences understood | Multiple pathways and feedbacks |
| Formulation of risk | Risk = P (Hazard) x Vulnerability | Risk = P (Vulnerability) e.g., critical threshold exceedance |
| Approach | Exploratory | Normative |

al., 2001; McCarthy et al., 2001; Metz et al., 2001; available at: http://www.ipcc.ch/), the main points of which are summarised hereafter.

Based on the most recent information, mainly from simulations of GCMs, it is believed that the average global temperature of the earth will be between 1.4°C to 5.8°C warmer than present by the end of the 21$^{st}$ century. Moreover, there is increasing evidence that the warming of the earth over the past 50 years is attributable to increased greenhouse gases resulting from human activities.

### 5.4.2.1.   IPCC Special Report on Emission Scenarios

The estimate of the range of temperature change at the end of the 21$^{st}$ century is based on results from climate models forced with scenarios of increasing greenhouse gases and aerosols, developed for the TAR (Nakicenovic and Swart, 2000). These scenarios, in turn, were based on four "storylines" of what the future of the world might be from the point of view of demographic, technological, political, social and economic developments (Box 5-1). Forty different scenarios were developed from those storylines. In addition to producing very different outcomes in terms of climate, the range of possible developments paths will also produce different adaptive capacities (TPs 6 and 7).

Across all Special Report on Emission Scenarios (SRES), the range of atmospheric $CO_2$ would reach levels between 540 ppm to 970 ppm by the end of the present century. There are also significant ranges of change across the scenarios for the other greenhouse gases such as methane and nitrous oxide. The trajectory of sulphate aerosols also varies considerably across the scenarios with some steadily decreasing and others with an initial increase, but then decreasing by the second half of the century.

---

**Box 5-1. SRES scenario storylines**

A1      Characterised by very rapid economic growth, global population peaking in mid-century, and then declining, and rapid introduction of new, efficient technologies. Three different subgroups in the A1 storyline are defined that present alternative changes in technology: fossil intensive (A1FI), non-fossil (A1T) and balanced across sources (A1B).

A2      Characterised by heterogeneity. Self-reliance and local identities are emphasised. Population increases continuously. Economic development is regionally oriented, and economic and technological growth is relatively slow compared to other storylines.

B1      A convergent world, having the population growth of the A1 storyline. Economic structures change rapidly toward a service and information economy, clean and resource-efficient technologies are introduced, with emphases on social and environmental sustainability.

B2      Local solutions to economic, social and environmental sustainability are emphasised. Global population grows continuously, but at rate lower than that of A2.

## 5.4.2.2. Projected climate changes

Based on atmosphere-ocean GCM (AOGCM) results, the IPCC determined that global annual average precipitation would increase from about 1.2% to 6.8% in the last 30 years of the 21$^{st}$ century, across the A2 and B2 scenarios. Global sea level is expected to increase by between 0.09 and 0.88 m by the end of the 21$^{st}$ century, based on the full range of the SRES scenarios. Regional increases in sea level rise show large variations between models.

Uncertainties in the responses of mean climate change, including variability, increase as one goes to finer levels of assessment (perception) than the global scale, especially for changes in regional precipitation. However, some specific regional changes are considered likely. It is believed that land temperatures will warm faster than the global average and oceans will warm more slowly. Polar regions are expected to experience greater increases in temperature than will tropical regions, and will also experience increases in precipitation in most seasons.

Based on a regional analysis of results of nine AOGCMs that used both the A2 and B2 emissions scenarios, more detailed common regional changes in temperature and precipitation were determined in the IPCC report (Giorgi and Hewitson, 2001). However, these results are more uncertain than those described in the previous paragraph. Large warming will occur during the winter in all high northern latitude regions, as well as in Tibet, whereas it is indicated to take place during the summer in the Mediterranean basin, as well as in northern and central Asia. Increases in precipitation are anticipated over northern mid-latitudes and tropical African regions in the boreal winter. Increases in precipitation are also seen in the boreal summer in South Asia (e.g., India), East Asia (i.e., central China), and Tibet. Consistent decreases in winter precipitation are seen over Central America in the boreal winter (December–February) and over Australia and southern Africa in the austral winter (June–August). Changes in precipitation tend to be larger in the A2 scenario, compared to the B2. In other regions of the world and/or seasons, there was a lack of consistency in the changes in precipitation across the models and scenarios and no clear signal could be determined. More details on these results can be found in Giorgi and Hewitson (2001) and Giorgi et al. (2001).

The IPCC also assessed possible future changes in extreme events. These estimates are particularly important since vulnerability to extreme events is usually high in human society, and our need to adapt to them is high. It is now believed that extreme high temperatures will increase, as will high-intensity precipitation events. Low temperature extremes would decrease. Mid-continental areas will likely experience greater drought in the summer. Unfortunately, little is known regarding how intense hurricanes or mid-latitude storms will change. There is some evidence that, on average, more El Niño-like conditions would be seen (TP4, Annex A.4.2 provides a summary).

## 5.4.3. Conducting sensitivity experiments

To obtain a first-order sense of how possible climate changes may affect different impacts and because of the degree of uncertainty in climate change, particularly at the regional scale, sensitivity experiments are a good means of exploring how impacts may respond to climate change. These make use of incremental changes in climate, e.g., applying a 1°, 2°, and then 3°C increase in temperature; and/or 5%, 10%, 15% increase/decrease in precipitation, and so on. These can be constructed as quantitative data sets for use as input to quantitative impact models (e.g., crop and hydrologic models; Risbey, 1998; Mehrotra, 1999) or applied to mental models (i.e., thought experiments) constructed with stakeholders.

Sensitivity experiments can produce important information on the basic sensitivity and vulnerability of the particular system and aid in the establishment of critical climate thresholds in the system (levels at which serious damage occurs). It is often recommended that such incremental changes be used early in a project so as to better understand the response of the system to climate shifts in a systematic way and to establish thresholds (e.g., Mearns and Hulme, 2001). The use of incremental changes should be limited to such explorations because they do not necessarily produce internally consistent and plausible scenarios of change. It is also possible to assess sensitivity to changes in climate variability, especially if it is difficult to develop scenarios for those changes from climate model data (e.g., assessing plausible but artificial changes in daily rainfall as part of flood modelling).

## 5.4.4. Selecting planning and policy horizons

Planning horizons will affect how far into the future a risk assessment may be projected. Planning horizons relate to the lifetime of decision-making associated with a particular activity – how far into the future is it planned? Is climate change likely to occur with this planning horizon? Do current planning decisions assume the continuation of historical conditions? How do we incorporate climate change into long-term planning?

The same activity can be affected by several planning horizons used by different stakeholders (e.g., financial, urban planning and engineering horizons for infrastructure). For example, in a water resource or catchment-based assessment, the planning life of water storages may be 50+ years, but planning for supply may only be 5–15 years (Figure 5-4). A risk assessment may then want to create scenarios based on two time horizons such as 2020 and 2050 to accommodate both water policy and infrastructure horizons respectively.

The policy horizon is related to the period of time over which a particular policy is planned to be implemented. This may not be on the same time scale as a planning horizon. For instance, the infrastructure affecting an activity will have an engineering life of many decades, but the policy horizon governing the operation of that infrastructure may be much shorter. Most natural resource policy is implemented over periods of 5 to 15 years. Such policies may be reviewed or updated over time but

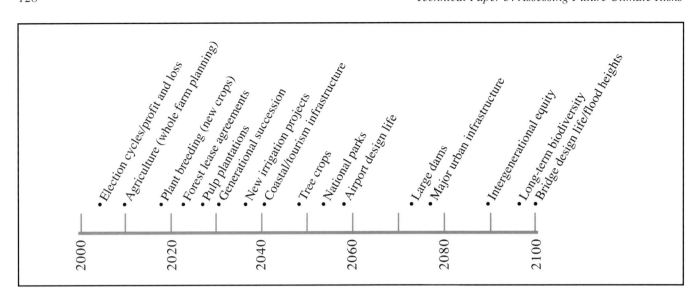

**Figure 5-4:** A representative section of planning horizons relevant to climate risk assessments. Few of these planning horizons are fixed. They cover a range of time and some (e.g., long-term biodiversity) will extend long beyond 2100.

are often expected to manage resources over a much longer planning horizon. Risk assessment may be extended over the longer planning horizon, but adaptations developed to manage those risks are likely to be applied over shorter-term policy horizons (e.g., a long-term strategic outlook is often used to inform shorter-term adaptations). These longer-term outlooks are important, because to ignore strategic objectives in favour of exclusively short-term management may lead to incremental changes accumulating in unintended or irreversible outcomes. If the existing planning horizon does not extend beyond the policy horizon, assessment of the potential risks under climate change may be used to alert policy-makers to the value of taking a longer-term view.

Planning and policy horizons influence the choice of climate scenario. Scenarios may represent a time slice in the future (e.g., 2020 or 2050), or project a pathway from the present into the future. The planning horizon may extend further than the policy horizon but knowledge of possible risks will influence the path taken, in policy terms, of reaching that planning horizon in good shape. If climate scenarios far into the future are chosen, but policy needs are much more immediate, several time-slices over the short to long-term may be used to bridge the distance between policy and planning horizons. A tension exists between the long-term needs of sustainable development and the short-term needs of economic and policy development. However, if adaptations can serve both shorter-term policy needs and long-term strategic objectives, the likelihood of achieving sustained benefits is maximised (as it is if both short- and long-term climate risks are managed). If adaptation is incremental, then policy horizons can be updated using adaptive management, by reviewing shorter-term actions in the light of new information about longer-term outcomes. If irreversible changes with significant consequences are possible, or if retrofitting infrastructure at some future time is likely to be too expensive, then adaptation may need to anticipate long-term changes almost immediately.

### 5.4.5.   *Constructing climate scenarios*

The major methods for constructing climate scenarios utilise results from climate model simulations. While there are other means (Table 5-1; Carter et al., 1994; Mearns and Hulme, 2001), climate model results provide the user with internally consistent and plausible scenarios of the future that are sufficiently detailed for use with quantitative impacts models.

### 5.4.5.1.   *Introduction to climate modelling*

Climate projections are produced by mathematical representations of the earth's climatic system using GCMs. These models are as physically representative as possible within the limitations of scientific knowledge, the ability to represent physical phenomena on an appropriate scale, and computer capacity. They link the atmosphere, ocean, land, and biosphere both vertically and horizontally in a series of three-dimensional grid boxes that partition the earth into layers and grids. The scale and thus the number of those boxes are limited by the computer power available to carry out the necessary computations. GCMs have grid box resolutions in the order of 100 km to 500 km on a side, while Regional Climate Models (RCMs) have a resolution between 5 km and 100 km. Regional climate models have a limited domain of higher resolution allowing large-scale simulations to be run, and may be nested in a GCM or as a zone of high resolution of grid squares within a lower resolution GCM.

The current generation of GCM is the *coupled GCM*, or AOGCM, that links a three-dimensional representation of the ocean to the atmosphere. In these experiments, the enhanced greenhouse effect is simulated by gradually increasing the radiative forcing equivalent to historical increases in greenhouse gases and sulphate aerosols to 1990 or 2000, then simulating the response to greenhouse gas and aerosol scenarios to 2100 or beyond. Although climate models are run on many time-steps per day usually, only daily and monthly data is saved. Monthly data is saved for many variables in the atmosphere and ocean, whereas daily data is generally saved for surface variables important for the diagnosis of results and for impact studies. Due to the large amounts of data saved and stored, monthly data is usually preferred to the use of daily data.

### 5.4.5.2. *Uncertainties of future climate*

The uncertainties affecting climate change are biophysical and socio-economic. Biophysical uncertainties are those dealt with in climate models and include interactions between the oceans, atmosphere and biosphere. Socio-economic uncertainties include the economy, technology, population and society. These uncertainties interact, e.g., where greenhouse gas emissions alter the climate and biosphere, which then affect human systems. Accurately forecasting the rest of this century's climate is not possible because we cannot accurately predict the necessary socio-economic drivers in terms of greenhouse gas futures – we can only produce a large range of possible outcomes. The uncertainties in the climate models themselves also contribute to this inability.

While there are many uncertainties in climate change, this section reviews only some of the major ones that impacts researchers can most likely take account of in their work.

The uncertainties in technological, political and economic futures are integrated in the production of emissions scenarios. Hence, the different emissions scenarios can be said to summarize a range of those uncertainties. The major uncertainties in climate system responses are represented by the different climate models that respond differently to the different emissions scenarios. These are the two summary types of uncertainty that are most available for consideration in impacts (and hence) adaptation research. Uncertainties also tend to propagate as one progresses through an assessment and as one moves from the global to local scale (Figure 5-5). Risk assessments need to account for these uncertainties as much as is practical. (A brief summary is in Box 5-2; IPCC-TGCIA, 1999; Carter and La Rovere, 2001; Mearns and Hulme, 2001).

Progress is rapidly being made in quantifying the uncertainties of climate change. These efforts have lead to recent papers quantifying the near future (i.e., next 20 years) using a combination of climate observations and climate model results (Allen et al., 2000; Forest et al., 2002). Moreover, attempts to assign probabilities to longer-term future climate have also been made (e.g., Schneider, 2001; Wigley and Raper, 2001). More recently progress has been made in determining the reliability of climate model results (Giorgi and Mearns, 2001, and assigning probabilities to climate change on a regional scale Giorgi and Mearns, 2003; Tebaldi et al., 2003). However, these works should be viewed as providing subjective examples as opposed to objective probabilities of long-term future climate. Box 5-2 summarises how climate scenarios can be used to manage uncertainty.

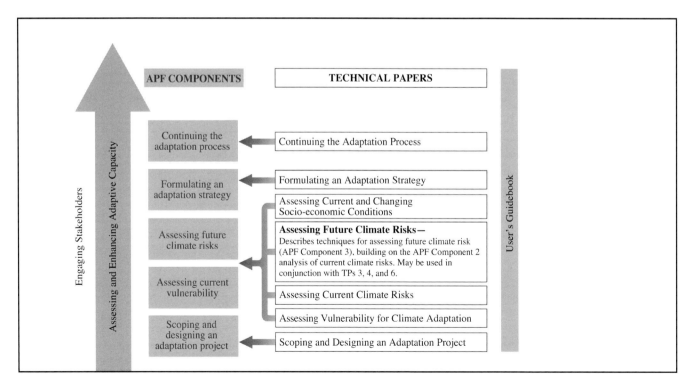

**Figure 5-5:** The relationship between (upper diagram) ranges of uncertainty cascading through an assessment, and (lower diagram) individual scenarios, S1 to S4, and resultant ranges of uncertainty. These diagrams are sourced from Jones (2000) and Schneider and Kuntz-Duriseti (2002).

## Box 5-2: Assessing likelihoods of climate change

Within the resources available to an assessment, the choice of how many and what kind of scenarios are needed has to balance the concerns between precision and the ability to explore key uncertainties. For instance, daily rainfall data from climate models is very imprecise and may need to be downscaled to obtain plausible values and distributions, but this task is resource intensive and may limit the number of scenarios that can be produced. The trade-off is between producing plausible scenarios that properly represent the data needed to simulate impacts, and exploring the major uncertainties that will affect an assessment's outcomes. This box outlines some strategies for assessing uncertainty and likelihoods. The IPCC Data Distribution Centre has both data and supporting material, as do a number of climate modelling centres. Even if only a limited number of climate scenarios are used, it would be valuable to scope the range of projected climate changes for the area in question before constructing those scenarios.

*Single scenario*

A single scenario can be used as a plausible outcome or to illustrate a storyline that tests various options in an environment of high uncertainty. It can be located within a range of uncertainty (e.g., low, median or high warming) or may be used to give a specific realisation to a generally accepted direction of change (e.g., increase in extreme rainfall). The downside is that a single scenario is often taken (erroneously) as a prediction.

*Two scenarios*

Two different scenarios will overcome the possibility of a single scenario being seen as a prediction. Strategies are to sample a range of uncertainty by choosing extreme outliers, or just to illustrate two distinctly different possibilities (as in the U.S. National Assessment).

*Several scenarios*

Undertaken to explore one or more ranges of uncertainty (e.g., greenhouse gas emissions, climate sensitivity, regional climate change). Three scenarios are sometimes discouraged to prevent users from gravitating towards the central estimate.

*Range of outcomes*

Constructing a range of uncertainty bounded by high and low estimates of the outcomes (e.g., global warming as expressed by the IPCC). This limits the uncertainty by identifying outcomes that are not likely, but on the other hand, can identify large ranges in impacts that make it difficult to design adaptation policy. Figure 5-5 shows how scenarios are related to a well-calibrated range of uncertainty (e.g., global warming, regional temperature, rainfall or sea level rise).

*Relating likelihood to global warming and sea-level rise*

It is possible to quantify likely impacts and the consequences of those impacts for systems affected by variables that can be linked closely to global warming, such as mean temperature and sea-level rise. For example, low-lying land in any given region will be the first to be affected by sea level rise and elevated land will be the last. This allows relative likelihoods to be attributed across a range of coastal areas, where the lowest levels of coast are the most likely to be inundated, and the highest are the least likely. This distribution is conditional and depends on factors such as trend in land movement, regional variability in mean sea-level rise, and changes in patterns of surge events. However, where mean sea-level rise is a significant driver of change, then the IPCC (2001) range of change will give a guide as to likelihood, and damage sea-level rise relationships will provide a guide as to consequences. Any section of coast proven vulnerable below the IPCC minimum projected sea-level rise will almost certainly be affected, the median part of the range is moderately likely to occur and the upper part of the range is unlikely to be reached. The same principle holds for systems strongly affected by temperature including coral reefs, tropical montane systems, permafrost regions and where biological thresholds are close to their upper temperature limits. Those impacts vulnerable to small levels of warming will be the first and most likely to be affected. If the direction of rainfall change is either overwhelmingly wetter or drier, then this principle can also apply to hydrological systems.

This principle is much more difficult to apply for variables that may either increase or decrease (e.g., rainfall in many regions), where systems are subject to complex interactions between variables, or where systems are driven largely by changes to variability rather than by accompanying changes to means. This covers many biological, health and hydrological systems.

***Combining ranges and probability distribution functions***

Recent efforts are beginning to quantify risk in terms of applying prior distributions to input ranges of uncertainty. These methods are in their early stages of development but where they have been applied (in Australia), policy makers have responded positively.

*5.4.5.3. Current climate data*

Current climate data is generally necessary in developing climate scenarios because errors in the reproduction of current climate by AOGCMs are still quite large. In general detailed climate data on a daily time scale are most easily acquired from the meteorological service of the relevant country. Monthly long-term datasets are available for the entire world on some web sites and institutes, such as the IPCC Data Distribution Centre, described in the next section. The way in which climate data is used to construct climate scenarios is described in later sections.

*5.4.5.4. Climate model output*

There are many sources of climate model output from future climate experiments. Different climate modelling centres provide their data upon request, and many have web sites from which one can download climate data.

The most complete repository of climate model data is the IPCC Data Distribution Centre web site, which was created to provide up-to-date climate and related scenarios for impacts researchers. The DDC is the main product of the Task Group on Climate Impact Assessment of the IPCC. At this site, GCM results for nine different AOGCMS are available using two of the SRES emissions scenarios (A2 and B2). Additional climate model results will be made available in the near future. Data for the major climate variables of interest for impacts work (e.g., temperature, precipitation, solar radiation) on a monthly timescale are made available. There are also data on the socioeconomic scenarios that were used in the formation of the emissions scenarios, as well as guidance material on how to develop scenarios and how to use them.

Observed climate data on a monthly timescale for the world is also available. Over time, results from many climate models for three additional SRES emissions scenarios will also become available. The web site is: http://ipcc-ddc.cru.uea.ac.uk/

*5.4.5.5. Methods of constructing scenarios*

There are various ways of constructing climate scenarios (reviews of methods in Carter et al., 1994, and Mearns and Hulme, 2001). These include climate model-based approaches, temporal and spatial analogues, expert judgement and incremental scenarios for sensitivity studies as discussed above. Table 5-2 presents an overview of the methods with their main advantages and disadvantages. The most common means is by using results from AOGCM simulations in combination with climatological observations. The classic method entails determining the change in climate, and using this change to perturb observed climate data. In the case of results from transient runs of AOGCMs, this is accomplished by taking the average of a series of simulated years of the current climate (1961–1990) and the same for a series of simulated future years (2071–2100), taking the difference of the future minus the current simulations, and then appending these differences (generically referred to as "deltas") to the observed data set. Quantitative impact models can be run using the actual observed data for present conditions and the "changed" observed data set to represent the future. In this manner, the errors in the climate model simulations do not directly affect the impact model results. In the case of presenting information on changes in climate to stakeholders, results from the simulations can also be discussed with them.

To account for uncertainties in future climate, it is recommended that results from multiple AOGCMs forced with multiple emissions scenarios be used.

*5.4.5.6. Using and communicating single-event and frequency-based probabilities*

A project may want to quantify likelihoods, or levels of confidence in outcomes developed using climate scenarios and communicate these to stakeholders. If no guidance as to likelihood of the outcomes of an assessment is provided to stakeholders, they may attach their own assumptions in an ad hoc manner, perhaps jumping to the wrong conclusions (Schneider, 2001). Therefore, we may want to qualify or even quantify the outcomes or to attach confidence levels to the conclusions as recommended by Moss and Schneider (2000). There are two aspects of probabilities that need to be considered before doing this:

1) What type of probabilities are you representing in your scenarios? They may be represented implicitly – so be mindful of such implicit assumptions.

**Table 5-2:** The role of various types of climate scenarios and an evaluation of their advantages and disadvantages according to the five criteria described in the table endnotes. Note that, in some applications, a combination of methods may be used (e.g., regional modelling and a weather generator). From Mearns and Hulme (2001).

| Scenario type or tool | Description/use | Advantages[a] | Disadvantages[a] |
|---|---|---|---|
| **Incremental** | • Testing system sensitivity<br>• Identifying key climate threshold | • Easy to design and apply (5)<br>• Allows impact response surfaces to be created (3) | • Potential for creating unrealistic scenarios (1, 2)<br>• Not directly related to greenhouse gas forcing (1) |
| **Analogue** | | | |
| Palaeoclimatic | • Characterising warmer periods in past | • A physically plausible changed climate that really did occur in the past of a magnitude similar to that predicted for ~2100 (2) | • Variables may be poorly resolved in space and time (3, 5)<br>• Not related to greenhouse gas forcing (1) |
| Instrumental | • Exploring vulnerabilities and some adaptive capacities | • Physically realistic changes (2)<br>• Can contain a rich mixture of well-resolved, internally consistent, variables (3)<br>• Data readily available (5) | • Not necessarily related to greenhouse gas forcing (1)<br>• Magnitude of the climate change usually quite small (1)<br>• No appropriate analogues may be available (5) |
| Spatial | • Extrapolating climate/ecosystem relationships<br>• Pedagogic | • May contain a rich mixture of well-resolved variables (3) | • Not related to greenhouse gas forcing (1, 4)<br>• Often physically implausible (2)<br>• No appropriate analogues may be available (5) |
| **Climate model-based** | | | |
| Direct AOGCM outputs | • Starting point for most climate scenarios<br>• Large-scale response to anthropogenic forcing | • Information derived from the most comprehensive, physically-based models (1, 2)<br>• Long integrations (1)<br>• Data readily available (5)<br>Many variables (potentially) available (3) | • Spatial information is poorly resolved (3)<br>• Daily characteristics may be unrealistic except for very large regions (3)<br>• Computationally expensive to derive multiple scenarios (4, 5)<br>• Large control run biases may be a concern for use in certain regions (2) |
| High resolution/stretched grid (AGCM) | • Providing high-resolution information at global/continental scales | • Provides highly resolved information (3)<br>• Information is derived from physically-based models (2)<br>• Many variables available (3)<br>• Globally consistent and allows for feedbacks (1,2) | • Computationally expensive to derive multiple scenarios (4, 5)<br>• Problems in maintaining viable parameterisations across scales (1,2)<br>• High resolution is dependent on SSTs and sea ice margins from driving model (AOGCM) (2)<br>• Dependent on (usually biased) inputs from driving AOGCM (2) |

| Scenario type or tool | Description/Use | Advantages[a] | Disadvantages[a] |
|---|---|---|---|
| Regional models | • Providing high spatial/ temporal resolution information | • Provides very highly resolved information (spatial and temporal) (3)<br>• Information is derived from physically-based models (2)<br>• Many variables available (3)<br>• Better representation of some weather extremes than in GCMs (2, 4) | • Computationally expensive, and thus few multiple scenarios (4, 5)<br>• Lack of two-way nesting may raise concern regarding completeness (2)<br>• Dependent on (usually biased) inputs from driving AOGCM (2) |
| Statistical downscaling | • Providing point/high spatial resolution information | • Can generate information on high resolution grids, or non-uniform regions (3)<br>• Potential for some techniques to address a diverse range of variables (3)<br>• Variables are (probably) internally consistent (2)<br>• Computationally (relatively) inexpensive (5)<br>• Suitable for locations with limited computational resources (5)<br>• Rapid application to multiple GCMs (4) | • Assumes constancy of empirical relationships in the future (1, 2)<br>• Demands access to daily observational surface and/or upper air data that spans range of variability (5)<br>• Not many variables produced for some techniques (3, 5)<br>• Dependent on (usually biased) inputs from driving AOGCM (2) |
| Climate scenario generators | • Integrated assessments<br>• Exploring uncertainties<br>• Pedagogic | • May allow for sequential quantification of uncertainty (4)<br>• Provides "integrated" scenarios (1)<br>• Multiple scenarios easy to derive (4) | • Usually rely on linear pattern scaling methods (1)<br>• Poor representation of temporal variability (3)<br>• Low spatial resolution (3) |
| **Weather generators** | • Generating baseline climate time-series<br>• Altering higher order moments of climate<br>• Statistical downscaling | • Generates long sequences of daily or sub-daily climate (2, 3)<br>• Variables are usually internally consistent (2)<br>• Can incorporate altered frequency/intensity of ENSO events (3) | • Poor representation of low frequency climate variability (2, 4)<br>• Limited representation of extremes (2, 3, 4)<br>• Requires access to long observational weather series (5)<br>• In the absence of conditioning, assumes constant statistical characteristics (1, 2) |
| **Expert judgment** | • Exploring probability and risk<br>• Integrating current thinking on changes in climate | • May allow for a "consensus" (4)<br>• Has the potential to integrate a very broad range of relevant information (1, 3, 4)<br>• Uncertainties can be readily represented (4) | • Subjectivity may introduce bias (2)<br>• A representative survey of experts may be difficult to implement (5) |

Numbers in parentheses under advantages and disadvantages indicate that they are relevant to the criteria described. The five criteria are: (1) *Consistency* at regional level with global projections; (2) *Physical plausibility and realism*, such that changes in different climatic variables are mutually consistent and credible, and spatial and temporal patterns of change are realistic; (3) *Appropriateness* of information for impact assessments (i.e., resolution, time horizon, variables); (4) *Representativeness* of the potential range of future regional climate change; and (5) *Accessibility* for use in impact assessments.

2) How do your stakeholders understand likelihood and probability? This understanding may or may not be compatible with the management of climate uncertainties, so a common understanding may need to be developed as part of an assessment.

Regarding the first aspect; there are two major types of probability that may be represented when dealing with climate risks. These can be divided into frequency-based and single-event uncertainties. Frequency-based uncertainties concern recurrent phenomena such as those that comprise climate variability and extremes (e.g., a flood, drought, or tropical cyclone). This type has a known or unknown statistical distribution that describes a series of events in terms of frequency and magnitude. The quantification of single-event uncertainties aims to determine the likelihood of a single event occurring within a given period (i.e., what is the likelihood of an El Niño event occurring next year or of global warming exceeding 3°C by 2100?).

Most climate hazards are described by frequency-based probabilities such as return periods or as a given frequency per unit time, including those contributing to the assessment of current climate risks, as described in TP4. These uncertainties are usually assessed using historical data and statistical and dynamical relationships based on that data. People familiar with weather and climate are most used to this type of uncertainty. Even if they are not well-versed in statistics, people understand that the more extreme events generally occur less frequently and that the more extreme events have the larger consequence. Risk assessment requires weighing up these two factors of frequency and magnitude. Return events such as the 1-in-100 year flood, likelihood of a specific extreme temperature, probability of a given severity of drought, cyclone frequency and magnitude are all examples (Table 4-1, TP4). Many criteria for assessing exceedance are also built on frequentist uncertainties (e.g., a given sequence of hot days >35°C and both thresholds described in Annex A.5.1).

Part of the scenario-building task involves deciding how explicitly these uncertainties need to be represented. If historical climate variability is used as a basis and the mean changed, then the implicit assumption is that the variability around the mean remains unchanged. Changing mean climate as a response to global warming requires the management of single event uncertainties.

Single-event uncertainties represent an event that may or may not occur (e.g., collapse of the West Antarctic Ice-Sheet), or an event with a range of potential outcomes where only one outcome is possible, (e.g., global warming in °C by 2100). Questions such as "How much will the earth warm by 2050?" or "What is the direction and magnitude of rainfall change in my region under global warming?" are examples. Many single-event uncertainties associated with climate change are without precedent, and have no prior statistical history from which a probability distribution can be constructed.

The uncertainties surrounding variables such as mean global warming and regional changes in average temperature, rainfall

and other such factors are single-event uncertainties. That is, only one outcome is possible. This is why such uncertainties are generally expressed as scenarios and ranges of change instead of forecasts with central estimates. Care must be taken when communicating such ranges because a range of rainfall change constructed from several GCM of -15% to +15% does not mean that zero rainfall change is the most likely outcome. If most of the GCM analysed simulate some change in mean this may suggest that zero rainfall change is very unlikely.

Many scenarios will combine both frequency-based and single-event uncertainties. Care will need to be taken to track both implicit and explicit assumptions in scenarios and to ensure that stakeholders understand how different uncertainties are being applied. If stakeholders can see how their existing understanding about climate and risk is incorporated into scenarios, then they will have a better chance of understanding how climate change uncertainties have been managed.

Figure 5-6 features different combinations of these two types of uncertainty in probabilistic terms:

- Graph "a" shows a normal distribution for a single variable shown as a distribution around the mean with nominal thresholds or risk criteria shown. This is a two-sided distribution.
- Graph "b" is a cumulative probability distribution that may be one sided, as for daily rainfall, or a cumulative representation of a probability distribution similar to the one on the left. These are typical of the types of frequency-based probabilities discussed in TP4.
- Graph "c" represents a change in variance with no change in mean.
- Graph "d" indicates multiple scenarios with changing means but fixed variance. This is the type of climate scenario where historical climate variability is scaled by a change in mean to estimate the impacts of different degrees of warming.
- Graph "e" exhibits a change in both mean and variance for a single scenario.
- Graph "f" displays changes in both variance and mean and is the most complex to produce and interpret.

Assessments that are considering the types of analyses illustrated in Figure 5-6 are encouraged to undertake a sensitivity analysis first, to quantify the impact for a given level of change. If changes in variance are likely to be dominated by changes to the mean (e.g., as in Figure 5-6, Graph d) then do not attempt producing scenarios for altered climate variability – use the historical variance. If changes in variance are important (e.g., where heavy rainfall is critical), then variability may be the most important factor.

By comparing scenarios to each other and situating them within broader ranges of uncertainty, it is possible to build up a picture of relative likelihoods. For example, if different climate models produce a consistent change in regional climate of warmer, wetter or drier conditions, then this change may seem

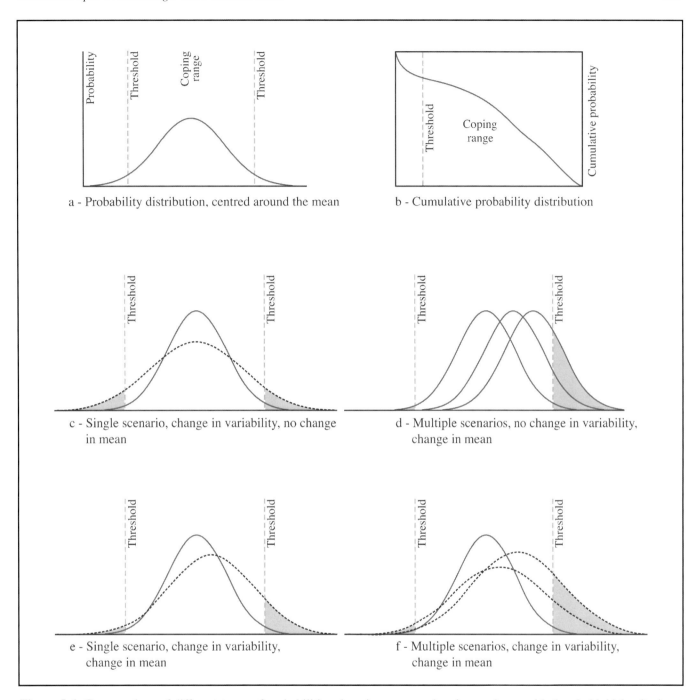

a - Probability distribution, centred around the mean

b - Cumulative probability distribution

c - Single scenario, change in variability, no change in mean

d - Multiple scenarios, no change in variability, change in mean

e - Single scenario, change in variability, change in mean

f - Multiple scenarios, change in variability, change in mean

**Figure 5-6:** Constructions of different types of probabilities changing mean and variance, shown with thresholds/risk criteria to demonstrate how different representations of probability in scenario construction can be used to estimate change in risk.

more likely. Critical thresholds linked to small magnitudes of global warming will be more likely to be exceeded than those that manifest under larger magnitudes of global warming. The same situation exists for sea level rise, low lying areas will be those most at risk from inundation and surge.

### 5.4.6. Conducting climate change risk assessments

The conventional seven-step method has been to apply climate change scenarios, either to perturb a baseline climate, or directly to impact models, to see how much impacts may change (Carter et al., 1994; Carter and Parry, 1998). Adaptation options are then assessed to reduce those impacts. Types of assessments and their needs have multiplied since that method was first formulated (Carter et al., 1994) creating a demand for a variety of assessment techniques. For that reason, Figure 5-3 is a generic procedure that can be populated by many different analytic techniques, including those used in the seven-step method. These techniques can range from qualitative analysis (e.g., partitioning the outcomes into low, medium and high risk) to highly advanced numerical techniques (probabilities calculated using statistical and/or modelling techniques).

Qualitative methods can use conceptual models incorporating elements of climate change (see TP4 for the development of conceptual models under current climate), informed by broad projections of global or regional climate change as being representative of "typical" climate scenarios. Narrative approaches may develop several plausible storylines of how climate may change, encouraging stakeholders to investigate how they would personally cope with such changes, suggesting adaptation options to manage potential risks. At the very least, this process will sensitise stakeholder groups to the issues surrounding adaptation to climate change. Hybrid approaches using existing quantitative models with qualitative assessments of future climate and socio-economic outlooks can also be instructive. The development of integrated scenarios, where consistent climate and socio-economic scenarios may also be addressed in a qualitative or semi-quantitative way, can also be used to promote a dialogue with stakeholders. See TP6 for issues relating to the alignment of SRES climate and greenhouse gas emission scenarios at a local or regional scale. Morgan et al. (2002) contains a rich assortment of techniques that can be used in risk communication. Willows and Connell (2003) also contains a range of useful methods.

Most risk assessments undertaken in developing countries are generally qualitative or semi-quantitative, but requests for quantitative information by policy makers require an improved capacity to quantify outcomes. Many of the established methods will still be used but will increasingly be modified for particular styles of assessment. For methods on how to create and apply climate scenarios, the IPCC-TGCIA guidelines (1999) and seven-step method of impact assessment (Carter et al., 1994; Carter and Parry 1998; UNEP, 1998) users are referred to existing guides.

Four assessments of current climate risk are featured in TP4. Of those, Box 4-2 has a future Component but is largely an assessment of current risk together with a brief assessment of possible future changes to determine whether managing current risk would also reduce future risks. This next example is similar but opens up the question of how to follow up once short-term adaptations are put in place.

Box 5-3 details an example of a risk assessment investigating a natural hazard (drought). The analysis shows that current climate risks are severe; climate variability, and therefore drought risk, is increasing. Projections from three GCM show that rainfall is likely to decrease and temperature (and by extension evaporation) will increase. A vulnerability study shows that drought currently causes armed conflict. This risk has been communicated to the government and stakeholders who have negotiated a series of adaptations.

In this case, adaptation was badly needed to prevent recurring shocks that were causing famines, requiring years of recovery. Once basic protection against climate hazards is achieved, the emphasis can shift to adapting to increase productivity and protection of the natural resource. This requires longer-term planning horizons, gradually moving the emphasis of assessments from current risks towards future climate risks. Permanent water points create their own environmental stresses, population growth will continue and further drying is projected to increase climate risks. Risk assessments that explicitly formulate the likelihood of continuing climate hazards, and those that investigate the vulnerability of local populations to climate will clearly be of value in ensuring a growing population can continue to reduce their exposure to environmental risks in a changing climate.

---

**Box 5-3: Drought risk assessment in Uganda**

*Location:* The Ugandan cattle corridor, running from the northeast to the southwest of the country, is a semi-arid area populated by over 41% and 60% of Uganda's human and cattle population respectively. The Karamoja region in the northeast of the cattle corridor is a nomadic pastoralist region covering approximately 24,000 km$^2$ (10% of the country). It has an average annual rainfall of 745 mm, ranging between 450 mm during severe drought years to 1000 mm during wet years.

*Impacts:* Droughts are increasing in frequency resulting in loss of water supply and pastures. Cattle keepers are forced to move livestock to other areas, resulting in cattle rustling, intertribal fighting and overall environmental insecurity. A recent study identified this area as one where environmental degradation, particularly drought, has caused armed conflict.

*Traditional adaptation:* Nomadism and migration are the major adaptive measures. Population growth is placing pressure on nomadic lifestyles while migration has been the catalyst for armed conflict and warfare. Warfare has moved from using bows and spears to automatic machineguns and rifles, threatening regional and national security.

*Risk analysis:* An initial vulnerability assessment under climate change using three GCM was carried out. It concluded that a doubling of $CO_2$ would increase the temperature by 2–4°C and decrease rainfall by 10–20% (>1 mm day$^{-1}$). The annual rainfall variability of the area has been increasing over the last 3 decades and is expected to increase further due to climate change.

*Adaptation measures:* Through a wide stakeholder consultation, the government has agreed to construct valley dams and tanks (surface water reservoirs) to supply stock during drought years. Eight reservoirs have been constructed of the 58 planned. The risk to drought impacts has decreased and the coping range increased, with available water for most drought years. However, land degradation is occurring near the reservoirs and water supply has periodically been contaminated. (Source: S. Magezi)

Although the APF stresses the need to assess current vulnerability and adaptation as part of planning for the future, current levels of adaptation need to be assessed for their adequacy in managing a changing climate. Box 5-4 shows an assessment that looks at possible changes to agricultural production in India. It uses an approach that accounts for current adaptations in agriculture, as expressed as farm-level net revenue aggregated to state and national scale (Kumar and Parikh, 2001). This assessment suggests two things: 1) that developing countries face possible decreases in agricultural production compared to gains in developed countries using similar assessment techniques and 2) that current adaptations may be insufficient to manage losses under climate change.

The advantage of this approach is that it factors current adaptation into the assessment, and includes climate variability, albeit as it affects mean net revenue. The disadvantage is that the effects of $CO_2$ are not included as they would be in a more conventional crop modelling exercise. However, crop models generally do not simulate adaptations all that well, although a new generation of models such as the Agricultural Production Systems sIMulator (APSIM) (Keating et al., 2003) are beginning to do so. Both the method in Box 5-4, and crop modelling approaches, have distinct advantages that can be used to illustrate different aspects of risk. When different methods agree, some added confidence can be attached to the results.

Annex A.5.1 summarises a risk assessment of water supply that uses both a natural hazards and vulnerability-based approach to assessing risk in a catchment in eastern Australia. This assessment applied multiple climate scenarios to an existing rainfall-runoff and river management model to determine changes in mean annual water supply, irrigation allocations and environmental flows. A relationship between changes in rainfall, potential evaporation and water supply allowed conditional probability distributions of possible outcomes to be created. A natural hazards-based approach concluded that storage, irrigation and environmental flows would most likely change by 0% to -15% by 2030 from a total range of possibilities of +10% to -35%.

A complementary vulnerability-based assessment utilised two thresholds that represented a serious risk within the catchment. The first was a failure of irrigation supply to exceed 50% of the allocation levels five years running and the second that breeding of colonial water birds in a RAMSAR-gazetted wetland failed ten years running. It was found that the risk of exceeding this threshold depended on long-term rainfall variability in addition to climate change. If rainfall variability was "normal", the probability of exceeding critical thresholds was negligible by 2030. However, if rainfall variability was in a drought-dominated phase, then the chance of exceeding the critical thresholds was about one in three. This catchment has been designated as fully to over-allocated in a recent audit (NLWRA, 2002), so adaptation to climate change is now seen as a necessary part of ongoing water reform, and investigations are ongoing.

Few risk assessments under climate change have so far utilised vulnerability-based approaches in a quantitative manner. However, a rich literature assessing qualitative approaches and vulnerability, to current climate suggests that significant development in this area is possible (TP3). Probabilistic approaches that apply a natural hazards approach in a "top down" manner, applying climate change scenarios to impact models to determine vulnerability are also being developed. Bottom-up approaches, where local criteria for risk denoting critical thresholds are constructed, then assessed for likelihood of exceedance are few, but this method has the potential to manage some (but not all) of the limitations of the natural hazards-based approach.

### 5.4.7. Managing climate risks

The main purpose of risk assessment is to determine the need for risk management (the reduction of risk). Adaptation to climate change reduces risk by altering human and environmental responses to climate hazards. (The hazards themselves are altered by the mitigation of greenhouse gases). Adaptation will increase the breadth of the coping range allowing successively

---

**Box 5-4: Sensitivity of agricultural production in India to climate change**

This study estimated the relationship between farm-level net revenue and climate variables in India using cross-sectional evidence (Kumar and Parikh, 2001). It used an economic approach expressed as farm-level net revenue. A number of variables including temperature, rainfall, soil, technology, fertiliser and altitude were used to estimate a regression relationship with economic data from the yields of twenty crops across India. Temperature and rainfall of January, April, July and October are converted into anomalies, along with crop prices. Data for the 271 districts was from the decade 1970–1980; climate data was from the 1960–1980 time period. The response functions that explain the variation in price across districts therefore incorporate climate variability and adaptation to the mean climate and variability for the 10-year period that baseline climate data were available.

A "best-guess" climate change scenario was used to estimate possible changes due to climate. A rise of 2°C and an increase in rainfall of 7% was used as an illustrative scenario to determine how a mid-range or "best guess" climate change might affect Indian agriculture. The decrease in total economic yield was approximately 8%, being largest in the northern states. The eastern states registered increases. The impacts were larger than for those estimated in the United States using similar models, presumably due to India's warmer temperatures and lower levels of technology.

larger and/or frequent climate hazards to be managed. For example, the provision of a reliable water supply or food aid to dryland farming communities will mean they can manage more severe and frequent droughts – to a point (Box 5-3 and TP4, Box 4-2). If an assessment system can quantify a change in critical thresholds, then it will be possible to quantify the benefits of adaptation under climate change, and to create the conditions by which a cost-benefit analysis may be carried out (TP8).

## 5.5.  Conclusions

The major purpose of assessing climate change risk within the APF is to help prioritise possible adaptations that may be feasible. Some measures, such as no-regrets options, or generic measures that will provide adaptation benefits in a broad range of plausible circumstances, will prove to be better than others. This applies to the development of adaptive capacity in particular (TP7). A detailed knowledge of both current and future hazards, and how they may affect societies, can help provide guidance for adaptation, even if a modelling system that quantifies these links cannot be constructed.

Again, given the levels of uncertainty that accompany assessments of future climate risks, teams will need determine how much information is needed in order to make decisions on adaptation policy. Projects should not over-deliver, but if policy makers have significant demands, projects can inform them of the resources needed to meet those demands, including the resources needed to develop assessment methods. There are some recipes available, but continuing exploration of a relatively new area of assessment will be needed.

## References

**Allen**, M.R., Stott, P.A., Mitchell, J.F.B., Schnur, R. and Delworth, T.L. (2000). Quantifying the uncertainty in forecasts of anthropogenic climate change, *Nature*, **407**, 617–620.

**Carter**, T.R., Parry, M.L., Harasawa, H. and Nishioka, S. (1994). *IPCC Technical Guidelines for Assessing Climate Change Impacts and Adaptations*, London: University College and Japan: Centre for Global Environmental Research. 59 pp.

**Carter**, T.R. and Parry, M. (1998). *Climate Impact and Adaptation Assessment: A Guide to the IPCC Approach*, London: Earthscan.

**Carter**, T.R. and La Rovere, E.L. (2001). Developing and applying scenarios. In: McCarthy, J.J., Canziani, O.F., Leary, N.A., Dokken, D.J. and White, K.S. eds., *Climate Change 2001: Impacts, Adaptation, and Vulnerability*, Contribution of Working Group II to the Third Assessment Report of the Intergovernmental Panel on Climate Change, Cambridge: Cambridge University Press, 145–190.

**Forest**, C.E., Stone, P.H., Sokolov, A.P., Allen, M.R. and Webster, M.D. (2002). Quantifying uncertainties in climate system properties with the use of recent climate observations. *Science*, **295** 113–117.

**Giorgi**, F. and Mearns, L. O. (2001). Calculation of best estimate, confidence, and uncertainty in regional climate changes from AOGCM simulations via the "reliability ensemble averaging (REA)" Method. *Journal of Climate*, **15**, 1141–1158.

**Giorgi**, F. and Mearns, L. O. (2003). Probability of regional climate change calculated using the Reliability Ensemble Averaging method. *Geophysical Research Letters*, **30**, 1629–1633.

**Giorgi**, F. and Hewitson, B. (2001). Regional climate information – evaluation

and projections. In Houghton, J.T., Ding, Y., Griggs, D.J., Noguer, M., Van Der Linden, P.J. and Xioaosu, D. eds., *Climate Change 2001: The Scientific Basis*, Contribution of Working Group I to the Third Assessment Report of the Intergovernmental Panel on Climate Change, Cambridge: Cambridge University Press, 583 –683.

**Giorgi**, F., Whetton, P.H., Jones, R., Christenen, J.H., Mearns, L.O., Hewitson, B., von Storch, H., Francisco, R. and Jack, C. (2001). Emerging patterns of simulated regional climate changes for the 21$^{st}$ century due to anthropogenic forcings, *Geophysical Research Letters*, **29**, 3317-3321.

**Houghton**, J.T., Ding, Y., Griggs, D.J., Noguer, M., Van Der Linden, P.J. and Xioaosu, D. eds. (2001). *Climate Change 2001: The Scientific Basis*, Contribution of Working Group I to the Third Assessment Report of the Intergovernmental Panel on Climate Change, Cambridge: Cambridge University Press.

**IPCC** (2001) Summary for Policy-makers, in Houghton, J.T., Ding, Y., Griggs, D.J., Noguer, M., Van Der Linden, P.J. and Xioaosu, D. eds. *Climate Change 2001: The Scientific Basis*, Contribution of Working Group I to the Third Assessment Report of the Intergovernmental Panel on Climate Change, Cambridge University Press, Cambridge, 1–20.

**IPCC-TGCIA** (1999). *Guidelines on the Use of Scenario Data for Climate Impact and Adaptation Assessment*. Version 1. Prepared by Carter, T. R., Hulme, M. and Lal, M. Intergovernmental Panel on Climate Change, Task Group on Scenarios for Climate Impact Assessment.

**Jones**, R.N. (2000). Managing uncertainty in climate change projections – issues for impact assessment. *Climatic Change*, **45**, 403–419.

**Jones**, R.N. (2001). An environmental risk assessment/management framework for climate change impact assessments. *Natural Hazards*, **23**, 197–230.

**Jones**, R.N. and Page, C.M. (2001). Assessing the risk of climate change on the water resources of the Macquarie River Catchment, in Ghassemi, F., Whetton, P., Little, R. and Littleboy, M. eds., *Integrating Models for Natural Resources Management across Disciplines, issues and scales* (Part 2), Modsim 2001 International Congress on Modelling and Simulation, Modelling and Simulation Society of Australia and New Zealand, Canberra, 673–678.

**Jones**, R.N., Lim, B. and Burton, I. (2003). Using risk assessment methods to inform adaptation, *Climatic Change* (submitted).

**Keating**, B.A., Carberry, P.S., Hammer, G.L., Probert, M.E., Robertson, M.J., Holzworth, D., Huth, N.I., Hargreaves, J.N.G., Meinke, H., Hochman, Z., McLean, G., Verburg, K., Snow, V., Dimes, J.P., Silburn, M., Wang, E., Brown, S., Bristow, K.L., Asseng, S., Chapman, S., McCown, R.L., Freebairn, D.M. and Smith, C.J. (2003) An overview of APSIM, a model designed for farming systems simulation. *European Journal of Agronomy*, **18**, 267–288.

**Kumar**, K.S.K. and Parikh, J. (2001). Indian agriculture and climate sensitivity, *Global Environmental Change*, **11**, 147–154.

**McCarthy**, J.J., Canziani, O.F., Leary, N.A., Dokken, D.J. and White, K.S. eds., (2001). *Climate Change 2001: Impacts, Adaptation, and Vulnerability*, Contribution of Working Group II to the Third Assessment Report of the Intergovernmental Panel on Climate Change, Cambridge: Cambridge University Press.

**Mearns**, L.O. and Hulme, M. (2001). Climate scenario development. In Houghton, J.T., Ding, Y., Griggs, D.J., Noguer, M., Van Der Linden, P.J. and Xioaosu, D. eds. *Climate Change 2001: The Scientific Basis* Contribution of Working Group I to the Third Assessment Report of the Intergovernmental Panel on Climate Change, Cambridge: Cambridge University Press, 739–768.

**Mehrotra**, R. (1999). Sensitivity of Runoff, Soil Moisture and Reservoir Design to Climate Change in Central Indian River Basins. *Climatic Change*, **42**, 725–757.

**Metz**, B., Davidson, O., Swart, R. and Pan, J. (eds.) (2001). *Climate Change 2001: Mitigation*. Contribution of Working Group III to the Third Assessment Report of the Intergovernmental Panel on Climate Change, Cambridge: Cambridge University Press.

**Morgan**, M.G. and Henrion, M. (1990). *Uncertainty: A Guide to Dealing with Uncertainty in Quantitative Risk and Policy Analysis*. Cambridge: Cambridge University Press.

**Morgan**, M.G., Fischhoff, B., Bostrom, A. and Atman, C.J. (2001). *Risk Communication: a mental models approach*, Cambridge: Cambridge University Press.

**Moss**, R.H. and Schneider, S.H. (2000). *Towards Consistent Assessment and*

*Reporting of Uncertainties in the IPCC TAR: Initial Recommendations for Discussion by Authors.* New Delhi: TERI.

**Nakicenovic**, N. and Swart, R. eds. (2000). *Emissions Scenarios*: Special Report of the Intergovernmental Panel on Climate Change. Cambridge United Kingdom: Cambridge University Press,

**NLWRA** (2002). *Australia's Natural Resources: 1997–2002 and Beyond.* National Land & Water Resources Audit, Turner ACT

**Risbey**, J.S. (1998). Sensitivities of water supply planning decisions to stream-flow and climate scenario uncertainties. *Water Policy*, **1**, 321–340.

**Schneider**, S.H. (2001). What is "dangerous" climate change? *Nature*, **411**, 17–19.

**Schneider**, S.H. and Kuntz-Duriseti, K. (2002). Uncertainty and Climate Change Policy, in Schneider, S.H., A. Rosencranz, and J.O. Niles, eds., *Climate Change Policy: A Survey*, Island Press, Washington D.C., 53–88.

**Smit**, B. and Pilifosova, O. (2001). Adaptation to climate change in the context of sustainable development and equity. In McCarthy, J.J., Canziani, O.F., Leary, N.A., Dokken, D.J. and White, K.S. eds. *Climate Change 2001: Impacts, Adaptation, and Vulnerability*, Contribution of Working Group II to the Third Assessment Report of the Intergovernmental Panel on Climate Change, Cambridge: Cambridge University Press.

**Tebaldi**, C. , Smith, R., Nychka, D., and Mearns, L.O. (2003). Quantifying uncertainty in projections of regional climate change: A Bayesian approach to the analysis of multimodel ensembles. Submitted to *Journal of Climate*.

**UNEP** (1998). *Handbook on Methods for Climate Change Impact Assessment and Adaptation Strategies*, Version 2.0, Feenstra, J.F., Burton, I., Smith, J.B. and Tol, R.S.J. eds. United Nations Environment Programme, Vrije Universiteit Amsterdam, Institute for Environmental Studies, http://www.vu.nl/english/o_o/instituten/IVM/research/climatechange/Handbook.htm

**Visser** H., Folkert R.J.M., Hoekstra J. and de Wolff J.J. (2000) Identifying key sources of uncertainty in climate change projections, *Climatic Change*, **45**, 421–457.

**Wigley**, T.M.L. and Raper, S.C.B. (2001). Interpretation of high projections for global-mean warming. *Science*, **293**, 451–454.

**Willows**, R.I. and Connell, R.K. (Eds.) (2003). *Climate adaptation: Risk, uncertainty and decision-making*. UKCIP Technical Report. UKCIP, Oxford. http://www.ukcip.org.uk/risk_uncert/risk_uncert.html.

**Yohe**, G. and Tol, R.S.J. (2002). Indicators for social and economic coping capacity – moving toward a working definition of adaptive capacity. *Global Environmental Change*, **12**, 25–40.

# ANNEX

### Annex A.5.1.  Climate change risk assessment utilising probabilities and critical thresholds

This annex describes a recent assessment that quantifies likely changes and assesses critical thresholds for an Australian catchment (Jones and Page, 2001). The modelling system coupled a climate scenario generator to a rainfall-runoff and river management model. Regional changes to potential evaporation (Ep) and precipitation (P) were used to perturb daily records of P and Ep from 1890–1996. The historical time series includes a drought-dominated (dry) period (1890–1947) and a flood-dominated (wet) period (1948–1996) allowing different modes of decadal rainfall variability to also be assessed. Three outputs were considered for risk assessment: storage in the Burrendong Dam (the major water storage), environmental flows to the Macquarie Marshes (nesting events for the breeding of colonial water birds), and proportion of irrigation allocations met over time.

*Quantifying outcomes*

Fifty-six simulations were run using a range of scenarios exploring the IPCC (2001) range of global warming, and regional changes in P and Ep from nine climate models. These models were then used to create the following transfer function:

$$\delta flow = a \yen (atan (\delta Ep / \delta P) - b)$$

where $\delta Ep$ and $\delta P$ were measured in mm yr$^{-1}$, $\delta flow$ is mean annual flow in GL yr$^{-1}$ and percent, atan is the inverse tan function, and a and b are constants. The results have an r$^2$ value of 0.98 (suggesting that 98% of the results fall within one standard deviation of the uncertainty contained within the relationship) and a standard error ranging from 1% to 2%.

According to the central limit theorem of statistics, if multiple ranges of uncertainty are combined, then the central tendencies are favoured at the expense of the extremes (e.g., Wigley and Raper, 2001). Three ranges of uncertainty contributed to the analysis: global warming and regional $\delta P$ and $\delta Ep$. Monte Carlo methods (repeated random sampling) were used to sample the IPCC (2001) range of global warming for 2030 and 2070. These were then used to scale a range of change per °C of global warming on a quarterly basis for P, sampling Ep using the above transfer function to estimate possible changes in mean annual water supply. The quarterly changes for P and Ep were then totalled to determine annual $\delta P$ and $\delta Ep$.

The following assumptions were applied:

- The range of global warming in 2030 was 0.55°–1.27°C with a uniform distribution. The range of change in 2070 was 1.16°–3.02°C.
- Changes in P were taken from the full range of change for each quarter from the sample of nine climate models.

- Changes in P for each quarter were assumed to be independent of each other (seasonally dependent changes between seasons could not be found).
- The difference between samples in any consecutive quarter could not exceed the largest difference observed in the sample of nine climate models.
- Ep was partially dependent on P ($\delta Ep = 5.75 - 0.53\delta P$, standard error = 2.00, randomly sampled using a Gaussian distribution, units in percent change).

Figure A-5-1-1 shows the results for 2030 where the probability distribution is tallied from wettest (best) to driest (worst) outcomes. Although there is an increased flood risk with increases, the drier outcomes are considered worse in terms of lost productivity and environmental function. The driest and wettest outcomes are less likely than the central outcomes where the line is steepest. The extremes of the range are about +10% to –30% in 2030 and about +25% to –60% in 2070, but the most likely outcomes range from about 0% to –15% in 2030 and 0% to –35% in 2070.

*Critical thresholds*

Two critical thresholds for the system were established:

1. Bird breeding events in the Macquarie Marshes, taken as 10 consecutive years of inflows below 350 GL.
2. Irrigation allocations falling below a level of 50% for five consecutive years.

Both thresholds are a measure of accumulated stress rather than a single extreme event. From the sample of runs described above, both thresholds were exceeded if mean annual flows fell below 10% in a drought-dominated climate, 20% in a normal climate and 30% in a flood-dominated climate.

*Uncertainty analysis*

Uncertainty analysis was carried out to understand how each of the Component uncertainties contributed to the range of outcomes. Three ranges of input uncertainty, global warming and local changes in P and Ep, were assessed by keeping each input constant within a Monte Carlo assessment, while allowing the others free play, consistent with Visser et al. (2000). Global warming was held at 0.91°C in 2030 and 2.09°C in 2070. $\delta P$ was taken as the average of the nine models in percent change per °C global warming for each quarter. $\delta Ep$ was linearly regressed from $\delta P$, omitting the sampling of a standard deviation. In both 2030 and 2070, $\delta P$ provides almost two-thirds of the total uncertainty, global warming about 25% and $\delta Ep$ just over 10% (Table A-5-1-1).

***Table A-5-1-1:*** *Results of uncertainty analysis for water storage in 2030 and 2070. The ranges shown are in percent change from mean annual storage.*

| 2030 | Limits of range | Range | Contribution to uncertainty |
|------|-----------------|-------|------------------------------|
| All | +10.3 to – 28.4 | 38.7 | |
| Constant global warming | +7.7 to –21.4 | 29.1 | 25% |
| Constant P | –1.9 to –15.9 | 14.0 | 64% |
| Constant Ep | 7.2 to –26.7 | 33.9 | 12% |
| | | | 101% |
| **2070** | | | |
| All | +23.8 to –60.1 | 83.9 | |
| Constant global warming | +17.3 to –45.8 | 63.1 | 25% |
| Constant P | –4.6 to –34.0 | 29.4 | 65% |
| Constant Ep | 16.3 to –57.7 | 74.0 | 12% |
| | | | 102% |

### Bayesian analysis

Bayesian analysis involves testing of input assumptions on the resulting probabilities. The tests are as follows:

1. Sampling intervals for δP and δEp were altered from quarterly to six-monthly and annually to determine whether the sampling interval affected the results. Figure A-5-1-2 shows the results as they affect the probability distribution of changes to mean annual Burrendong storage in 2030. Also shown are the original individual scenario runs, which are treated as having equal probability. The resultant probability distributions for six-monthly and annual sampling produce higher flows, but the results do not change by more than 10% from the original distribution in most cases.

2. The next test was to determine this impact of a non-uniform distribution of global warming, compared to the uniform distribution originally used. Wigley and Raper's (2001) non-linear distributions for global warming in 2030 and 2070 – based on input uncertainties for emissions scenarios, radiative forcing, atmospheric greenhouse gas modelling and climate sensitivity – were substituted for a uniform distribution. This has little effect on the results (Figures A-5-1-3 and A-5-1-4), which is consistent with global warming forming only 25% of the input uncertainties. Only very large changes in the range or distribution of global warming would be expected to significantly affect the outcome.

3. The distributions of rainfall change were altered by applying cubic polynomial regressions to the range provided by the nine models, counting the lowest and highest sample as the 10th and 90th percentile, respectively, thereby extending the range of rainfall change. These were added to the non-linear distributions for global warming (Figures A-5-1-3 and A-5-1-4). Although the total ranges have increased by 2% and 31% in 2030 and 20% and 55% in 2070 for the "W&R warming" and "All" cases, the distributions remain similar for the major part of the range.

These results show that the "most likely" parts of the ranges are not greatly expanded by increasing the ranges of uncertainty by the amounts here. The input ranges of uncertainty for rainfall for the Macquarie catchment are about ±4% per degree of global warming. These would have to be expanded considerably to alter the risk to water supply.

### Impact on policy

Previously, water managers in Australia were influenced by the uncertainty in rainfall change that indicated increases and decreases (in the Macquarie catchment, the range is about ±4% per degree of global warming), transferring this outcome to similar uncertainties in flow. The identification of seasonal decreases of rainfall

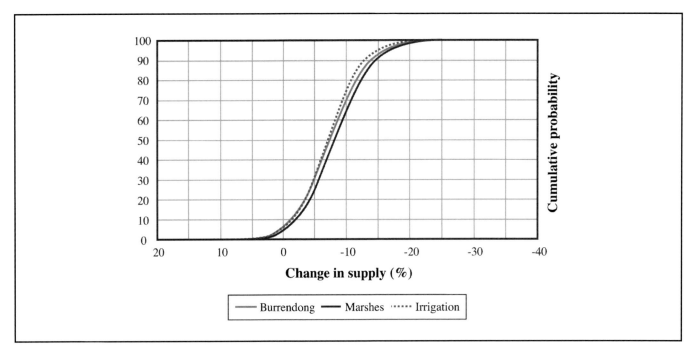

**Figure A-5-1-1:** Probability distribution for changes to mean annual Burrendong Dam storage, Macquarie Marsh inflows and irrigation allocations, based on Monte Carlo sampling of input ranges of global warming, δP and δEp in 2030.

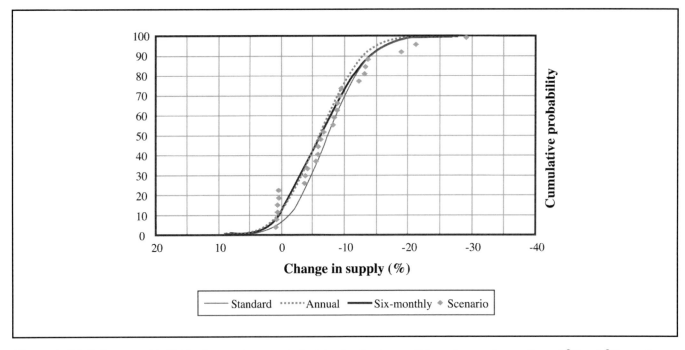

**Figure A-5-1-2:** Impact of individual scenarios, quarterly (standard), six-monthly and annual sampling of δP and δEp on the probability distribution for changes to mean annual Burrendong storage in 2030.

in the winter-spring period in all the climate models investigated, construction of potential evaporation scenarios and this work, has contributed to a change in attitude.

This risk assessment has already contributed to policy that is overseeing the development of environmental flow regimes for the Murray River. The finding that water availability is likely to decrease, and that critical thresholds may be crossed under a drought-dominated climate, has been sensitised by a series of dry years and findings that allocations in the catchment being investigated were above sustainable levels. It is now being speculated that the decrease in rainfall may be similar to decadal shifts experienced in southwest Western Australia and in the Sahel. Further work is investigating whether current water policy measures and changes being planned are sufficient to manage the risks that have been identified.

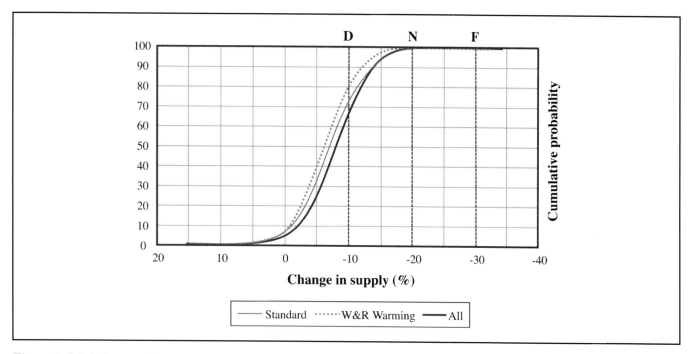

**Figure A-5-1-3:** Impact of uniform sampling, non-linear sampling of global warming (Wigley and Raper, 2001) and non-linear sampling of rainfall change (All) on the probability distribution for changes to mean annual Burrendong storage in 2030. Critical thresholds under a drought-dominated climate (D), flood-dominated climate (F) and normal climate (N) are also shown.

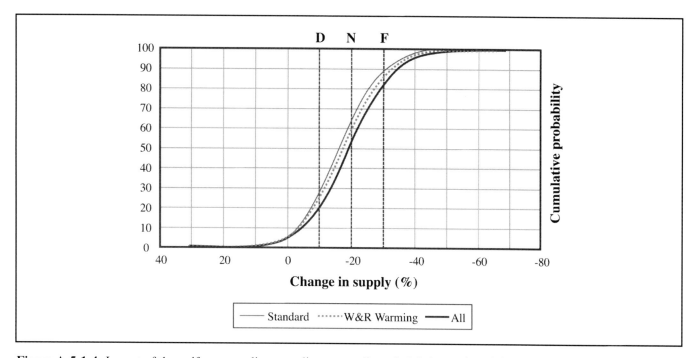

**Figure A-5-1-4:** Impact of the uniform sampling, non-linear sampling of global warming (Wigley and Raper, 2001) and non-linear sampling of rainfall change (All) on the probability distribution for changes to mean annual Burrendong storage in 2070. Critical thresholds under a drought-dominated climate (D), flood-dominated climate (F) and normal climate (N) are also shown.

# 6

# Assessing Current and Changing Socio-Economic Conditions

ELIZABETH L. MALONE[1] AND EMILIO L. LA ROVERE[2]

Reviewers

*Suruchi Bhawal[3], Henk Bosch[4], Hubert E. Meena[5], Moussa Cissé[6], Roger Jones[7], Ulka Kelkar[3], Khandaker Mainuddin[8], Mohan Munasinghe[9], Atiq Rahman[8], Samir Safi[10], Barry Smit[11], and Gina Ziervogel[12]*

[1] Pacific Northwest National Laboratory, Washington, DC, United States
[2] Centre for Integrated Studies on Climate Change and the Environment, Rio de Janeiro, Brazil
[3] The Energy and Resources Institute, New Delhi, India
[4] Government Support Group for Energy and Environment, The Hague, The Netherlands
[5] The Centre for Energy, Environment, Science & Technology, Dar Es Salaam, Tanzania
[6] ENDA Tiers Monde, Dakar, Senegal
[7] Commonwealth Scientific & Industrial Research Organisation, Atmospheric Research, Aspendale, Australia
[8] Bangladesh Centre for Advanced Studies, Dhaka, Bangladesh
[9] Munasinghe Institute for Development, Colombo, Sri Lanka
[10] Lebanese University, Beirut, Lebanon
[11] University of Guelph, Guelph, Canada
[12] Stockholm Environment Institute Oxford Office, Oxford, United Kingdom

# CONTENTS

## 6.1.    Introduction

Understanding the socio-economic pattern(s) of any system(s) is essential for adapting to climate change. Vulnerability to climate change depends on the interactions between changing socio-economic conditions and climate hazards; the feasibility of its adaptation options requires socio-economic analyses of the underlying barriers and opportunities. Therefore, socio-economic conditions must be described in enough detail to evaluate the merits of policy options.

Earlier approaches for assessing vulnerability and adaptation made simplified assumptions, which limited the usefulness of the proposed adaptations. At worst, climate change impacts were projected on a static society, without accounting for changes in key socio-economic drivers of human development. In other assessments (Pepper et al., 1992; Nakicenovic et al., 2000), impact predictions were carried out with a very limited set of socio-economic indicators (such as population, GDP per capita, and land-use change and technological improvement) using computer-based models. For global models, this minimalist treatment is appropriate. But at smaller scales, where adaptation actually takes place, more detail is needed about the residents, and how they live and work in communities. Government policies – including taxes and regulations – encourage certain economic and social activities and discourage others. The culture of societies, their forms of social solidarity and organisation, are all important factors in shaping adaptation policy.

The challenge is to develop adaptation strategies appropriate to the societies of the future. To achieve this goal, first, the relationship between current and future climate and changing socio-economic conditions must be explicit. Second, projected socio-economic conditions and their implications for vulnerability of systems should be explored. Adopting this approach increases the realism of the analysis.

To support this type of analysis, this Technical Paper (TP) provides guidance in three areas:

* characterising socio-economic conditions and drivers with indicators;
* relating these indicators to vulnerability and climate analyses;
* integrating adaptation to climate change into sustainable development objectives.

This paper emphasises qualitative or mixed quantitative/qualitative approaches. Its application will produce either a qualitative or quantitative description of current and future socio-economic conditions for the priority system. Specific outputs may include (1) a general overview of historical socio-economic conditions, (2) detailed description of current conditions, (3) and a set of alternative "storylines" describing future socio-economic prospects in the context of potential climate change impacts.

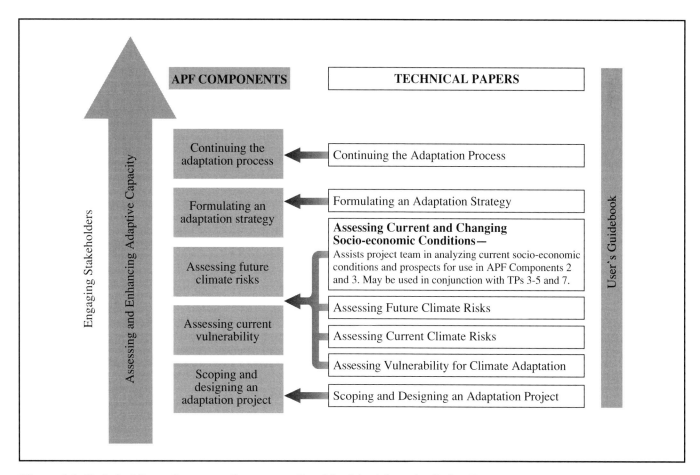

**Figure 6-1:** Technical Paper 6 supports Components 2 and 3 of the Adaptation Policy Framework

## 6.2.    Relationship to the Adaptation Policy Framework as a whole

This paper relates to Components 2 and 3 of the Adaptation Policy Framework (APF) process (Figure 6-1). It assumes that the APF users have designed and scoped a project using Component 1. At this point, the team will use concepts from this paper to analyse current socio-economic conditions and prospects within the identified priority system(s). Components 2 and 3 of the APF provide the basis for developing and implementing coherent adaptation strategies, policies and measures (TPs 8 and 9).

Depending upon the methodological choices made under Component 1, this paper may be used in combination with TPs 3, 4, 5 and 7 to varying degrees. Essentially, the extent of the team's socio-economic analysis will be dictated by the degree to which they have incorporated other APF analyses, including those for vulnerability (TP3), climate risks (TPs 4 and 5), and adaptive capacity (TP7).

In other words, an analysis of socio-economic conditions and prospects can be conducted as a stand-alone exercise – this would constitute the policy-based approach – or as part of a vulnerability assessment (TP3). As a stand-alone exercise, the project would use the guidance outlined here to assess the potential efficacy of an existing or proposed policy (or strategy or measure) in a scenario of future climate change. Depending on the project requirements, it is possible to develop an adaptation strategy, using this policy-based approach and this TP as a resource. Used as part of a vulnerability assessment, the analysis of socio-economic conditions outlined here would provide indicators for this larger assessment. This analysis can, in turn, be integrated with the results from the climate risk analysis (TPs 4 and 5) for APF Components 2 and 3. (The information on current socio-economic conditions is similar to that required for United Nations Framework Convention on Climate Change (UNFCCC) National Communications.)

## 6.3.    Key concepts

The concepts presented below are central for characterising socio-economic conditions. These concepts are also outlined in the APF User's Guidebook.

**Indicators** – Since socio-economic conditions and prospects are intangible and cannot be measured directly, analysts use indicators, i.e., parameters that characterise these abstract concepts. For example, although social welfare is important, it cannot be measured directly; often, GDP per capita is used as an indicator. GDP per capita is a flawed indicator for either welfare or growth, since it neglects a range of important values,

from the ability of a household to meet basic needs to the use/depletion of natural resources.[1] As a measure of economic productivity, however, GDP can be observed, measured, and compared across areas. When using indicators, project teams should ensure that informal activities are taken into account. Such activities are not captured in official statistical data, yet are so important to livelihoods in developing countries.

**Qualitative and quantitative analysis** – Qualitative and quantitative approaches are mutually dependent. Quantitative analysis rests on judgmental, qualitative assumptions about how the world works, what are suitable categories for data, what constitute good data, and the validity of scientific procedures. For future projections, the role of qualitative assumptions is even more marked. Qualitative research, on the other hand, if it is to make sense of the world at all, must weigh and measure, and judge what is important, and what are critical variables in human development. Whether or not numbers are used, these are essentially quantitative tasks.

The question is not "Shall we use a qualitative or quantitative approach?" but "How can we use both to answer the question usefully?" This approach means including policy-makers and other stakeholders in the process, debating the starting assumptions, and being willing to re-examine categories, assumptions, and data as the analysis proceeds.

**Scenarios** – A scenario represents a plausible and simplified description of how the future may develop, based on a coherent and internally consistent set of assumptions about driving forces and key relationships. Scenarios may be derived from projections, but are often based on additional information from other sources, sometimes combined with a narrative storyline.

**Storylines** – Storylines are qualitative, holistic pictures of the general structures and values of society. Storylines can be developed at any scale – from the global, to the regional, national, or local levels. They describe conditions that might be produced by human choices about economic and social policy, reproduction, occupations, and energy/technology use. Storylines are useful tools for policy makers to "vision" alternative future worlds.

## 6.4.    Guidance on characterising current and changing socio-economic conditions

In this chapter, users will find guidance for characterising current socio-economic conditions (i.e., developing an adaptation baseline), and projected conditions (scenarios or prospects) in their priority system, with three variants: no policies regarding adaptation; and two alternatives for adaptation policy.

This effort can range from a qualitative description to a full-

---

[1] GDP per capita neglects the value of unpaid work, people's satisfaction with their occupations, and many other aspects of welfare. As it does not represent income or real wages, the measure of GDP per capita does not capture a household's ability to meet its needs. The additions provided in various versions of "green" GDP compensate for some of the shortcomings.

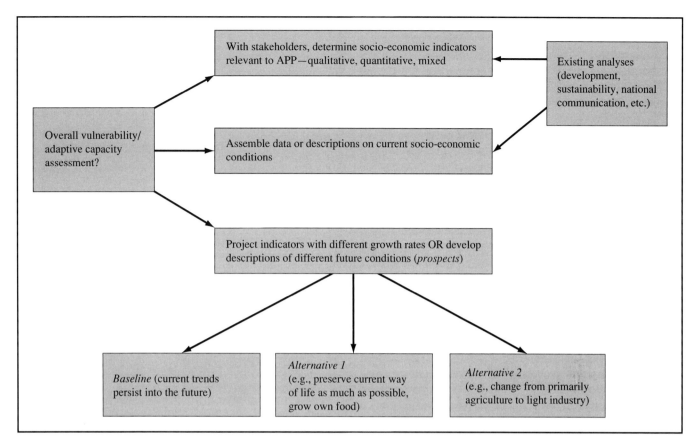

**Figure 6-2:** Overview of developing socio-economic conditions and prospects

blown assessment, based on resource-intensive, model-based processes. The output may be summarised as a brief (five to eight page) section of the full adaptation report, or as an extensive report that includes model results. The "example" boxes in the text show how this information may be summarised.

The description should accomplish the following (Figure 6-2):

- Analyse current socio-economic conditions, including current natural resource management practices, describing changes in the last 10 to 20 years (50 years, if possible) due to climate change, including variability. This analysis constitutes the adaptation baseline.
- Develop qualitative storylines, and quantitative or qualitative future scenarios: construct a reference scenario without adaptation, scenarios with past and current adaptation measures, and scenarios with additional adaptation policies and measures.
- Ensure consistency among global, regional, national, sub-national and local scenarios.
- Analyse socio-economic prospects, taking into account the lower and upper parts of the economic cycle.
- Analyse vulnerability to climate change (TP3), considering the cyclical, sometimes random fluctuations found in different sectors and regions.

### 6.4.1. Setting up study boundaries

In keeping with the APF process, this guidance assumes that a socio-economic analysis will be focused on a priority system (TP1). Setting system boundaries is important for adaptation issues. However, it is unlikely that a system will be self-contained; most likely, the priority area will be "key" for many reasons.[2] This priority system may be connected to the national or international economy, many people may be affected by it, and it may contain an important natural or cultural resource. Thus, the system must account for links among the elements, both inside and outside the locale or sector. Such elements include trade, kinship, migration, culture, transportation, communication, etc.

As a starting point, the team should review existing documents and modify them as needed. Examples include development plans, poverty reduction strategies, and sustainability assessments. In developing countries, most of these plans exist.

### 6.4.2. Using indicators

Desirable indicators fulfil three criteria: (1) summarise, quantify and simplify relevant information; (2) capture phenomena

---

[2] The terms "priority area" and "priority system" are used interchangeably here to refer to the area of focus of the adaptation project. (See TP1 for guidance on identifying a priority system.)

of interest; and (3) communicate relevant information. They may be qualitative, quantitative, or both. If quantitative scenarios of the future relevant to climate change vulnerability and adaptive capacity are desired, the process involves choosing relevant indicators, collecting appropriate data, and estimating future values for those proxies (Malone et. al., 2002).

### 6.4.3.  *Characterising socio-economic conditions today*

Together, the adaptation project team and stakeholders select the indicators and/or descriptions that are most relevant to the area, sector, and people that are being analysed. The suggestions below are not prescriptive; characterising the socio-economic conditions could use any combination or none of the indicators discussed. For example, one indicator may stand for several others in a specific place. Also, stakeholder knowledge may be more important than any quantitative data.

In developing an adaptation baseline, the starting point is an overview of the socio-economic elements that make the selected priority area important. These elements are likely to include the significance of the area for: supporting its population; producing food and other goods for consumption; natural resources such as forests, fisheries, and tourism; and facilitating (or inhibiting) trade and markets. Box 6-1 provides an overview of such information for a coastal region in China. In any adaptation process, the overview should be tailored to suit the priority system or area. This example relies on quantitative data (statistics), although qualitative data may be as good as or better.

It is likely that this priority system was chosen because its important assets and economic activities have been systematically impacted by climate hazards, and that this maladaptive trend has increased its vulnerability. Its socio-economic elements should be described (e.g., people and infrastructure at-risk from floods; or hunger, disease and internal migration consequences of drought). Recent experience, both events and responses, should be summarised, e.g., several good harvests may have encouraged more extensive (i.e., expansion of) agriculture. The overall assessment may include biophysical information as well as socio-economic information.

### 6.4.4.  *Exploring specific characteristics*

This phase of the analysis focuses on socio-economic elements most relevant to current conditions. (This analysis corresponds to the assessment of adaptive capacity discussed in TP7.) For convenience, the elements – or indicators – are divided into five categories: demographic analysis, economic analysis, natural resource use, governance and development policies, and culture. For all categories, the description should be more detailed than simply trends in population growth and GDP per capita over the past two to five decades. If the information is available, a set of appropriate indicators could be established for each category. Whenever possible, both quantitative and qualitative approaches should be used.

The availability of data and data quality, the level of detail, and the selection of specific indicators are matters for the individual teams and their stakeholder groups to consider and decide. Data at specific time and spatial scales may not be available. However, many countries carry out periodic population and agricultural censuses and household income and expenditure surveys for development planning. If the quality of these data are adequate, they may be used for the APF process. Even in developed countries the data set will never be perfect.

Table 6-1 shows an example of an indicator set for water resources. These indicators represent just a small sampling of the many possibilities. Although these indicators have been divided into categories, there are significant linkages between them.

#### 6.4.4.1.  *Demographic analysis*

Demographic characteristics are essential for an analysis of socio-economic conditions. Since gathering every available statistic would be impractical, key demographic indicators should be selected. The objective is to assess the socio-economic vulnerability of people in the priority system.

The number of people living in the priority system is a starting point, but the population's well-being also depends on how they are distributed in the area (in terms of urbanisation, for instance, or the number of hectares per farm household), the

---

**Box 6-1: Example of brief overview of current socio-economic conditions using geopolitical, demographic, and economic data**

The coastal region (1.27 million km²) of the People's Republic of China (hereafter referred to as China) includes Tianjin and Shanghai D.C.; Liaoning, Hebei, Shandong, Jiangsu, Zhejian, Fujian, Taiwan, Guangdong, Hainan provinces; and Guangxi Zhuangzu Zizhiqu. This region accounts for 12.24% of the total land area of the country, yet it supports 40.2% of China's population and contributes 55% of the country's gross agricultural and industrial output. The narrow 40- to 50-km-wide portion of this region along the coastline includes 44 coastal cities with prefecture status, 35 coastal cities with county status, and 111 coastal counties or districts from two D.C. and nine provinces. Although this coastal zone makes up 2.9% of the area of the country, its population constitutes 13.43% of the total, which makes the population density 4.7 times the average for all of China. The region's total social output value is 28.8% of the total, and the output value per unit is 9.9 times the average for the country (adapted from Yang, 1996, pp 265-266).

*Table 6-1: Example set of indicators for water resources*

| Demographic indicators | • Access to clean water and sanitation<br>• Withdrawals as a % of available water<br>• % shares of total use (household, industry, agriculture) and rate of increase in uses |
|---|---|
| Economic indicators | • Presence or absence of water markets<br>• Contribution of water to products (e.g., irrigation to agricultural products)<br>• Amount/kinds of water infrastructure (reservoirs, dams, etc.) |
| Governance and policy indicators | • Treaties or agreements regarding available water resources<br>• % of water resources not under regional control<br>• Development plans for area (e.g., population growth, agricultural development and water-use implications) |
| Cultural indicators | • Cultural meaning and recreational uses of rivers/lakes (sacred or forbidden uses)<br>• % unpolluted stream and beach kilometres (and nature of protection) |

Partial Source: Moldan and Billharz, 1997

land tenure regime, the rate of population growth (e.g., fertility trends and death rates), the age distribution (e.g., "How many working-aged people?" "What is the dependency ratio?"), the workforce versus unemployment levels, health characteristics, and male/female education levels. Such demographic characteristics are key to the priority area's vulnerability and adaptive capacity. For example, the presence of young children raises unique education and health issues.

The next step is to relate the demographics of the priority area to national-level information. "What are the differences between the priority area or sector and the country?" Relevant socio-economic changes include rural-to-urban migration, epidemic disease, and fluctuating educational levels. The differences between, first, *levels of change*, and, second, *rates of change*, in the priority area, and the country as a whole, will yield insights about vulnerabilities at both scales.

The example of analyses in Box 6-2 is incomplete, but it is intended to suggest elements of focus and indicators for use in an economic analysis.

Examples of **potential indicators for use in demographic analysis** include the following: population size, age structure, population density, location/urbanisation, migration, education (e.g., literacy rate), fuel used by households (e.g., firewood), housing with electricity, rate of poverty and extreme poverty, health characteristics (e.g., infant mortality), food security (e.g., dietary needs, composition and costs, local basic diet, food sources, availability and accessibility).

### 6.4.4.2. Economic analysis

An economic profile of the people who live in the area – their types of employment activity – is an important element of current socio-economic conditions. "What are the principal ways people make their livings, and what is the share of each activity in the priority area or sector's overall economy?" "Are shifts being seen, e.g., in the types of crops planted or livestock raised?" "Is off-farm employment an increasing trend?" "What is the unemployment rate?" Such questions can inform the eco-

---

**Box 6-2: Example of demographic analysis: Urbanisation, education and health**

Within the Sudano-Sahel region, urbanisation is increasing rapidly and is expected to continue in the near future. These migration trends impose a burden on the existing education and health systems in the region, and increase people's vulnerability. To this extent, progress already made in the region on education is being seriously threatened by the deteriorating economic trends, which are caused to a large extent by droughts, heavy external debt and political instability. The priority area is already affected by these factors, and access to health services has not kept up with population increases (adapted from Wang'ati, 1996, pp 76-77).

To describe changes in education and health, the following indicators may be used: income per capita and its distribution, the number of school-age children enrolled in schools, access to food and health care, and the average life expectancy at birth. However, accurate censuses are rare in the region and this type of data is often limited.

nomic analysis. For these surveys, the household may be a more appropriate unit than the individual.

The principal economic activities of the priority system can be captured by the following development patterns, policies, and associated indicators.

**Monetary policies**

- *Market participation:* Adaptation choices are profoundly affected by national policies, free trade agreements, and the extent of participation in domestic and international markets. For example, the well-being of subsistence farmers is directly dependent upon weather, while the well-being of farmers growing cash crops is highly dependent upon market prices. The impacts of privatisation policies should be identified (here and/or in the governance category) as they significantly affect the economic vulnerability of rural farmers.

- *Public and private investment:* The level of investment in economic activity, such as manufacturing and other business enterprises, can provide powerful indicators of economic conditions. Investments provide employment and marketable products.

- *Income:* If the economy is largely monetised, some measure of income or wages, along with income inequality, may capture important information. If there is a large informal or subsistence economy, better ways to characterise well-being include wealth, assets, or consumption. One of the accepted measures of the poverty rate may also be used (e.g.,, $1 or $2 per day, or a nationally-defined poverty level).

- *Savings:* Similarly, national savings and borrowing can be used to finance reconstruction following a climate disaster. However, the creation of national debt could reduce economic growth, and further exacerbate poverty and vulnerability to future climatic disasters.

**Industrial and infrastructure policies**

- *Industrialisation:* The extent of industrialisation and diffusion of associated technologies are related to market activity. One facet of industrialisation is the presence (or absence) of modern farming methods and cultivars.

- *Infrastructure:* The extent of infrastructure, such as roads, rail and air transport, electricity generation, communications, irrigation districts, dams, and buildings, are important for economic development. One indicator might be the portion of the public budget that is dedicated to social infrastructure, such as schools and hospitals.

**Labour policy**

- *Labour:* In many countries, the secondary and tertiary economic sectors are the most important in terms of the labour market. Moreover, in many developing countries, a large informal economy generates most of the labour demand.

- *Migration:* In developing countries, members of the rural household units often migrate abroad and provide additional incomes to their families (e.g., Sri Lanka, Kerala). In some countries, these sources of income represent an important percentage of the national budget. Domestic migration – seasonal, for example – may be a current strategy to cope with climate variability.

**Agricultural policy**

- *Food security:* In countries where food security is a major issue, it is critical to evaluate the implications of a changing climate on agricultural production. "What are the dietary needs and how are they met by domestic production and/or imports?" (See also the food security example in TP3.)

- *Land tenure:* The land tenure regime and the extension of household plots are especially useful in characterising rural economic conditions.

**Environmental policy**

- *Environmental impacts:* The environmental sustainability of economic activities should be identified. This assessment will reveal the extent to which current development processes (e.g., industrialisation, international free-trade agreements, privatisation) are facilitating adaptation or promoting maladaptation.

The examples in Boxes 6-3 and 6-4 are incomplete, but are intended to suggest areas of focus and indicators that could be

---

**Box 6-3: Example of general economic analysis for an urban area**

Mexico City is divided into two broad industrial sectors: the means of production and consumer goods. In 1970, the first sector represented 27.4% of all industry, including machinery, tools, and raw materials for other industries. Consumer goods comprised 73% of all industry, including immediate and durable goods. Also, the city holds an enormous share of Mexico's major financial exchanges, private businesses, and central offices. In 1980, 4.9% of the active population was employed in the primary sector, 41.4% in industrial activities, and 53.7% in services (adapted from Guillermo Aguilar, et al., 1995).

---

**Box 6-4: Example of household analysis for an agricultural area**

In 1979 the average area of land cultivated by households studied in the Kosi Hills of Nepal was about one hectare. The 43% of households with access to less than a half-hectare of cultivable land were only able to produce about one-half of their own food-grain requirements and were short of food for a few months each year. Household members thus depend on off-farm seasonal employment opportunities and the health of employable household members. In practice, these households are likely to be in debt, causing them to adopt strategies that provide quick returns without necessarily giving them the best possible yield from their endeavours. For example, they may accept local, low-wage, casual work rather than travel for several days to find buyers for their hand-made crafts (adapted from Nabarro et al., 1990, pp 68-69).

---

used in an economic analysis. In these examples, the APF users will need to ask, "What are the implications for climate change adaptation?"

### 6.4.4.3. Natural resource use

The priority system most likely has natural resources that are used by the population in various cultural and economic activities. An assessment of these resources and their uses can reveal a lot about a population's vulnerability to climate change. If the area is largely dependent upon agriculture for food security and/or income, the quality and amount of land available for crops and/or livestock is critically important to understanding the climate risks that people face. Similarly, water quality and availability – the amount and timing of precipitation, surface water, and groundwater – are important and potentially limiting factors for a population's agriculture, industry, sanitation, and consumption, and can thus be central to vulnerability.

Other natural resources may also be important. Resources such as minerals; forest products; abundant sun, wind, or water; scenery; and biodiversity can reduce a population's dependence on agriculture or its exposure to agricultural risks and, in this way, reduce its vulnerability to climate risks. However, the exploitation or use of natural resources may also damage other natural resource systems. A natural resources assessment should consider current and potential uses, along with current and potential negative consequences.

The examples in Boxes 6-5 and 6-6 are incomplete but are intended to suggest areas of focus and indicators that could be used in a natural resources analysis.

### 6.4.4.4. Analysis of governance and policy

Economic development and environmental policies provide both constraints and opportunities for adaptation, as noted above. This analysis consists of three major steps: (1) evaluating existing policies and programmes; (2) detailing the planning and policy-making processes for the priority system; and (3) assessing adaptive capacity to implement policies and programmes.

First, specific policies and programmes should be evaluated for their potential to advance sustainable development and adaptation to climate change. The anticipated environmental consequences should also be specified. State reforms, such as privatisation and liberalisation of trade, are especially important.

Additional relevant policies include:

Examples of **potential indicators for use in natural resource analysis** include the extent of natural resources, current uses and state of health/degradation (e.g., water quality and quantity, forest cover, deforestation rates, expansion/abandonment of agricultural lands, soil degradation or desertification), and the potential for further and different uses (considering sustainability).

- The government makes policy choices about economic development—whether or not to encourage domestic markets and international trade, to develop supporting institutions such as banks and intellectual property protections, or to focus on increasing GDP. Domestic economic policies are used to reduce the negative effects of transitions to privatisation and liberalised trade through emergency relief programmes, job training, insurance, the establishment of reserves, etc.

---

**Box 6-5: Example of innovative resource development**

In Port Antonio, a town in Jamaica, wild fish stocks have been depleted and the beach has been treated as a dump. However, oyster farming is addressing both of these issues. Oyster production includes both collecting spat-on pieces of old tires strung on fishing line in government nurseries on the southern coast, and cultivation on the north coast which has no natural stock but ideal growing conditions. Pressure on wild fish stocks is being reduced and impetus is being provided for protecting coastal marine habitat – not only for oysters, but also for thousands of other marine species (adapted from Bourke, 1995).

---

---

**Box 6-6: Example of oil extraction impacts on ecosystems**

With rich biodiversity and immense oil reserves, the Niger Delta is an important source of biological and economic wealth. Since the 1950s, Nigeria has exported large quantities of oil from the southern region of the Delta known as the "oil belt". However, the extraction and production of oil has caused environmental damage in this extremely sensitive ecosystem. Oil spills have destroyed freshwater ecosystems, fouled farmland, killed wildlife and endangered human life. In addition, canals built to support oil pipelines have impacted the hydrology of the Niger Delta, creating a scarcity of water, and channelling pollutants back into the ecosystem (adapted from *PECS News*, Spring 2000).

---

- Economic development and internal welfare policies, such as these are most relevant to adaptation. They may, however, be considered low priority when compared with other policies, e.g., trade agreements, border or internal security proposals, or existing government support legislation. Policies such as land set-asides and tenure reforms are critical for natural resource management.

- Policy choices also greatly affect the internal well-being of a country's citizens; a government may choose to emphasise poverty reduction, preservation of traditional cultures, development of endogenous technologies, provision of funding for research programmes, and extension of education and health services.

Second, once the most relevant policies for the priority system have been identified, the project team should outline the planning and policy-making process. The policy analysis (laws, standards, regulations, etc.) will be relevant to selecting pathways for implementing alternative adaptation. Ultimately, the adaptation choice may be determined by the path of least resistance.

Finally, the capacity of government institutions to carry out current policies and development programmes should be assessed (TP7). The team should identify the relevant agencies and actors, and their roles and effectiveness. "What agencies and other

**Examples of potential indicators for use in governance and policy analysis** include environmental trends and policies, the extent of integration of economic and environmental policies, and the planned state reforms (e.g., privatisation, current and planned free-trade agreements).

actors are involved?" "Is it a participatory or top-down approach?" "Who makes the decisions?" "Are there ways to alert the policy makers who implement policy changes?"

In an urban area, relevant policies might include those to improve slums (e.g., sanitation, housing, electricity supply, local security); capacity development may include training people to enter the workforce; investing in public schools, hospitals, roads or clinics; and combating air pollution and the urban heat island.

In an agricultural area, policies could include research into drought-resistant cultivars and other technological options (e.g., irrigation, dams); capacity development may involve implementing land reform and environmental policies (e.g., regulations, laws, standards, incentives), and providing off-farm employment.

The examples in Boxes 6-7 and 6-8 are incomplete but are intended to suggest areas of focus and indicators that could be used in an analysis of governance and policy.

*6.4.4.5.  Cultural analysis*

Culture can be expressed as "the way we do things here". Cultural values include the way families are defined, and their obligations toward one another, their relationship with nature (e.g., the maize culture in Meso America), the role and forms of governance, and technology diffusion. To a large extent, culture dictates social behaviour. It is a powerful force that can enable certain activities, and constrain others.

Cultural values have significant bearing on climate change adaptation. For example, where a strong culture of mutual self-

---

**Box 6-7: Example of production-focused policies**

The focus of production programmes in Nepal's mountain areas is primarily on resource-use intensification and extraction. These considerations also guide public interventions related to infrastructure, development and investment. The focus is on mountain niches such as irrigation and hydropower, mining, tourism, and horticultural production. These are largely guided by external demand and the revenue needs of the state. They result in a high rate of resource extraction, reduced diversification of resource-centred activities, and negative side-effects on fragile mountain resources. Moreover, the negative effects of intensification and over-extraction are accentuated by the absence of any measures to control or regulate the demand on mountain resources (adapted from Jodha, 1995, pp 167-170).

---

**Box 6-8: Example of general development policies**

Recent Brazilian development efforts, following an industrialisation and urbanisation model, have favoured certain areas of Brazil and imposed conditions on the northeast and the Amazon. Such economic development has not addressed the agrarian problem, nor has it encouraged a search for new organisations of the agricultural economy—concentration of agrarian landholdings, the exploitation of rural labour, and the appropriation by large landowners or other elites of the economic surplus generated by farmers or the landless. Sustainable development in this context has been difficult to achieve (adapted from Bitoun et al., 1996, p 145).

---

**Examples of potential indicators for use in cultural analysis** include cultural values and traditions relevant for adaptation especially education, knowledge development, technical assistance, endogenous technological development, local research, communication and public awareness.

help and co-operation exists, adaptation strategies may benefit from tapping into and building upon this social capital. As reflected by changing lifestyles, culture is greatly influenced by globalisation.

The example in Box 6-9 is an incomplete analysis but is intended to suggest areas of focus and indicators that could be used in a cultural analysis.

### 6.4.5. *Characterising current adaptations*

Current adaptations for coping with today's climate constitute the *adaptation baseline*. This baseline is a comprehensive description of adaptations that are in place to cope with current climate. The baseline may be both qualitative and quantitative, but should be operationally defined with a limited set of parameters (indicators). It also represents the analytical starting point for an adaptation project that uses the policy-based approach (both the User's Guidebook and TP1 explain the four major project approaches; Section 6.1 also briefly discusses the policy-based approach).

As economic and social conditions change, an area's climate can take on different meanings. For example, when traditional inter-cropping is practiced, the timing of rainfall is not as significant as when certain crops are grown exclusively. When rivers are the major form of transport, keeping the flow at a certain level is essential. But, if other means of transportation replace the rivers, water can become available for other uses, and the timing of precipitation becomes less significant. Current adaptations also represent an opportunity to address maladaptation to current climate.

Ideally, socio-economic conditions should describe historical changes over the past 10-20 years. However, depending on the recent political or socio-economic history of each country, timeframes could be 50 or even more years in the past.

The analysis is not intended to be comprehensive; in most cases, a narrative description will suffice, augmented by quantitative data if available. The examples in Boxes 6-10 and 6-11 suggest two types of recent adaptations to climate.

### 6.4.6. *Characterising changing socio-economic conditions*

At this point, project teams will have gathered enough information about the present and past to assess future socio-economic conditions. There are two remaining tasks. The first is to develop alternative "storylines" of the future for an appropriate time period (probably between 20 and 50 years into the future – see User's Guidebook and TP5 Section 5.4.4, *Selection Planning and policy horizons* for assistance in determining an appropriate time period). The second task is to make projections about how socio-economic conditions – indicators, if this approach has been taken – will change in the future under the alternative storylines. If the indicators chosen are qualitative, the description of socio-economic prospects will also be qualitative.

---

**Box 6-9: Example of cultural analysis**

Fonogram, a West Bengal village, has a strong tradition of mutual support among the poorest in the village, based on the informal system of loans and an often strongly-voiced animosity toward the rich. Villagers go from house to house asking for *khud* and *bhater fan,* as well as for building materials. The main kind of loan made between the poorest households is in the form of small amounts of money or foodstuffs. Other forms of mutual support include looking after children or livestock. But an important distinction is made between a "loan" between poor people that is seen as an expression of support, friendship and solidarity, and a loan begged from either another poor person or someone better off, which involves a subordinate relationship (adapted from Beck, 1990, pp 28-29).

---

**Box 6-10: Example of recent adaptation**

In the 1930s, there was doubt that Kenya's Machakos district could feed itself. However, the district has not only produced enough food under current climatic conditions, but has evolved a complex farming system that appears sustainable. Abolition of the colonial economy, which restricted some cash crops to the coloniser, has led to widespread growth of such crops. Also, there has been significant, ongoing intensification of food crops for both subsistence and the national market. The increase in production has been accomplished under a complex land-use system, in which management of vegetation is practiced systematically for both production and conservation purposes. Machakos farmers have organised extensive co-operatives (adapted from Wang'ati, 1996). A farming system that is well-adapted to current climate will probably be resilient to mean changes in future climate.

---

### 6.4.6.1. Developing storylines

In order to examine future adaptation to climate change, analysts construct accounts of what the future may be like. For this purpose, the Special Report on Emissions Scenarios (SRES) (Nakicenovic et al., 2000) developed "storylines"—i.e., coherent pictures of the future within which certain trends make sense. These narrative descriptions provide very general accounts based on two dimensions: the extent of sustainable development and economic development at local or global levels. These storylines allow for integrated analysis and for identification of key systems.

This guideline uses multiple storylines to characterise three alternative futures for the priority system (population). This approach accounts for a wide variety of possible futures and the large uncertainties involved in such projections (TP5 provides a discussion of quantitative and qualitative approaches to uncertainty in developing scenarios).

(1) The first storyline is a *reference scenario*, which does not consider climate change. The "current socio-economic conditions" (already discussed) are projected into the future. For example, if deforestation is currently taking place, the reference scenario prospects are for continued deforestation.

(2) and (3) are two significantly different projections in which development will proceed, taking climate change into account through adaptation policies. One set of policies may attempt to preserve current economic activities and socio-economic conditions using technologies (regulations for buildings to resist damage from storms or sea level rise, for instance); another set of policies alternatively could emphasise different crops or a reduction of agricultural activity. These policies should be described in

the storylines. Figure 6-3 illustrates this approach.

Given the wide spectrum of possible development paths open to countries, the choice and assumptions of the reference scenario is thus crucial. Further, it may be useful to repeat the comparison of reference scenarios with two storylines to test the sensitivity of adaptive policies to different reference scenarios, and provide a range for their possible outcomes. Several strategies can be used, with increasing levels of complexity, from extrapolation (1) to the desirable integrated analysis (4).

**(1) Extrapolation**

An approach to define a reference scenario is the extrapolation of historical trends. Quantitative scenarios can be calculated through the adjustment of linear, exponential or logarithmic curves to time series, describing the behaviour of a given variable in the past and its projection into the future. A qualitative analysis of the priority area will describe how "more of the same" will play out in the future.

**(2) Perpetuation of short-term trends**

The current conditions and trends expected to prevail in the near future may be assumed to continue in the medium and long term. The implementation of current government policies can be assumed to continue in the same direction, for example. The external constraints coming from regional or global economic conditions and agreements should be taken into consideration; in many cases, they are strong drivers for national policies. This approach can be helpful to assess the outcomes of a given course of action evolving over time.

---

**Box 6-11: Example of greater use of climate-sensitive agricultural resources**

In the last 20 years, the Egyptian government has promoted agricultural expansion into the New Lands (located in desert regions) and reclamation of the New-Old Lands (long-used areas now salinised or waterlogged). Crops have been selected according to soil and water limitations in each region to produce the best yields and qualities (adapted from El-Shaer et al., 1996). In this way, agricultural policy has maximised agricultural production by selecting crops that are well adapted to current climatic conditions.

---

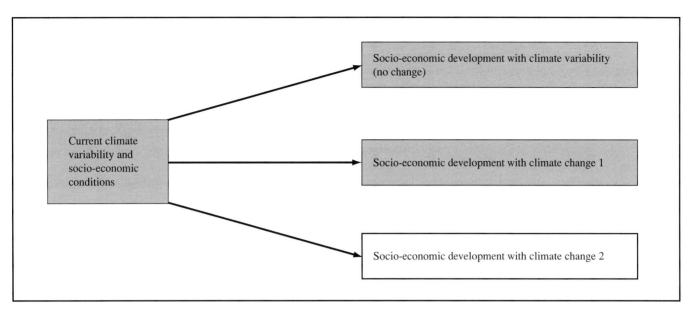

**Figure 6-3:** Schematic view of the multiple storylines approach

**(3) Analogues using key indicators or other countries as prospective futures**[3]

This approach requires an in-depth qualitative analysis in order to fully consider the national circumstances and drivers, including the social organisation and the nature of the development process. One or several analogue areas may be identified that are similar to the priority area in terms of natural resources, economic activities, history, culture, and governance but that have successfully addressed some of the issues currently facing the priority area. (Note: Similarities in demographic and economic characteristics will probably not be sufficient for a good analogue.) For example, an area that has made a successful transition from subsistence agriculture to high-yielding crops for markets may provide a "with adaptation policy" analogue that suggests, e.g., protection of smallholders, local technical support, and institutional mechanisms of cooperation as important dimensions of the transition. The use of analogues implies a rather normative approach to the design of storylines; i.e.,, the analogue is treated as a "norm" against which the priority area is compared.

**(4) Integrated analysis**

Recognising that accuracy is not possible when projecting the future, this approach aims for ownership and believability of the storyline built. A flexible design accounts for the interplay of trends, trade-offs among policy mixes, making use of subjective judgments to depict a feasible unfolding of the future. Here the approach is to ensure the pertinence of the storyline, through the involvement of key stakeholders in its design.

Within any given storyline, it is important to emphasise the transition pathway to the future as much as the final outcome at the end of the time horizon. This is particularly true for vulner-

ability and adaptation. The effects of extreme events, both in the ecological and socio-economic dimensions, are usually more relevant than the average conditions. It is unusual to observe continuously smooth growth or decline in socio-economic variables, particularly in the developing world. Up-and-down, stop-and-go patterns are found in many sectors and regions. Vulnerability and adaptation analyses can draw useful lessons from risk analysis and the design of contingency plans. For example, coping with floods or droughts tends to be far more painful in recession times than during high economic growth periods.

It is very useful to conduct the analysis of adaptive capacity separately, in the lower and upper parts of the cycles, corresponding to the behaviour of socio-economic variables in each storyline. The ups and downs of economic conditions should be accounted for as well. Some sectors or geographic regions could experience recession or even sustained decline. Even if the national economy enjoys economic growth, many groups may be marginalised and may receive no benefit from this growth. Therefore, resources for adaptation to climate impacts would not necessarily be available.

*6.4.6.2. Projections of socio-economic changes*

For any of the storylines, the projection of the socio-economic context is of utmost importance to the climatic vulnerability of the priority area. The impacts of climate change will depend, not only on the magnitude of the change, but also on the adaptive capacity of the ecosystems, including the local social structures and organisation.

The same five categories described in the overview of the socio-economic conditions today (i.e., demographic, economic, natur-

---

[3] UNFCCC Secretariat (2004) *Compendium on methods and tools to evaluate impacts of, vulnerability and adaptation to climate change,* for additional information on the use of analogues.

---

**Box 6-12: Example of three scenarios (no climate change)**

Korzeniewicz and Smith (1999) discuss three qualitative scenarios for Latin American countries that they term the "low-road," "middle-road," and "high-road" scenarios. In the low-road scenario, power remains concentrated in the state and high-status groups, high levels of inequality persist, and poverty is likely to rise. This scenario is "often accompanied by a lack of transparency, a deterioration of accountability, and widespread corruption among office-holders (features that become major obstacles to sustained economic growth)" (Korzeniewicz and Smith, p 21). The middle-road scenario is characterised by market reforms and sustained economic growth in a stable democratic regime. Although significant power remains with currently dominant groups, there are also consistent decreases in unemployment and poverty, increases in transparency and accountability, and efforts to combat corruption and clientelism. In the high-road scenario, a country exhibits strong economic growth, movement toward equality in income and wealth, and advances toward democracy and accountability.

---

al resource, government/policy, and cultural analysis) could be used to develop projections of socio-economic conditions. As illustrated in the previous section, the relationships among these different categories can be explored to evaluate overall progress towards sustainable development for the priority area.

The examples in Boxes 6-12 to 6-14 are provided to illustrate possible approaches to the projection of socio-economic changes in vulnerability and adaptation analyses.

---

**Box 6-13: Example of four scenarios for East Anglia consistent with SRES Scenarios**

*World markets (A1)*

Responsibility for action at enterprise level under market forces. Fast growing sectors: health care, leisure, financial. Declining sectors: manufacturing, agriculture. Annual country GDP growth: high (% see region; modify for country or location). Global carbon emissions: medium increase (cf. 1990 levels).
Weak international climate regime. Voluntary reduction of emissions. Emissions trading through markets

*Provincial enterprise (A2)*

Responsibility for action at individual level. Fast growing sectors: private health care, defence, maintenance services. Declining sectors: high-tech specialised services, finance. Annual GDP increases moderate. Global carbon emissions: high increase (cf. 1990 levels).
Very weak climate regime. Increased emissions. No controls. Voluntary action.

*Global sustainability (B1)*

Responsibility for action at state level, dictated by international government. Fast growing sectors: renewable energy, business services, clean technology. Declining sectors: fossil-fuel based and resource-intensive systems. High GDP growth. Global carbon emissions: low increase (cf. 1990 levels).
Strong international climate regime. Stringent reduction of emissions. Regulatory approach.

*Local stewardship (B2)*

Responsibility for action at collective level, supportive governmental framework. Fast growing sectors: small-scale manufacture and agriculture, local enterprises. Declining sectors: retailing, leisure and tourism. Low annual GDP increases. Global carbon emissions: medium low increase (cf. 1990 levels)
Strong/weak climate regime. Uneven emission controls. Fragmented regulatory approach.

Note: Annex A.6.1 has a summary description of SRES scenarios A1, A2, B1 and B2
Source: Lorenzoni et al., 2000

---

**Box 6-14: Example of socio-economic scenarios of the future for Egypt**

Strzepek et al. (2001) developed model-based socio-economic scenarios of the future. They then integrated the scenarios with climate scenarios and developed alternative futures with different adaptation strategies. The details of the representative socio-economic scenarios are given below.

| | **Population** | **Non-agricultural productivity** | **Agricultural productivity** | **Investment efficiency** | **Terms of trade** |
|---|---|---|---|---|---|
| **Scenario A** | Low | High | Low | High | High |
| **Scenario B** | Low | High | Low | Low | Low |
| **Scenario C** | Higher | High | High | High | High |
| **Scenario D** | Higher | High | Low | High | High |
| **Scenario E** | Higher | High | Low | Low | Low |
| **Scenario F** | Higher | Low | High | Low | Low |

---

## 6.5. Conclusions

This paper has provided guidance on how to analyse current and prospective socio-economic conditions in the context of the APF. In designing adaptation strategies, stakeholders will guide integrated quantitative and qualitative approaches to achieve:

- Coherent descriptions of the socio-economic conditions relevant to current adaptation and adaptive capacity to climate change
- Development of two or more storylines that provide the outlines of socio-economic prospects in the context of future climate change impacts
- Stakeholder participation in both defining current socio-economic conditions and prospects.

Project teams could use the following example outline as a checklist:

1. General overview of recent historical socio-economic conditions
2. Stakeholder input and selection of indicators for analysis
3. Current conditions (adaptation baseline)
   a. Demographic analysis
   b. Economic analysis
   c. Natural resource assessment
   d. Governance/policy-based analysis
   e. Cultural analysis
4. Prospects
   a. Three storylines (constructing a reference scenario with adaptation to current climate, two significantly different alternatives)
   b. Demographic prospects
   c. Economic prospects
   d. Prospects for natural resource use
   e. Governance/policy prospects
   f. Cultural prospects

# References

**Beck**, Tony. (1990). *Survival strategies and power amongst the poorest in a West Bengal village. IDS Bulletin* 20, 23-32.

**Bitoun**, Jan, Leonardo Guimarães Neto and Tania Bacelar de Araújo. (1996). Amazonia and the Northeast: the Brazilian tropics and sustainable development. in *Climate Variability, Climate Change and Social Vulnerability in the Semi-arid Tropics*, Jesse C. Ribot, Antonio Rocha Magalhães and Stahis S. Panagides (eds). Cambridge: Cambridge University Press, pp. 129-146.

**Douglas**, Mary, Des Gasper, Steven Ney and Michael Thompson. (1998). Human needs and wants. In *Human Choice and Climate Change, Volume 1: The Societal Framework*, Steve Rayner and Elizabeth L. Malone, eds. Columbus, OH: Battelle Press, pp.195-263.

**Fourke**, Sean. (1995). Fishing for a Better Tomorrow. *Nature Conservancy* January-February, 18-19.

**El-Shaer**, M.H., H.M. Eid, C. Rosenzweig, A. Iglesias and D. Hillel. (1996). Agricultural adaptation to climate change in Egypt. in: *Adapting to Climate Change: Assessments and Issues*, Joel B. Smith, Neeloo Bhatti, Gennady Menzhulin, Ron Benioff, Mikhail I. Budyko, Max Campos, Bubu Jallow and Frank Rijsberman eds., New York: Springer-Verlag, pp. 109-127.

**Guillermo** Aguilar, Adriár, Exequiel Ezcurra, Teresa García, Marsa Mazri Hiriart and Irene Pisanty (1995). The basin of Mexico. in *Regions at Risk: Comparisons of Threatened Environments*, Jeanne X. Kasperson, Roger E. Kasperson and B.L. Turner II. Tokyo: United Nations University Press, pp. 304-366.

**Hulme**, M., T. Jiang and T. Wigley. (1995). SCENGEN: A Climate Change SCENario GENerator. Software User Manual, Version 1.0. Climatic Research Unit, University of East Anglia, Norwich, United Kingdom, and WWF International, Gland, Switzerland.

**Jodha**, N.S. (1995). The Nepal middle mountains. in *Regions at Risk: Comparisons of Threatened Environments*, Jeanne X. Kasperson, Roger E. Kasperson and B.L. Turner II. Tokyo: United Nations University Press, pp.140-185.

**Korzeniewicz**, Roberto Patricio and William C. Smith. (1999). *Growth, Poverty and Inequality in Latin America: Searching for the High Road*. Rights vs. Efficiency Paper #7, Institute for Latin American and Iberian Studies at Columbia University. http://www.ciaonet.org/wps/smw01/

**Lorenzoni**, I., A. Jordan, M. Hulme, R.K. Turner, and T. O'Riordan. (2000). A co-evolutionary approach to climate change impact assessment: Part I, Integrating socio-economic and climate change scenarios. *Global Environmental Change* **10**, 57-68.

**Malone**, Elizabeth L., Joel B. Smith, Antoinette L. Brenkert, Brian Hurd, Richard H. Moss, and Daniel Bouille (2002). *Developing Socioeconomic Scenarios for Use in Vulnerability and Adaptation Assessments*. UNDP-NCSP, New York.

**Moldan**, Bedrich and Suzanne Billharz. (1997). *Sustainability Indicators: A Report on the Project on Indicators of Sustainable Development*. Eds., Chichester: John Wiley & Sons.

**Moss**, R.H., Brenkert, A. and E.L. Malone. (2001). *Vulnerability Indicators*. Washington, DC: Pacific Northwest National Laboratory,

**Nabarro**, David, Claudia Cassels and Mahesh Pant (1990). *Coping Strategies of Households in the Hills of Nepal: Can Development Initiatives Help? IDS Bulletin* 20(2), 68-74.

**Nakicenovic** et al. (2000). *Special Report on Emissions Scenarios*. Cambridge: Cambridge University Press.

*PECS News* [Population, Environmental Change, and Security Newsletter] (Spring 2000). Washington, DC: Woodrow Wilson Center

**Pepper**, W.J., J. Leggett, R. Swart, J. Wasson, J. Edmonds and I. Mintzer (1992). *Emissions Scenarios for the IPCC: An Update—Assumptions, Methodology, and Results*. Intergovernmental Panel on Climate Change, Geneva.

**Strzepek**, K., David Yates, Gary Yohe, Richard Tol and Nicholas Mader. (2001). Constructing "not implausible" climate and economic scenarios for Egypt IN: *Integrated Assessment* 2 (3, 2001): 139-157

**Tol**, Richard S.J. (1998). Socio-economic scenarios. In *UNEP Handbook on Methods for Climate Change Impact Assessment and Adaptation Studies*, Jan F. Feenstra, Ian Burton, Joel B. Smith, and Richard S.J. Tol. eds. United Nations Environment Programme and Vrije Universiteit. http://www.vu.nl/english/o_o/instituten/IVM/research/climatechange/Handbook.htm

**Wang'ati**, Fredrick Joshua (1996). The impact of climate variation and sustainable development in the Sudano-Sahelian region. in *Climate Variability, Climate Change and Social Vulnerability in the Semi-arid Tropics*, Jesse C. Ribot, Antonio Rocha Magalhães and Stahis S. Panagides (eds). Cambridge: Cambridge University Press, pp.71-91.

**World Bank** (1998). World Development Indicators 1998 [CD-ROM]. International Bank for Reconstruction and Development/The World Bank, Washington, DC.

**Yang**, H. (1996) Potential effects of sea-level rise in the Pearl River Delta area: preliminary study results and a comprehensive adaptation strategy. in *Adapting to Climate Change: Assessments and Issues*, Joel B. Smith, Neeloo Bhatti, Gennady Menzhulin, Ron Benioff, Mikhail I. Budyko, Max Campos, Bubu Jallow and Frank Rijsberman eds. New York: Springer-Verlag, pp. 265-276.

# ANNEXES

## Annex A.6.1. Methodological guidance in using models to construct socio-economic scenarios

(Source: Extracted from Malone et al., 2002)

This guidance begins at the global-regional level to help the reader establish general directions for and limits to scenarios so they will (1) account for global factors that have been analysed and, in the case of the SRES scenarios (Nakicenovic et al., 2000), approved by the IPCC; and (2) be internally consistent as the scenarios "tier down" to national and sub-national levels.

### A.6.1.1. Using existing scenarios

Socio-economic scenarios for use in climate change analyses exist at global and regional (multi-national) levels; these can be adapted for use in more localised vulnerability analyses. Tol et al. (1998) provide information and references for five socio-economic scenarios generated by the World Bank, IPCC, and integrated assessment modelling groups.

Many projections of climate change have made use of the IPCC's IS92 scenarios (Pepper et al., 1992). More recent work focuses on the new set of IPCC reference (no intervention through specific climate policies) SRES scenarios (Nakicenovic et al., 2000). The authors of the SRES report define and elaborate the socio-economic scenarios now used by the IPCC to project various emissions pathways. An argument for using the SRES scenarios is that their outputs will be used as inputs into global climate models that will create estimates of change in global climate to be used in impacts assessment (Hulme et al., 1995). By reflecting the SRES scenarios, the socio-economic scenarios will be consistent with the climate change scenarios.

The SRES features alternative "storylines" about the future. The storylines are qualitative, holistic pictures of the general structures and values of global society. They describe conditions that might be produced by human choices about economic and social policy, reproduction, occupations, and energy/technology use. The paces of population growth and economic development are set within and partially explained by the alternative tendencies of policies to support forms of global governance or localised self-sufficiency. There are four storylines (Nakicenovic et al., 2000):

- The A1 storyline and scenario family describe a future world of very rapid economic growth, global population that peaks mid-century and declines thereafter, and rapid introduction of new and more efficient technologies. Major underlying themes are economic and cultural convergence and capacity building, with a substantial reduction in regional differences in per capita income. The A1 scenario family develops into three groups that describe alternative directions of

technological change in the energy system: fossil intensive (A1F1), non-fossil energy sources (A1T), and a balance across all sources.

- The A2 storyline and scenario family describe a very heterogeneous world. The underlying theme is self-reliance and preservation of local identities. Fertility patterns across regions converge very slowly, which results in continuously increasing global population. Economic development is primarily regionally oriented and per capita economic growth and technological change are more fragmented and slower than in other storylines.

- The B1 storyline and scenario family describe a convergent world with the same global population that peaks in mid-century and declines thereafter, as in the A1 storyline, but with rapid changes in economic structures toward a service and information economy, with reductions in material intensity, and the introduction of clean and resource-efficient technologies. The emphasis is on global solutions to economic, social, and environmental sustainability, including improved equity, but without additional climate initiatives.

- The B2 storyline and scenario family describe a world in which the emphasis is on local solutions to economic, social, and environmental sustainability. It is a world with continuously increasing global population at a rate lower than A2, intermediate levels of economic development, and less rapid and more diverse technological change than in the B1 and A1 storylines. While the scenario is also oriented toward environmental protection and social equity, it focuses on local and regional levels.

Note, however, that the SRES scenarios were developed for the specific purpose of projecting future emissions of greenhouse gases. This means that they are not ready-made answers to the problem of developing socio-economic scenarios for vulnerability and adaptation analyses. They are a good starting point for considering such important factors as population growth and composition, economic conditions, and technological change. They do not explicitly represent other social institutions, such as farming, labour organisations, or the ways in which a government provides for the welfare of its citizens.

### A.6.1.2. Adapting storylines and projections from SRES scenarios

This section will help the user choose the appropriate storylines, data, and projections for their socio-economic scenarios. A country or a region such as an urban area or watershed exhibits its own variety of linked environmental-social conditions, providing the challenge of representing these in the context of a global socio-economic scenario. A region may have fragile ecosystems; major pollution problems, particularly air and water; and growing population and economy. International differences may further complicate the situation. Future developments in society hinge on the types of choices that are made, so many paths to the future are possible.

In other words, a region has its own set of storylines, which can be derived from the SRES storylines and adapted to regional circumstances. A scenario developer should ask, "What does an A1 kind of world mean for this specific region, and how would the A1 characteristics be manifested here? "

Vulnerabilities will be very different if a country seeks rapid industrialisation, takes food imports for granted, seeks self-reliance in food production or chooses a path of agricultural export-led growth. Vulnerabilities will also be different if a country chooses to protect and support its farmers, or let them face the whims of the market and the weather on their own strength (Tol, 1998, pp 2-14).

A country's likely approach to these policy matters must be considered in developing a storyline that will determine many of the socio-economic characteristics. Then appropriate values for the SRES variables can be determined by proportional calculations, i.e., applying the SRES percentage increases in population and GDP from the appropriate scenarios to the existing data for the region under study.

Using the SRES data and projections, users can review data on population and GDP projections, at a minimum. (For further methodological guidance see Malone et al., 2002.)

### A.6.1.3.  Adding country-specific factors to the socio-economic scenario

This section discusses national-level factors and storylines that will delineate two or more directions for the future. The primary concern is to keep a country's future development choices consistent with potential global developments and the country's own current policy directions. Storylines of the future will help the user to decide the most influential elements of that future and to construct ways to represent—and, if possible to quantify—those elements.

Besides the variables adapted from SRES or other sources of socio-economic scenarios, additional data for scenarios to be used in vulnerability analyses should be gathered from the literature (studies done about the user's particular country) and relevant databases (e.g., World Bank, 1998) to describe the social, economic, and institutional context in which climate variability and change will take place in the user's country. The important factors for the country's social future must be represented in its socio-economic scenario.

These factors include national indicators of well-being. Users should add to population and GDP figures (for the present and projections into the future) elements that capture more dimen-

sions of overall development and the variations as well as the averages. It is possible to develop a specific and highly detailed set of indicators of national well-being. (For example, Douglas et al., 1998 contains descriptions of human needs, particularly Box 3.1.) Or the user can use the UNDP's Human Development Index (HDI) (World Bank, 1998). The HDI uses three indicators:

- life expectancy at birth
- literacy rates
- purchasing-power-adjusted GDP per capita (in logarithmic form).

The first two indicators reflect the supporting infrastructure for an individual's life. Life expectancy is a good indicator of public health, resulting from clean water, sewerage, medical practice, and nutritional status. Literacy indicates the spread of education and access to information. The third indicator, purchasing power, is relevant to the individual's ability to acquire goods and services.

The HDI rankings are given to countries on the human deprivation continuum (0 to 1) for each indicator; the average of the three indicators, subtracted from 1, provides the overall HDI.

Table A-6-1 demonstrates an approach midway between an elaborate set of country-specific indicators, and the three that comprise the HDI. This approach is multidimensional, with indicators for economic capacity, human and civic resources, and environmental capacity. Within each category a selection of proxy variables has been made, the relationship between the proxy and the category has been specified, and the functional relationship has been defined.

The discussions above should give the user a picture of the methodology that they can adapt to develop projections, again using the storylines they have selected to provide a basis for the determination of rates of change. For example, access to health care may increase more under the global solutions scenario than under the self-reliant scenario, since presumably a country would be able to obtain medical services and products on the global market more easily than developing them in-country. Conversely, a self-reliant scenario would indicate that the user's country would have more development of national programmes to address climatic and other extreme events.

Each choice the user makes of projected values must have an underlying rationale. Users should remember that a straight-line extrapolation will rarely be defensible. For example, a literacy rate cannot improve indefinitely, and increasing calories over the amount to ensure adequate nutrition actually decreases well-being. Also remember that the projections must be realistic; projected reductions in income inequality must be based on the potential of the national society to achieve them, a difficult goal for any country to attain. Finally, many of the proxies that can be identified may reinforce one another; increased GDP may have implications for educational advancement and technological change – another reason to be very selective in choosing proxies to use.

*Table A-6-1:* Country-level factors for use in socio-economic scenarios

| Category | Proxy variables | Proxy for: | Functional relationship |
|---|---|---|---|
| Economic capacity | GDP(market)/capita<br><br>Gini index | Distribution of access to markets, technology, and other resources useful for adaptation | Adaptive capacity ↑ as GDP/cap ↑<br><br>at present Gini held constant |
| Human and civic resources | Dependency ratio<br><br>Literacy | Social and economic resources available for adaptation after meeting other present needs<br><br>Human capital and adaptability of labour force | Adaptive capacity ↓ as dependency ↑<br><br>Adaptive capacity ↑ as literacy ↑ |
| Environmental capacity | Population density<br><br>$SO_2$/area<br><br>% land unmanaged | Population pressure and stresses on ecosystems<br><br>Air quality and other stresses on ecosystems<br><br>Landscape fragmentation and ease of ecosystem migration | Adaptive capacity ↓ as density ↑<br><br>Adaptive capacity ↓ as $SO_2$ ↑<br><br>Adaptive capacity [of the environment]↑ as % unmanaged land ↑ |

Source: Moss et al., 2001

These additional characteristics, along with the adapted SRES projections, will provide a more detailed picture of a country's socio-economic future. Within these constraints, the user can extend the analysis into important sectors in their country. (Malone et al., 2002 has additional methodological guidance.)

## Annex A.6.2. An example of the use of socio-economic analysis within vulnerability assessment

This annex outlines an example of a vulnerability assessment that draws on socio-economic analysis. (TP3 contains additional information on vulnerability assessment.)

### Income and employment

As a result of El Salvador's growing economy, per capita income in 1994 was US $1440, which places this Central American country in the middle-income group of all countries. Nevertheless, with reference to the reduction of extreme poverty and improving the quality of life for the population, especially in rural areas, there is an important gap between urban and rural per capita incomes.

Urban per capita income is approximately $2200 annually; however, the level of rural income is only $500. With this per capita income, the majority of the rural population cannot acquire a basket of basic nutritional goods for a family, the price of which was $1100 per year in rural areas. In urban areas this same basket of

goods cost $1512; nevertheless, minimum urban salaries represent $1550, which indicates that the most urban income goes to the acquisition of food.

This situation demonstrates overall that the majority of the population, whether rural or urban, is at risk of food insecurity. It has been pointed out that the average income in the rural sector cannot catch up with or cover the food requirements; at the same time, the average urban income is destined to satisfy only 90% of food requirements. In this way, food insecurity takes on a chronic and structural character.

The indicators used in El Salvador's First National Communication on Climate Change (February 2000) were population; per capita income, disaggregated into rural and urban per capita incomes; food prices; real wages; nutritional requirements; production of basic grains; and a calculated nutritional gap and consequent import needs.

(Source: El Salvador First National Communications to the UNFCCC; http://unfccc.int/resource/docs/natc/elsnc1e.pdf)

# 7

# Assessing and Enhancing Adaptive Capacity

NICK BROOKS[1] AND W. NEIL ADGER[1]

Contributing Authors
*Jon Barnett[2], Alastair Woodward[3], and Bo Lim[4]*

Reviewers
*Emma Archer[5], Mohammed Atikullah[6], Suruchi Bhawal[7], Henk Bosch[8], Hallie Eakin[9], Jose Furtado[10], Molly Hellmuth[11], Ulka Kelkar[7], Maynard Lugenja[12], Mohan Munasinghe[13], Anthony Nyong[14], Atiq Rahman[6], Samir Safi[15], Juan Pedro Searle Solar[16], Barry Smit[17], Juha Uitto[4], and Thomas J. Wilbanks[18]*

[1] Tyndall Centre for Climate Change Research, School of Environmental Sciences, University of East Anglia, Norwich, United Kingdom
[2] School of Anthropology, Geography and Environmental Studies, University of Melbourne, Australia
[3] Wellington School of Medicine, Wellington, New Zealand
[4] United Nations Development Programme – Global Environment Facility, New York, United States
[5] University of Cape Town, Rondebosch, South Africa
[6] Bangladesh Centre for Advanced Studies, Dhaka, Bangladesh
[7] The Energy and Resources Institute, New Delhi, India
[8] Government Support Group for Energy and Environment, The Netherlands
[9] Center for Atmospheric Sciences, Universidad Nacional Autónoma de México, Mexico DF, Mexico
[10] Imperial College, London, United Kingdom
[11] UNEP Collaborating Center on Energy and Environment, Roskilde, Denmark
[12] The Centre for Energy, Environment, Science & Technology, Dar Es Salaam, Tanzania
[13] Munasinghe Institute for Development, Colombo, Sri Lanka
[14] University of Jos, Jos, Nigeria
[15] Lebanese University, Faculty of Sciences II, Beirut, Lebanon
[16] Comisión Nacional Del Medio Ambiente, Santiago, Chile
[17] University of Guelph, Guelph, Canada
[18] Oak Ridge National Laboratory, Oak Ridge, United States

# CONTENTS

## 7.1. Introduction

This Technical Paper (TP) addresses the assessment and enhancement of adaptive capacity of both social and physical systems, so that these systems may cope better with climate change, including variability. Users will find guidance on a range of important activities, including the development of adaptive capacity for priority groups, the development of adaptive capacity indicators, and identification and assessment of key adaptation options. After outlining the relationship of this paper to other Adaptation Policy Framework (APF) TPs, the authors explain the key concepts of hazards, systems and adaptive capacity. In addition to listing the determinants of adaptive capacity and discussing the uses of indicators, this paper addresses the nature of current and future hazards, and – based on the five APF Components – outlines guidance on assessing and enhancing the capacity of systems (and populations) to adapt to these hazards. Examples and links to resources are provided throughout the text and in the Annex.

## 7.2. Relationship to the Adaptation Policy Framework as a whole

Since a distinguishing feature of the APF is its focus on adaptive capacity, this TP relates to all five Components of the APF process (Figure 7-1). In other words, the enhancement of adap-

tive capacity should be considered at all stages of the adaptation process.

- Component 1 (TP1), *Scoping and designing an adaptation project:* TP7 recommends assessing adaptive capacity in terms of the capacity of particular systems and groups to adapt to specific types of hazards. The question of defining systems and identifying hazards (i.e., "who adapts and to what?") is explored through Component 1. This question should inform the design of any adaptation strategy.

- Components 2 and 3 (TPs 3-6), *Assessing current vulnerability and Assessing future climate risks:* Vulnerability assessments must form the basis for strategies to enhance adaptive capacity. Similarly, the nature of adaptive capacity and appropriate adaptation strategies is partly determined by the nature of the hazards to which systems must adapt; factors relating to development, economic well-being, health and education status are important determinants of adaptive capacity.

- Component 4 (TP8), *Formulating an adaptation strategy:* Identifying existing adaptive capacity and developing strategies for enhancing capacity are essential prerequisites for designing and implementing adaptation strategies.

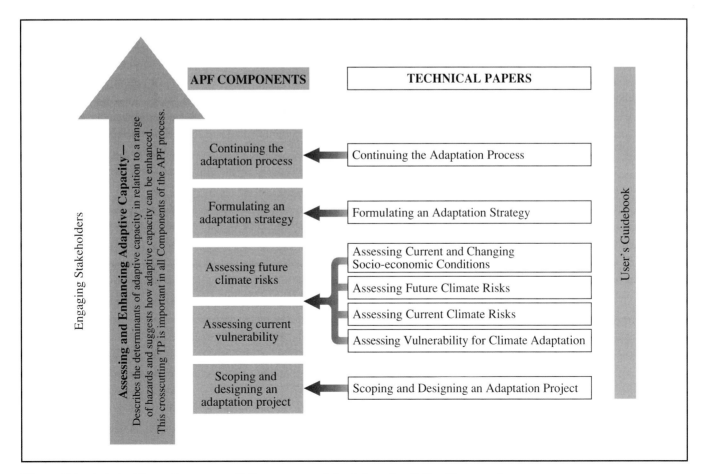

**Figure 7-1:** Technical Paper 7 supports all Components of the Adaptation Policy Framework

- Component 5 (TP9), *Continuing the adaptation process:* The processes of reviewing, monitoring and evaluating are important in maintaining levels of adaptive capacity. These processes can collectively identify where capacity development has succeeded or failed, and the extent to which it has been translated into actual adaptation.

- All Components (TP2), *Engaging stakeholders in the adaptation process:* Engaging stakeholders is the APF's other cross-cutting activity. Strategies to enhance adaptive capacity should engage stakeholders at all stages if they are to be successful and equitable.

## 7.3.    Key concepts

The glossary contains short definitions of terms used throughout the APF, whereas extended definitions of important concepts for this paper are described here.

### 7.3.1.    *Adaptive capacity*

*Adaptive capacity* is the property of a system to adjust its characteristics or behaviour, in order to expand its coping range under existing climate variability, or future climate conditions. In practical terms, adaptive capacity is the ability to design and implement effective adaptation strategies, or to react to evolving hazards and stresses so as to reduce the likelihood of the occurrence and/or the magnitude of harmful outcomes resulting from climate-related hazards. The adaptation process requires the capacity to learn from previous experiences to cope with current climate, and to apply these lessons to cope with future climate, including surprises.

The expression of adaptive capacity as actions that lead to adaptation can serve to enhance a system's coping capacity and increase its coping range (TPs 4 and 5) – thereby reducing its vulnerability to climate hazards (TP3). The adaptive capacity inherent in a system represents the set of resources available for adaptation, as well as the ability or capacity of that system to use these resources effectively in the pursuit of adaptation. Such resources may be natural, financial, institutional or human, and might include access to ecosystems, information, expertise, and social networks. However, the realisation of this capacity (i.e., actual adaptation) may be frustrated by outside factors; these external barriers, therefore, must also be addressed. At the local level, such barriers may take the form of national regulations or economic policies that hinder the freedom of individuals and communities to act, or make certain adaptation strategies unviable. However, many models of capacity development (UNDP-GEF, 2003) consider regulatory and policy framework to be internal to the system.

*Capacity development* refers to the process of enhancing adaptive capacity, and is discussed as a key Component of adaptation. The role of capacity development is to expand the coping range and

strengthen the coping capacity of a priority system with respect to certain climate hazards, and thus to build the capacity of the system to adapt to climate change, including variability. Many social service agencies view capacity development as a change-management process (UNDP-GEF, 2003) within a governance framework; in this case, as defined by the determinants of adaptive capacity (TP9). As such, adaptive capacity development is viewed as a central goal of most adaptation strategies.

### 7.3.2.    *Key Components of adaptive capacity*

Information on the nature and evolution of the climate hazards faced by a society – both historical climate data and data from scenarios of future climate change – is key to enhancing adaptive capacity.

On the other hand, information on socio-economic systems, including both past and possible future evolution, is important. Within these evolving socio-economic and developmental contexts, viable adaptation strategies can be designed. Adaptation and capacity development strategies must also be acceptable and realistic, so information on cultural and political contexts is also important.

The implementation of adaptation strategies requires resources, including financial capital, social capital (e.g., strong institutions, transparent decision-making systems, formal and informal networks that promote collective action), human resources (e.g., labour, skills, knowledge and expertise) and natural resources (e.g., land, water, raw materials, biodiversity). The types of resources required and their relative importance will depend on the context within which adaptation is pursued, on the nature of the hazards faced, and on the nature of the adaptation strategy.

Adaptation strategies will not be successful unless there is a willingness to adapt among those affected, as well as a degree of consensus regarding what types of actions are appropriate. Adaptive capacity, therefore, depends on the ability of a society to act collectively, and to resolve conflicts between its members – factors that are heavily influenced by governance.

Adaptive capacity can be undermined by a refusal to accept the risks associated with climate change, or by a refusal of key actors to accept responsibility for adaptation. Such refusals may be ideological in nature, or the consequence of vested interests denying the existence of risks associated with climate change. Large-scale structural economic factors and prevailing ideologies, therefore, play a vital role in determining which adaptations are feasible.

### 7.3.3.    *Scales of adaptation*

At the national or state level, governments and institutions will undertake a combination of **planned** and **reactive** adaptation, in which lessons learned from past hazard events are incorporated into forward-looking adaptation strategies. Climate projections will play a key role in planning for future climate change, facilitating anticipatory adaptation to new hazards and informing ongoing adaptation to familiar evolving hazards. Historical records will be of great value in identifying climate trends and "early warnings" of climate change. Clearly, climate information will be vital in planning adaptation strategies, and a system's capacity to adapt to climate change will be heavily influenced by its ability to collect and interpret such information.

Nonetheless, it must be recognised that adaptation will ultimately be a localised phenomenon. It will be driven by the need for people to adapt to the local manifestations and impacts of climate change, which will be mediated by geography and local physical, social, economic and political environments. Individuals tend to adapt in a reactive and often haphazard manner. At the local level, adaptation is a complex process that "emerges" as social systems reorganise themselves, in a largely unplanned fashion, through a series of responses to external stresses. Top-down, prescriptive strategies to undertake planned adaptation are therefore only a partial solution. Governments, non-governmental organisations (NGOs) and other bodies should address how they can enhance the capacity of systems (people) to adapt reactively and autonomously by creating enabling environments for adaptation. Such an approach must recognise that people will pursue adaptation strategies appropriate to their individual circumstances, and that adaptation may be unpredictable.

### 7.3.4.  Systems and hazards

Adaptive capacity is most easily perceived in terms of the capacity of a particular **system** to adapt so as to better cope with a particular climate **hazard** or set of hazards. A system may be a region, a community, a household, an economic sector, a business, a population group, or ecological system. Systems will be exposed to varying degrees to different climate hazards, defined in TP4 as events with the *potential* to cause harm. Hazards are physically defined here, and it is the interaction of a climate hazard (e.g., a drought, windstorm, or extreme rainfall event) with the properties of an exposed system – its sensitivity or socially-constructed vulnerability – that results in a particular outcome (TP3; Adger and Kelly, 1999; Brooks, 2003; Pelling and Uitto, 2001). Three principal hazard categories may be identified:

1.  **Discrete recurrent hazards** including simple and complex hazards (as described in TP4).
2.  **Changes in mean conditions** occurring over years or decades (e.g., continuous increases in mean temperature), or desiccation (e.g., such as that experienced in the Sahel over the final decades of the 20th century).
3.  **Singular or unique hazards** such as shifts in climatic regimes associated with changes in ocean circulation; the paleo-climatic record provides many examples of abrupt climate change events associated with the onset of new climatic conditions that prevailed for centuries or millennia (Roberts, 1998; Cullen et al., 2000; Adger and Brooks, 2003).

Climate change will likely be associated with all three categories of hazard, although the manifestations of climate change will vary geographically and over time. In the short term, perhaps the most likely changes will be in the frequency and severity of familiar recurrent hazards. The capacity to adjust to such changes in frequency and severity – and to support systems so that they can adapt to the altered levels of hazard – will be critical.

Changes in mean climate conditions will likely to be associated with changes in extremes. But adaptation to gradual change will be necessary in some cases, e.g., in certain agricultural systems where gradually increasing evapo-transpiration rates affect water demands. Gradual changes in mean conditions may ultimately result in the breaching of critical thresholds, beyond which a system's ability to cope is compromised (TP5).

### 7.3.5.  Ecological systems

Much of the discussion in this TP refers to human systems and the role of human behaviour in mediating adaptive capacity. However, practitioners may also be concerned with the adaptive capacity of ecological systems, or coupled social-ecological systems. For unmanaged ecological systems, adaptive capacity will depend on factors such as biodiversity and migration potential. In a system with high biodiversity there may be more potential for species to occupy new niches created by changed environmental conditions or the loss of other species, although the loss of keystone species may have dramatic implications for the survival of ecosystems. Ecosystems that are geographically constrained will be less able to adapt to change than those that have space to migrate with shifts in climatic zones. Migration of ecosystems in response to shifts in climatic zones will also be limited by the growth rates of their constituent flora; rapid shifts in climatic zones may exceed rates at which such systems can migrate in response to an expansion of favourable climatic conditions.

Adaptation in ecosystems may be promoted by human actions, such as the creation of migration corridors through urban or agricultural areas, and the avoidance of fragmentation. It may also be possible to relocate certain species, and even whole ecosystems, to areas that are more favourable to their survival under changed climatic conditions. Adaptive capacity may also be enhanced by the reduction of non-climatic stresses related to factors such as pollution and resource exploitation; the promotion of sustainable development is thus likely to enhance the adaptive capacity of ecosystems. However, it should be recognised that most ecosystems are managed to some extent, and an approach that views sustainable development in terms of coupled ecological and social systems is likely to be more fruitful than one that attempts to separate "human" and "natural" systems in most instances.

### 7.3.6.  Risk frameworks for adaptation

The impacts of a climate hazard on an exposed system are mediated by that system's vulnerability (TP3). The determinants of vulnerability will depend on how a system is defined – and where its boundaries are drawn – but may include social, economic, political, cultural, environmental and geographic factors. The risk posed to a system may be viewed as a function of the nature of the hazard faced and system's vulnerability (Brooks, 2003). The vulnerability of a system to climate change will be inversely related to the capacity of that system to respond and adapt to change over time; a description of a system's vulnerability to climate change (i.e., vulnerability integrated over time) will therefore require a knowledge of that system's adaptive capacity, in contrast to a description of the instantaneous vulnerability of a system at a given time, e.g., the time of onset of a short-lived hazard event. Risk may be measured probabilistically, in terms of the likelihood of a particular outcome (outcome risk) or the likelihood of a particular hazard event (event risk) (Sarewitz et al., 2003). Alternatively, risk may be measured in terms of indicators of outcome, e.g., the number of people killed, injured or displaced, or the economic losses resulting from climate hazards over a particular period. The purpose of capacity development and adaptation strategies is ultimately to reduce risk, or to prevent the exacerbation of risk in the face of increasing hazards. Risk indicators are therefore useful in terms of assessing the success of strategies designed to enhance adaptive capacity.

### 7.3.7. *Indicators of adaptive capacity*

Indicators of risk say little about the processes that make systems and populations vulnerable and determine whether they can adapt to evolving climate hazards. Indicators of adaptive capacity, however, are more difficult to identify than indicators of risk, as adaptive capacity is not directly measurable. Recognising this difficulty, UNDP-GEF (2003) uses a score card (subjective) approach for assessing changes in capacity attributable to a project.

Capacity development projects should consider the role of external or contextual factors that affect systems, but are outside of their control, as well as internal factors operating within systems that may be directly addressed through interventions to enhance adaptive capacity. Whether a factor is internal or external depends on the scale of the system in question. For example, national level data used to develop adaptive capacity indicators could represent internal factors if the scale of analysis is national and external factors if the scale is local. In the project context, the team needs to make a judgment as to whether the factors are internal or external to the system boundary.

At the national level, adaptive capacity is strongly related to factors such as health, literacy and governance (Brooks et al., 2004). These, in turn, are related to economic development, although the nature of these relationships is complex and the subject of debate. Health, literacy, governance and economic

wealth are representative of a country's overall development status; they are determined, to a large extent by the national development context, and thus contribute to the context within which sub-national scale systems must adapt. It might be well beyond the scope of most adaptive capacity development projects to affect national economic development, national governance, and the investment of central government in health and literacy. Capacity development projects might choose to address such factors at the local scale where they can be particularly effective in developing the capacity of highly vulnerable communities.[1]

If capacity development projects choose to operate at sub-national scales, they should address a range of factors that are important at the local level. The factors that represent adaptive capacity will be determined to a certain extent by the nature of the hazard(s) faced and by the characteristics of the system or population in question (e.g., the types of livelihoods that sustain the communities in question). For example, the factors that determine whether small-scale rural farmers can adapt to drought will not be the same as the factors that determine whether wealthy owners of waterfront properties can adapt to flooding, although there may be some common factors (e.g., the availability of information).

It is therefore not possible to provide a list of "off-the-shelf" indicators to capture universal determinants of adaptive capacity that are useful at the project level. Appropriate indicators for assessing adaptive capacity must be tailored to each case. These may be identified by asking the following nine questions. (The four key questions for the identification of adaptive capacity indicators are in bold; the other questions should have been addressed in the previous TPs. Annex A.7.1 contains sample responses to these questions.)

1. What is the nature of the system/population being assessed?
2. What are the principal hazards faced by this system/population?
3. What are the major impacts of these hazards and which elements/groups of the system/population are most vulnerable to these hazards? (See TP3 for vulnerability mapping/assessment.)
4. Why are these elements/groups particularly vulnerable? (See TP3 for how vulnerability is constructed.)
5. What measures would reduce the vulnerability of these elements/groups?
6. **What are the factors that determine whether these measures are taken?**
7. **Can we assess these factors in order to measure the capacity of the system population to implement these measures?**
8. **What are the external and internal barriers to the implementation of these measures?**

---

[1] However, capacity development efforts must also be sustainable in the sense that the benefits of a project can last beyond project completion. While desirable in themselves, efforts to improve health and literacy, for example, may provide only temporary adaptation benefits where there is a lack of state-supported infrastructure to provide continuity after a project has finished. Teams must therefore judge for themselves which factors may be effectively addressed and which should be viewed as providing the context or limits within which the project must be carried out.

---

**Box 7-1: Identifying indicators to assess adaptive capacity and barriers to adaptation to flooding**

Using the question-based approach outlined in Section 7.3.6, a team might identify the groups most vulnerable to flooding in a particular community or region. They conclude that vulnerability might be reduced by a combination of relocating certain groups to less exposed areas, and introducing and enforcing stricter building codes.

Indicators of capacity to adapt through these measures might capture awareness of flood risks, willingness of people to move, availability and affordability of housing in less exposed areas, and ability of local authorities to impose financial penalties on developers building in flood-prone areas or failing to incorporate measures to make new buildings more resilient. In certain developing countries where people build their own dwellings, the affordability and availability of the materials required to build more flood-resistant housing will be an indicator of their capacity to adapt, as will a knowledge of appropriate building design. A combination of quantitative and qualitative indicators would be required to assess the above factors (TP6).

External barriers to adaptation might include the lack of new land available for relocation, or limitations placed on local authorities by central government, preventing the introduction and enforcement of building regulations. (Insufficient financial resources and certain social factors might also prevent the enforcement of regulations.) Population density might be a quantitative indicator of such barriers, and political autonomy (most likely a qualitative indicator, perhaps based on results of surveys of local decision makers).

Internal barriers to adaptation might be the unwillingness of people to move away from flood-prone areas (due to the nature of their livelihoods), the high prices of land or property, or a lack of awareness of the risk of flooding under anticipated changes in climate. The latter two barriers might be addressed through the provision of social housing, loans or grants, and awareness-raising (education). The first barrier might be mitigated by supporting alternative livelihoods that do not require proximity to flood-prone areas. In this circumstance, team members must closely examine the impacts on the local economy and on food security. In a society where literacy rates are low, awareness-raising would be best pursued through non-printed media; the developmental context influences the nature of the capacity development activities.

---

**9.  How can capacity constraints be removed from key barriers to adaptation?**

Indicators may also be developed to assess the extent of external and internal barriers (Box 7-1).

Indicators might be used to map the geographic and social differentiation of adaptive capacity within a region or community, e.g., examining the variation in capacity at the household level, based on factors such as income and dependency ratio. Alternatively, indicators representing aggregations at the regional level might be used to compare capacity across different regions and to monitor its evolution over time. Regional-level indicators might include overall population density, transport network density, regional income and inequality, the nature of economic activity, etc. The development of local-level indicators will benefit from stakeholder participation: local people are generally the best equipped to identify factors that facilitate and constrain their own adaptation. In the project context, pragmatism is paramount when choosing a set of key indicators (TP1 contains criteria for selecting indicators).

Indicators may be quantitative, representing a measurable quantity, such as population density or average income, or qualitative, representing factors such as the principal type of economic activity in a region, or people's perceptions of risk. TP6 discusses both quantitative and qualitative approaches.

**7.4.  Guidance on enhancing adaptive capacity**

Since enhancing adaptive capacity is a process that cuts across all adaptation activities, the sections below provide guidance about each of the other Components. The process of enhancing adaptive capacity will be relevant to all projects, regardless of the approach. However, for projects using the *adaptive capacity approach* – distinguished by the identification of capacity development as its primary objective – it is possible to structure an assessment around the guidance provided below and in other TPs. (TP1 Section 1.4.4. contains information on selecting an approach).

### 7.4.1.  *Component 1: Scoping and designing an adaptation project*

*What is the adaptive capacity priority of the project, and what is the specific capacity enhancement goal?*

The nature of a project that enhances adaptive capacity will depend on the nature of the system or systems targeted by the project (TP1). A project might target the general apparatus of government to raise awareness of the need for adaptation and for mainstreaming adaptation issues into the policy process at all levels of government. However, most projects will be less ambitious in scope, targeting specific systems, regions or population groups that are at greatest risk from climate change, and/or sectors that are particularly important to a national economy. A project should start by identifying the priority system, the existing and/or potential hazards that threaten the system, and the timescales over which these hazards are likely to unfold. Priority systems, regions and populations might be identified on the basis of risk associated with existing climate hazards (Section 7.4.2), or with potential future hazards, as identified using climate change scenarios (Section 7.4.3).

Once the system and risks have been identified, the project team should consider the project's adaptation objective (TP1). For example, is the objective to make economic or agricultural systems more resilient, to reduce mortality from climate-related disasters, to prepare for specific, anticipated future manifestations of climate change, etc.? The aim of a capacity development project should be to increase the ability of systems to adapt, and of individuals and groups to design and implement adaptations. A capacity development project might be broken down into the following activities:

- identify a range of adaptations;
- prioritise adaptations based on their efficacy, feasibility and acceptability;
- remove barriers to adaptation;
- identify who is to act for planned adaptations.

Once these elements have been addressed, the team should be able to implement specific adaptation strategies. These might be single, large-scale planned projects, or multiple, diverse responses – the latter would be undertaken in a more ad-hoc, reactive way by individual agents. The role of "autonomous" adaptation should not be neglected; in past societies, adaptation to environmental variability and change has largely emerged in an unplanned manner as individuals responded in a variety of ways to change as it happened.

A real-world example set of questions about the early stages of the scoping process is provided in Box 7-2.

### 7.4.2.  *Component 2: Assessing current vulnerability*

*What adaptive capacity already exists to reduce current vulnerability to recurrent hazards?*

In many countries, vulnerability to existing hazards is significant. In such cases, capacity development projects should seek to enhance the ability of systems and populations to cope with these hazards. Failure to address existing hazards will undermine longer-term adaptation strategies, as damage from present-day climate extremes can reduce economic and social development and undermine a country's resource base. Furthermore, in the short to medium term, climate risk is likely to be associated with hazards similar to those of recent record, although with varying frequency and severity over time. Enhancing its capacity to cope with and adapt to such hazards will enhance coping and adaptive capacity with respect to near-term climate change. Table 7-1 provides examples of measures in place to respond to different types of current hazards.

For projects using the adaptive-capacity approach, it is possible to develop an adaptive capacity baseline. Since there are few clear, quantitative indicators of adaptive capacity, this baseline will generally be constructed from qualitative indicators. (TP6 contains a discussion about the selection and use of qualitative indicators, or for use of a score card approach, see UNDP-GEF, 2003).

Capacity development for adaptation to existing climate hazards will be most effective when it is carefully targeted at the systems and populations most at-risk from climate hazards, where risk is a function of both vulnerability (TPs 3 and 4) and exposure to hazard (TPs 4 and 5). Combined hazard-vulnerability mapping projects can be of particular use, as these identify regions and groups with high vulnerability, as well as "hot spots" (i.e., elevated socially-determined vulnerability and climate hazard; TP3, Annex A.3-5). Information from mapping projects can also identify which types of hazard should be addressed in terms of capacity development projects. Prioritisation may also be undertaken

---

**Box 7-2: Adaptation guidance for local authorities in the United Kingdom**

The United Kingdom Climate Impacts Programme (UKCIP) offers guidance to local authorities in adapting to climate change (UKCIP, 2003, p.1). It encourages local authorities to ask themselves the following questions:

- Do you know how climate change could impact your area?
- Do your current policies, strategies and plans include provisions for the impacts of climate change?
- Can you identify and assess the risks from climate change to your services?
- Are developments with a lifetime of more than 20 years required to factor in climate change?
- Does your Emergency Planning Service take into account climate change?
- Are you addressing climate change in your local community strategy or community plan?
- Have you briefed your elected members on any key risks arising from climate variability and long-term climate change?

Addressing the above questions will significantly enhance institutional adaptive capacity at the local level. The report also lists potential impacts of climate change on local government authority services and potential adaptation responses, and is a useful template for other similar communities.

*Table 7-1:* *Types of current hazard and adaptation responses*

|  | **Familiar discrete recurrent hazards** | **Existing trends** |
|---|---|---|
| **Principal types of adaptation response** | • Combined reactive and anticipatory (planned and autonomous) | • Responsive (autonomous assisted/facilitated by policy) |
| **Examples of hazard** | • Floods, droughts, wind storms, heat waves, cold waves, extreme rainfall events, hail storms, dust storms | • Increased evapo-transpiration, long-term reductions in rainfall (e.g., Sahel), increases in minimum temperatures, rising water tables, salinisation of aquifers |
| **Who acts?** | • Government, planning bodies, communities, individuals | • Communities and individuals, planning bodies |
| **Measures to enhance adaptive capacity** | • Establish monitoring networks<br>• Assess historical data and case studies (identify successful and unsuccessful adaptations)<br>• Disseminate information on successful adaptations<br>• Develop short-range forecasting capacity<br>• Improve access to credit and insurance<br>• Encourage autonomous adaptation<br>• Prevent maladaptation through regulation<br>• Enforce environmental regulations<br>• Assess adaptation needs (including technological needs) through stakeholder engagement | • Establish monitoring networks<br>• Assess historical data and past/existing adaptations (identify successful and unsuccessful adaptations)<br>• Disseminate information on successful adaptations<br>• Develop long-range forecasting capacity<br>• Assess adaptation needs through stakeholder engagement<br>• Create "enabling environments" to encourage further adaptation |

on the basis of recent historical outcomes from climate hazards. (Box 7-3 has additional information on prioritisation using various data sources).

Case studies can also illuminate examples of "good practice" in terms of risk management (see the case study section), and lessons may be learned from examples of successful adaptation/vulnerability reduction from other contexts (e.g., from other countries). Box 7-4 briefly summarises an example of successful adaptation in the African Sahel.

### 7.4.3. Component 3: Assessing future climate risks

*What capacity will societies have to adapt to future hazards?*

Current socio-economic, political and environmental conditions, described (depending on the project approach) in terms of current vulnerability and existing adaptations, represent the project baseline (TP1, Section 1.4.3). Adaptive capacity will exist within current socio-economic, political and environmental contexts, as discussed in TP6. The capacity to adapt to a given set of hazards may be enhanced or reduced over time, depending on development pathways. The use of socio-economic scenarios to assess how vulnerability, and by extension adaptive capacity, may change over time under different development trajectories is also discussed in TP6.

Vulnerability to climate change over significant time periods (years to decades) is crucially dependent on the ability to adapt to the manifestations of climate change. The determinants of vulnerability and adaptive capacity will vary to some degree depending on the nature of the climate changes being experienced – e.g., agricultural adaptation to drought will be a very different process from adaptation of settlements to increased flooding; in reality, even the vulnerability approach to risk management will require some knowledge of what hazards are likely to be associated with future climate change. In the absence of detailed data from climate models and scenarios, it is not unreasonable to extrapolate from existing conditions. At least in the near term, climate change is likely to be associated with changes in the frequency and severity of historically familiar hazards. Consequently capacity development is likely to be most useful if it focuses on these hazards. Nonetheless, such a strategy should be augmented by efforts to gather information on potential climate change as projected by climate models, and also on recent observable climate trends which may act as "early warnings" of further changes to come.

The capacity to adapt to future climate hazards will be enhanced by the following measures:

- Develop an understanding of possible future climate hazards based on model projections and climate scenarios where these are available.

- Where the above are not available, focus on the types

### Box 7-3: Data sources and prioritisation of systems

The following sources can provide valuable information on hazards, vulnerability and current adaptations, and adaptive capacity at the sub-national level, assisting the identification of high priority systems, regions and populations:

- National vulnerability assessments
- National Adaptation Programmes of Action (NAPAs)
- Vulnerability and hazard assessment and mapping projects

If these sources are not available, prioritisation might be undertaken using records of climate-related disasters – if available – from national statistical agencies, government departments, NGOs, or research organisations. Data on climate disaster-related mortality, displacement, total economic impacts, and other adverse outcomes, can be useful in identifying the areas at greatest risk for climate change hazards. Where data is limited or unavailable within a country, project teams might wish to use the following international databases:

- Emergency Events Database (EM-DAT) (http://www.cred.be/emdat) contains data relating to a variety of disaster types, including those with a climatic Component, for most countries. See Brooks and Adger (2004) and Brooks et al. (2004a, b) for applications of EM-DAT to studies of climate risk and vulnerability.
- DesInventar database (http://www.desinventar.org/desinventar.html) contains sub-national data on disaster outcomes for selected countries in the Americas.

These data sources may be used to prioritise regions, systems and population groups for capacity development projects, based either on the distribution of adaptive capacity, or on the need for capacity development to improve outcomes from climate hazards. For example, in high-risk regions that exhibit persistently high negative outcomes (in terms of mortality, displacement, economic losses, etc.), the question-based approach outlined in Section 7.3.6 may be used to (a) identify determinants and indicators of adaptive capacity and (b) design capacity development and adaptation strategies. Adaptive capacity indicators and measures of outcomes from climate hazards can be used to monitor the success of these strategies. Indicator identification and monitoring of success will be greatly assisted by consultations with stakeholders: those affected by climate hazards will be best placed to identify the factors and processes that determine their capacity to adapt, and also to assess the success of strategies aimed at enhancing this capacity (Box 7-5).

### Box 7-4: Agricultural adaptation in the Sahel

During the final decades of the twentieth century, inhabitants in parts of the Sahel (northern Nigeria and parts of Niger) successfully adapted to both drought and economic liberalisation, as reported by Mortimore and Adams (2001). The devastating drought of the early 1970s led to substantial loss of human life, and also resulted in widespread loss of livelihoods, transforming sections of Sahelian societies. Nonetheless, since the 1970s, agricultural systems have been transformed through a process of autonomous adaptation. With the abolition of subsidies on farm inputs, and in the face of uncertainties in world markets, many farmers have moved away from export agriculture, instead exploiting local markets. Agricultural diversity has increased as more integrated systems of farm management have been adopted. Livestock numbers have increased, and artificial fertilisers have been replaced with animal manure. Soil and water conservation measures have been introduced. Household incomes have also diversified, with non-farm income increasing in importance.

Other countries and regions facing drought might look to such examples when addressing adaptation to existing climate hazards or future climate change. The Sahel can provide examples of the successful adaptation of agricultural systems to increasing aridity and rainfall variability in a semi-arid environment, conditions which might be faced by other regions in the future. The cases described by Mortimore and Adams (2001) and other authors demonstrate the importance of local (informal) markets. In communities in which state-fixed prices are too low to act as incentives for agricultural innovation, adaptation has not occurred, and people have instead migrated to cities. In government-sponsored efforts to promote food security, programmes should encourage agricultural adaptation by supporting local markets, rather than focusing on export agriculture. (See Annex A.7.1 for additional information.)

***Table 7-2:*** *Types of future hazard and adaptation responses*

| | **Future discrete recurrent hazards** | **Future trends** | **Future singular events** |
|---|---|---|---|
| **Principal types of adaptation response** | • Initially anticipatory (planned, policy driven); also reactive when hazards are realised | • Responsive and anticipatory (planned and autonomous) | • Anticipatory (planned, policy driven), reactive if/when events occur |
| **Examples of hazard** | • Floods, droughts, wind storms, heat waves, cold waves, extreme rainfall events, hail storms, dust storms | • Warming, cooling, desiccation, sea-level rise | • Changes in thermohaline circulation, ice-sheet collapse, glacial dam-bursts, abrupt warming/cooling, circulation shifts |
| **Who acts?** | • Government and planning bodies | • Communities, individuals, government and planning bodies | • Government and planning bodies |
| **Measures to enhance adaptive capacity** | • Establish monitoring networks<br>• Develop forecasting capacity<br>• Develop ability to assess climate model output<br>• Build resilience to existing hazards | • Establish monitoring networks<br>• Develop forecasting capacity<br>• Develop ability to assess climate model output<br>• Create enabling environments | • Participate in global climate monitoring programmes<br>• Develop ability to assess climate model output<br>• Develop contingency plans for dealing with impacts of singular events |

of hazards that are familiar from the recent historical record, while gathering more quantitative information on possible future climate hazards from modelling studies, scenarios and analysis of recent trends.

- Develop an observational capacity to identify trends that may constitute "early warnings" of climate change.
- Adopt a vulnerability-based approach to risk management that is nonetheless informed by a prioritisation of hazards based on the above considerations.
- Create an environment in which adaptation is possible by disseminating information about climate change and its potential consequences, and addressing uncertainty.
- Engage stakeholders to discuss and formulate strategies to increase the capacity to adapt to future climate change.

Table 7-2 frames these and other measures for enhancing adaptive capacity, in relation to the types of future hazards to which they can respond.

### 7.4.4. Component 4: Formulating an adaptation strategy

*What measures, policies and strategies enhance adaptive capacity and encourage autonomous adaptation?*

The aim of capacity development projects is to create resilient and flexible systems that will be better prepared to adapt autonomously (i.e., without external intervention). Capacity enhance-ment will also facilitate the efficient implementation of adaptation strategies by reducing obstacles and making people more receptive. These principles should be at the heart of methods to enhance adaptive capacity, which are a prerequisite to implementing adaptation strategies and measures (Section 7.4.5).

Capacity development strategies must be tailored to the systems where adaptation is to be promoted (identified in Components 1 to 3) and to the climatic, environmental, socio-economic and political contexts within which these systems exist, e.g.,:

- Nations that experience little damage from existing climate variability will wish to concentrate on enhancing the adaptive capacity of systems that are likely to be vulnerable to anticipated future hazards.
- If substantial uncertainty exists as to the nature of future hazards, the focus would be on enhancing the resilience of economically or culturally important systems; in such cases, projects will focus on the issues raised in Component 4.
- Countries that suffer frequent losses as a result of existing climate variability will wish to focus, at least initially, on enhancing the capacity of systems and populations to increase their coping range with respect to familiar hazards (focusing on Component 2). These countries will also need to consider how strategies that deal with current hazards may incorporate measures to deal with future risks.

---

**Box 7-5: The importance of awareness raising for capacity development**

Awareness raising is important as it helps stakeholders and decision makers recognise the need for adaptation, and promotes willingness to engage in the identification, prioritisation and implementation of adaptation options. Decision-makers and stakeholders need to understand the risks climate change poses to their society; people will not pursue potentially disruptive and expensive adaptation strategies unless they are convinced that they are necessary. Scepticism concerning the reality of climate change may need to be overcome through the dissemination of information relating to the science of climate variability and change, including considerations of uncertainty. There is a need for clear communication by scientists to decision makers and stakeholders about the nature of anticipated climatic changes and the risks they pose to society. Training in science communication, as well as funding of scientific research is desirable, as is the formation of databases of explanatory materials for use in public education and communication with policy makers and others.

Awareness raising will also be facilitated by the keeping of reliable, detailed meteorological records, which may be used to identify climatic variations and trends on multi-decadal timescales. Climate scenarios and socio-economic scenarios will also be useful for visualising the potential impacts of climate change and their implications for stakeholders. The development of seasonal forecasting ability will also enhance the capacity of those in climate sensitive sectors such as agriculture to adapt. Forecasts will become increasingly important in the event of increased interannual climatic variability, particularly where agriculture depends on the planting of crops to take advantage of a short-lived wet season. Uncertainty must be addressed explicitly in seasonal forecasts and in long-range climate change scenarios, and the dissemination of this information should be undertaken by an adequately staffed and funded meteorological or climate change unit. Dissemination might be via public service broadcasting, particularly where there is a large, widely dispersed rural population and where literacy rates are low. In such areas, access to information will be enhanced by measures such as the distribution of free or very low-cost wind-up radios.

---

The principal elements of the capacity development process are as follows (Yohe and Tol, 2002):

- Raise awareness of the risk associated with the hazard (Box 7-5).
- Identify a set of possible adaptation options, including those that may be undertaken by actors at a range of scales, from institutions and government to communities and individuals (discussed in TP8).
- Prioritise options based on their efficacy, feasibility and acceptability (discussed in TP8).
- Remove barriers to adaptation within the system being addressed (discussed in TP9).

Some adaptation options will involve considerable planning and co-ordination, while others may be undertaken on an ad hoc basis. These latter, "autonomous" adaptations can be encouraged by providing an economic, regulatory and policy environment in which people are likely to pursue these options, rather than through coercive measures. Examples might be (i) encourage agricultural diversification through grants, loans, subsidies on specific farm inputs, and support local markets, or (ii) provide incentives via local tax regimes for people to settle in less hazard-prone areas.

*How can we identify and prioritise adaptation and capacity development options?*

One of the most common needs is the capacity to design integrated policy packages that sufficiently identify trade-offs, synergies and conflicts among key sectors. An initial shortlist of options for adaptation/capacity development may be drawn up, based on considerations of what is appropriate and technically feasible within the existing socio-economic and political context. Involving stakeholders from the outset reduces conflict (TP2). The short-listed options can then be prioritised based on how likely they are to be effective (efficacy), how easy they are to implement (feasibility), and how acceptable they will be to those affected by them (acceptability). To a large extent, feasibility and acceptability might be based on considerations of cost, although non-financial criteria must also be considered (TP8). Prioritisation might therefore be performed using a multi-criteria analysis, or by seeking consensus among the stakeholders. Although the latter approach is less likely to lead to conflict, consensus might be difficult to achieve. Different interest groups will exhibit preferences for certain adaptation options, and the resolution of inter-group conflicts will be central to the adaptation process. Clearly, fostering dialogue and nurturing a culture of consensus may be important in enhancing adaptive capacity (Box 7-6). For practical examples of prioritisation of options, see Yohe and Tol (2002).

*What constraints might there be on adaptive capacity?*

A number of adaptations may be feasible and effective for a system or population that needs to increase its ability to cope with a climate hazard. However, for various reasons, these options may not be acceptable. In such cases, acceptability represents an important constraint on adaptive capacity. For example, building a dam to buffer a region against drought – by storing and providing water for domestic, industrial and agricultural use – may be unacceptable for social and ecological reasons. Its construction may displace people, destroy valued ecosystems, or inundate culturally important areas. Alternatively, it might be prohibitively expensive, or threaten the

---

**Box 7-6: Adaptive capacity and participatory decision-making**

Stakeholder involvement (TP2) in the identification and prioritisation of adaptation options is absolutely vital, since to be successful, adaptation measures must be acceptable to those who are to implement them. Where there is no consensus as to the feasibility and acceptability of these options, the capacity to adapt will be very limited, and what adaptation does occur will be constrained by conflict.

The origin of a capacity development initiative is an important factor in the commitment of decision makers and stakeholders. When the impetus for adaptation comes from, and is generally acceptable to, both the government and stakeholder communities, progress is likely. Alternatively, if the adaptation agenda is imposed by external groups – without local representation – community buy-in will be difficult to achieve. The role of external groups should be to support locally-driven initiatives for adaptation strategies. An opportune time to develop such initiatives is after crises (e.g., cyclones, droughts, or floods). At these times, political and social awareness of environmental change issues is high, and resistance to adaptation strategies is low.

The exclusion of poor and marginalised members of society from the decision-making process is likely to lead to further undermining of their socio-economic status that may, in turn, lead to social conflict and political instability. This is particularly likely if adaptation measures involve displacement. Further marginalisation may also lead to environmental degradation, as the extremely poor are forced to use resources in an unsustainable manner in order to survive. Strategies with such consequences are as likely to be maladaptive as they are to help adaptation. Adaptive capacity is strengthened by the existence of networks and mechanisms that encourage participation and prevent marginalisation.

In the relationship between society and the state, capacity development should take the form of engagement between civil society, in the form of stakeholder groups, and local and national government. Stakeholder representatives should come from all sections of society likely to be affected by climate change, or by the implementation of adaptive measures. Stakeholder groups with little or no historical power to influence decision-making should be represented, and the fact that adaptation may create "winners and losers" must be recognised. A wide variety of stakeholders should participate in adaptation policy formulation, and in the case where those who share concerns and interests regarding climate change have no framework for collective representation, they should be assisted in building such networks. People are far more likely to support adaptation strategies if they feel their views have been taken into account.

Decision makers might have to weigh the interests of those who will be physically displaced against those who stand to profit economically from the implementation of the adaptation measure. In such circumstances, adaptive capacity will be enhanced by the existence of formal mechanisms for addressing such conflicts of interest, and through the pursuit of conflict management strategies. Those who will be most adversely affected by an adaptation measure should have a greater input, in addition to offers of compensation.

TP2 provides guidance on stakeholder engagement; additional information is provided in the UNDP/GEF handbook listed in the references (UNDP/GEF, 2004).

---

security of communities downstream. It may also lead to reduced stream flow in neighbouring downstream countries, and become a source of potential political conflict. In such a case, acceptability represents the "weakest link" in terms of adaptive capacity. If building a dam were the most effective, or only, adaptation measure available, efforts might be made to remove the barriers to its implementation. Such efforts might involve the relocation of threatened settlements (perhaps augmented by financial compensation), ecosystems or heritage sites, or the negotiation of water management agreements with neighbouring countries. The first step towards enhancing adaptive capacity is identifying the "weakest link" of the system in terms of its capacity.

Alternatively, an adaptation measure may be effective and acceptable, but might not be feasible due to technological limitations. What is technically feasible for one country may

not be feasible for another. Similarly, cost might be the deciding factor, making certain measures feasible in wealthy countries but impossible in poor nations, again emphasising the importance of developing adaptation solutions that are appropriate to local circumstances, with input from stakeholders.

Capacity constraints might also originate from outside a country's borders. For example, options that require restructuring economic policy at the national level may be vetoed by creditor nations or international financial institutions, which often dominate the economic policies of highly indebted developing countries. These constraints are much more difficult to overcome. Even where a country has a significant degree of economic independence, those running capacity development projects at a sub-national scale are likely to have little influence over national economic policy. Their efforts will be better employed by promoting local measures to facilitate autonomous adaptation, particularly if they are concerned with a single locality or with a sector that does not make a large contribution to the national economy. (TP9 contains additional discussion on dealing with potential constraints.)

*What policy considerations are important in capacity development strategies?*

Policies aimed at enhancing adaptive capacity must achieve a balance between strong regulations to prevent maladaptation (e.g., steering development away from flood plains) and measures to encourage adaptive behaviour. Policies should provide individuals, communities and organisations with sufficient flexibility to pursue adaptation strategies appropriate to their circumstances. Restrictive policies must be carefully targeted to avoid undermining adaptive capacity. New policies should be assessed in terms of their potential impacts on adaptive capacity, particularly for groups and systems that already exhibit high vulnerability and/or exposure to climate hazards. The impacts of policies on systems and communities in sensitive ecosystems, such as coastal and riverine zones, should be given special attention. Policies designed to address issues at a regional scale can have unforeseen effects at local scales; cross-scale linkages should therefore be examined in a "policy impact assessment" process.

### 7.4.5.   Component 5: Continuing the adaptation process

*How can efforts to enhance adaptive capacity be sustained and improved over time?*

Once a strategy has been developed and barriers to adaptation addressed, adaptation measures can be implemented. Of all the APF Components, this is one of the most complex. It requires the capacity to recognise opportunities for mainstreaming adaptation into on-going processes. TP9 suggests actions that can be taken to facilitate adaptive capacity.

Adaptation measures must be ongoing, and strategies to encourage and facilitate adaptation should not be seen as "one-off" measures. For this reason, it is important that the adaptation strategies be assessed on a continual or regular basis. Reviewing, monitoring and evaluating the success of adaptation strategies is addressed in detail in TP9. The following questions are important to the adaptation learning process:

- Are the strategies working – i.e., are they as effective as anticipated at reducing vulnerability and/or effectively managing risk?
- Once implemented, are the adaptation strategies still viewed as acceptable – i.e., are there any unexpected negative consequences of these strategies that reduce their acceptability?
- Are the strategies as feasible as was anticipated – i.e., are there any previously unforeseen difficulties in their implementation?

- Has adaptive capacity really been increased?
- Are people more willing and better able to pursue autonomous adaptation?

Assessments of the success of adaptation strategies and capacity development programmes, and the modification of such strategies where necessary, will benefit from the following activities:

- Meteorological monitoring, which provides information on the evolution of hazards.
- Monitoring outcomes (mortality, morbidity, displacement, economic losses), which enables project teams to assess the success of adaptation strategies. Improvements in outcomes under conditions of constant or increasing hazards is indicative of effective adaptation; even where outcomes apparently do not improve, adaptation may be working if hazards are increasing in severity and/or frequency (TP9).
- Monitoring of vulnerability and adaptive capacity using indicators, which can yield direct information on the impacts of adaptation strategies, even in the absence of hazard events (e.g., where strategies are designed to increase resilience to, or prepare for, anticipated future hazards) (TP9).
- Stakeholder involvement in the assessment process, which can offer valuable feedback on whether adaptation and capacity development strategies are proving successful, as well as on any unforeseen consequences of these strategies (TPs 2 and 9).

Monitoring the success of adaptation and capacity development strategies is necessary, but not sufficient, to ensure that the adaptation process continues effectively. In addition, adaptation strategies must be flexible, and able to incorporate new information on climate hazards and on socio-economic and environmental systems. Given the high degree of uncertainty in both climate and socio-economic scenarios, it is highly probable that, as new information becomes available and our understanding of the climate system and processes of adaptation improves, existing strategies will need revision or updating. A flexible approach is required to prevent societies from becoming "locked in" to policies and procedures that may prove inappropriate in the mid- to long-term. A danger in large-scale, long-term projects is that political inertia and vested interests encourage their continuation, even if it becomes apparent that they are inappropriate, or that better alternatives are available. Adaptive capacity will be enhanced if accompanied by policies that require their future modification and revision. (TP9 provides additional discussion on continuing the adaptation process.)

### 7.5.   Conclusions

In its broadest context, the APF treats adaptive capacity as a change management process. In other words, adaptation will only occur if the system is able to adjust its characteristics or behaviour, so that its coping range is expanded under future climate, including variability. However, external barriers to adaptation often exist and the adaptation process does not automatically occur if capacity in the system is constrained. It follows that an adaptation project can be designed to catalyse a change process if the key capacity constraints are removed. In a given system, it is necessary to understand the Components of the change process in terms of: "Who needs to adapt?" "To which climate risks?" "What are the barriers to adaptation?" "What are the capacity constraints of the adaptation process?"

A prerequisite to enhancing adaptive capacity is the baseline analysis of adaptive capacity to cope with current climate. Because adaptive capacity cannot be directly measured, it is characterised by examining potential changes of the sensitivity of human and ecological systems to climate. A capacity assessment includes an examination of the willingness and resources necessary to adapt to climate hazards. An assessment should avoid the potential pitfall of trying to identify a comprehensive list of quantitative capacity indicators. It is more important to understand and to characterise the adaptation process in a pragmatic manner.

Following the guidance in this paper, project teams should be able to produce some of the following:

- A list of priority systems and target groups most in need of adaptive capacity development (TPs 1, 3 and 6).
- A set of qualitative indicators that characterise adaptive capacity within and between systems, population groups and regions (TPs 3 and 6).
- A shortlist of realistic options for adaptation and adaptive capacity development for a priority system/population facing a particular hazard or set of hazards (TP8).
- A set of preferred adaptive capacity development options based on considerations of feasibility, efficacy and acceptability, identified in consultation with stakeholders (TP8).
- A strategy for implementing the preferred adaptive capacity development options involving significant stakeholder involvement, frequent review of progress, and assessment of options for revision (TPs 2, 8 and 9).

## References

**Adger**, W.N. and Brooks, N., (2003). Does environmental change cause vulnerability to natural disasters? In Pelling ed., *Natural Disasters and Development in a Globalising World*, London: Routledge, pp. 19-42.

**Adger**, W.N. and Kelly, P.M. (1999). Social vulnerability to climate change and the architecture of entitlements, *Mitigation and Adaptation Strategies for Global Change*, **4**, 253-266.

**Brooks**, N. (2003). *Vulnerability, Risk and Adaptation: A Conceptual Framework*. Working Paper 38, Tyndall Centre for Climate Change Research, University of East Anglia, Norwich, United Kingdom. Available at: http://www.tyndall.ac.uk

**Brooks**, N. and Adger, W.N. (2004). Country level risk indicators from outcome data on climate-related disasters: an exploration of the Emergency Events Database. Tyndall Centre for Climate Change Research. Available from the author at nick.brooks@uea.ac.uk.

**Brooks**, N., Adger, W.N. and Kelly, P.M. (2004). The determinants of vulnerability and adaptive capacity at the national level and the implications for adaptation. Submitted to *Global Environmental Change*. Available from the author at nick.brooks@uea.ac.uk

**Cullen**, H.M., deMenocal, P. B., Hemming, S., Hemming, G., Brown, F.H., Guilderson, T. and Sirocko, F. (2000). Climate change and the collapse of the Akkadian empire: Evidence from the deep-sea, *Geology*, **28** (4), 379-382.

**Mortimore**, M. and Adams, W.M. (2001). Farmer adaptation, change and "crisis" in the Sahel, *Global Environmental Change*, **11**, 49-57.

**Pelling**, M. and Uitto, J. I., (2001). Small island developing states: natural disaster vulnerability and global change, *Environmental Hazards* **3**, 49–62.

**Roberts**, N. (1998). *The Holocene: An Environmental History*, Oxford, Blackwell

**Sarewitz**, D., Pielke, R., and Keykhah, M. (2003). Vulnerability and risk: Some thoughts from a political and policy perspective. *Risk Analysis*, **23**, 805-810.

**UKCIP** (2003). *Climate change and local communities – How prepared are you? An adaptation guide for local authorities in the UK*, United Kingdom Climate Impacts Programme. Available at: http://www.ukcip.org.uk/resources/publications/documents/Local_authority.pdf

**UNDP/GEF** (2003). *Capacity Development Indicators*, New York, US.

**UNDP/GEF** (2004). *Assessing Technology Needs to Address Climate Change*, New York, US. Available at: http://www.undp.org/cc/technology.htm

**Yohe**, G. and Tol, R.S.J. (2002). Indicators for social and economic coping capacity – moving towards a working definition of adaptive capacity, *Global Environmental Change*, **12**, 25-40.

# ANNEXES

## Annex A.7.1.  Capacity to adapt to drought in the Sahel

The question-based approach to identifying indicators of adaptive capacity is illustrated below using the example of drought in the African Sahel. The indicators are suggestions; the example is a general one and does not relate to any specific country or region. A combination of quantitative and qualitative indicators is suggested (qualitative indicators are identified in the text). Most indicators represent the local scale, but in some cases, national-level indicators, representing interactions across scales, are also identified. The example draws on the work of Mortimore and Adams (2001).

Note that the UNDP-GEF (2003) approach differs from the example below. In the latter, evidence that adaptation has occurred is required. A score card rather than specific indicators is used.

*What is the nature of the system/population being assessed?*

- Rural livelihoods, including small-scale farmers and pastoralists

*What are the principal hazards faced by this system/population?*

- Drought

*What are the major impacts of the hazard(s) and which elements of the system/population are most vulnerable to these hazards?*

- Food shortages, famine, loss of livelihoods, rural-urban migration, economic losses.
- Rural poor, isolated communities, small households, pastoralists.

*Why are these elements/groups particularly vulnerable?*

- Poor households are unable to afford food when production fails.
- Isolated communities are often inaccessible or overlooked in terms of aid distribution; opportunities to exploit local markets for income diversification and to seek temporary salaried work in urban centres are limited.
- Labour availability for agricultural tasks is determined by household size, age and sex of household members, and options for bringing in labour from outside the household.
- Once pastoralists lose their animals they are reliant on aid or forced to resort to begging, at least in the short term. Pastoralists are often marginalised by governments that prefer sedentary populations and favour settled agriculture.

*What adaptive measures would reduce the vulnerability of the above groups?*

- Agricultural innovation to promote resilience.

- Improved transport networks and accessibility of isolated communities.
- Development of local markets.
- Increased resource sharing (including labour).
- Recognition of and support for pastoral groups – availability of grazing, mobility. Shift to livelihoods based on animals better adapted to drought, e.g., from cattle to sheep goats, camels.

*What capacity exists to implement these measures?*

- Agricultural innovation requires financial and human resources, technical and/or traditional knowledge, availability of crop and livestock varieties for diversification. In the Sahel, farmers have opportunities to sell produce. People are more likely to invest in agriculture if they are secure in their tenure. *Indicators: household income and size, dependency ratio, biodiversity, prices of farm inputs and outputs, land ownership, economically-active population, knowledge of traditional farming practices (qualitative indicator).*
- Isolation can be tackled locally by strengthening links between communities, or by government, e.g., building roads. These require good community relations and public investment respectively. *Local level indicators: settlement density, road density, "social capital" indicators. National level indicators: political accountability and representation of region, financial and technical resources.*
- Local markets can be developed through subsidies and controls on imports and commodity prices, although these might be politically unacceptable. Deregulation and the removal of price controls – where prices of agricultural goods are artificially low – may also stimulate local agricultural and economic development. Transport networks will also facilitate local trade and exchange. *Local level indicators: price of agricultural outputs, road density. National level indicators: political representation, economic autonomy (e.g., linked to debt).*
- Resource sharing is most likely to occur where community relations are good and traditional social institutions are strong. *Indicators: indicators of community cohesion (e.g., crime rate).*
- The ability of pastoral groups to access pasture and water is, to a certain extent, determined by geography and the nature of the local or regional physical environment. However, their capacity to adapt by exploiting new areas or retreating to more productive areas is often limited by restrictions on their movement due to agricultural expansion, political marginalisation and the existence of national boundaries. A shift from cattle to other animals requires that the latter are available, affordable and culturally acceptable. *Local level indicators: rate of agricultural expansion, per cent of land area covered by rangeland, water availability*

*(e.g., well density), livestock prices, proximity to national borders. National level (qualitative) indicators: internal and external conflict, relations between pastoral groups and ruling groups.*

*What barriers are there to the implementation of these measures?*

Some constraints to the realisation of adaptive capacity have been mentioned above, where they are generally represented by the "national level" indicators. These national level indicators represent processes and factors that provide the broader political or economic context for local adaptation, and which may be viewed as external to the local systems in which adaptation occurs. Constraints on the realisation of adaptive capacity may result from economic policies that affect the price of farm inputs or outputs (e.g., imported foodstuffs that compete with farm output). These policies may be the result of conditions imposed on a country by creditor nations or international financial institutions. In such a case, adaptive capacity might be developed at the local level by recognising these economic barriers and developing alternative livelihood strategies. At the national level, capacity might be enhanced by a renegotiation of debt repayments or by a rethinking of relationships with international financial institutions. These financial interventions should have greater focus on regional co-operation and reduced emphasis on integration into the world economy, allowing the government to support local markets and livelihoods.

While isolated communities are likely to be vulnerable in terms of livelihood and food security, and lacking in adaptive capacity, their isolation might also mean that they are less adversely affected by factors such as cheap imports that undermine local markets. Multiple and opposing consequences of strategies to enhance adaptive capacity should be assessed; poorly-conceived strategies can undermine adaptive capacity if they have unforeseen consequences.

Further constraints on developing adaptive capacity might be the result of internal conflict (e.g., mitigating against long-term planning and/or investment and preventing regional co-operation). Conflict in neighbouring countries, which might result in border closures, may hinder the mobility of pastoralists. While nomadic groups are generally highly adapted to variable rainfall, anecdotal evidence suggests that their capacity to adapt in some Sahelian countries was constrained as a result of their displacement by sedentary agriculture, which expanded northward into marginal areas during the wet 1950s.

If constraints on adaptive capacity can be identified, capacity development and adaptation may be pursued. Such development and adaptation may either occur within the context of those constraints – recognising which options are realistic – or through a strategy that involves the removal of constraints where feasible and desirable. The latter strategy will often involve intervention at the governmental and international levels.

# 8

# Formulating an Adaptation Strategy

ISABELLE NIANG-DIOP[1] AND HENK BOSCH[2]

Contributing Authors
*Ian Burton[3], Shaheen Rafi Khan[4], Bo Lim[5], Nicole North[6], Joel Smith[7], Erika Spanger-Siegfried[8]*

Reviewers
*Mozaharul Alam[9], Anne Arquit Niederberger[10], Suruchi Bhawal[11], Moussa Cissé[12], Mohamed El Raey[13], Ulka Kelkar[11], Jyoti K. Parikh[14], Hubert E. Meena[15], Mohan Munasinghe[16], Atiq Rahman[9], Roland P. Rodts[17], Samir Safi[18], Juan-Pedro Searle Solar[19], Barry Smit[20], and Thomas J. Wilbanks[21]*

[1] Department of Geology, Faculty of Science, University Cheikh Anta Diop, Dakar, Senegal
[2] Government Support Group for Energy and Environment, The Hague, The Netherlands
[3] University of Toronto, Toronto, Canada
[4] Sustainable Development Policy Institute, Islamabad, Pakistan
[5] United Nations Development Programme – Global Environment Facility, New York, United States
[6] INFRAS, Zurich, Switzerland
[7] Stratus Consulting Inc., Boulder, United States
[8] Stockholm Environment Institute, Boston, United States
[9] Bangladesh Centre for Advanced Studies, Dhaka, Bangladesh
[10] Policy Solutions, Hoboken, United States
[11] The Energy and Resources Institute, New Delhi, India
[12] ENDA Tiers Monde, Dakar, Senegal
[13] University of Alexandria, Alexandria, Egypt
[14] Indira Gandhi Institute of Development Research, Mumbai, India
[15] The Centre for Energy, Environment, Science & Technology, Dar Es Salaam, Tanzania
[16] Munasinghe Institute for Development, Colombo, Sri Lanka
[17] Independent consultant, Ouderkerk a/d Ijssel, The Netherlands
[18] Lebanese University, Faculty of Sciences II, Beirut, Lebanon
[19] Comisión Nacional Del Medio Ambiente, Santiago, Chile
[20] University of Guelph, Guelph, Canada
[21] Oak Ridge National Laboratory, Oak Ridge, United States

# CONTENTS

## 8.1. Introduction

Adaptation to climate change is a process by which strategies to moderate, cope with and take advantage of the consequences of climate events are developed and implemented (IPCC, 2001). Governments and/or communities can proactively adapt (Smit et al., 2001) but, at this time, the public policy world is not yet motivated by climate change impacts. Policy making is dominated by competing priorities and interest groups, and decisions with their own schedules – often unrelated to climate change. Among the many examples are elections, natural disasters and fiscal crises. At the same time, adaptation itself is a long-term process, requiring sustained attention. Developing an adaptation strategy is not a simple "one-shot" deal; instead it is an iterative, continuous learning process. The objective of this Technical Paper (TP) is to help countries develop their own adaptation strategy by providing guidance on setting priorities.

In designing an adaptation strategy, teams can acknowledge the reality of how a policy is made and yet, in the interests of climate change adaptation, offer a clear vision of where it ought to go. The strategy itself involves working within the context and opportunities of the political structure, taking advantage of opportunities as they arise, but having a sense of priorities based on that vision.

This TP outlines the elements of adaptation strategies that are likely to be consistent across the range of environmental or climate contexts. The first activity is the synthesis of available information. Once this data is compiled, the second activity involves designing an adaptation strategy with consideration to objectives, indicators, and integration in national development plans and other synergies. The third activity involves the formulating the adaptation options for policies and measures. Selecting and prioritising the policies and measures, and then extending the analysis beyond a simple list, comprise the fourth element. Finally, the last activity is the formulation of an adaptation strategy for implementation.

An adaptation strategy will, in many ways, be a "living" document; the process will not end with developing the strategy. Instead, it will mark the beginning of a new phase in which lessons from implementing the approach are fed back into the strategy to improve it over time (TP9). This strategy should include flexibility mechanisms to address the climate "surprises" that will almost certainly occur in the future, and account for new technologies and findings in the field of climate change (Klein et al., 1999).

The following pages focus on adaptation at a national level. However adaptation strategies, policies and measures must

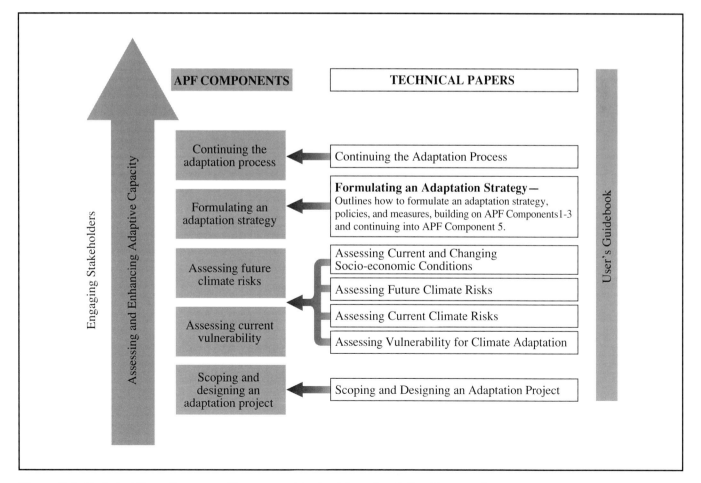

**Figure 8-1:** Technical Paper 8 supports Component 4 in the Adaptation Policy Framework

also be considered at the sub-regional and regional levels (e.g., when they are related to shared resources, as is the case with international rivers). Furthermore, this TP focuses on public and private planned – as opposed to autonomous – adaptation.[1]

The following sections describe the relationship of TP8 with the Adaptation Policy Framework (APF) as a whole (Section 8.2) and present key concepts (Section 8.3). Section 8.4 offers guidance on formulating an adaptation strategy. Tasks described here must be considered as indicative, since some of them may already have been completed. The annexes address methodological issues related to the process of developing an adaptation strategy.

## 8.2.    Relationship to the Adaptation Policy Framework as a whole

This TP builds on the first three Components of the APF process (Figure 8-1) and provides direct input to Component 4. At this stage it is assumed that the team has designed and scoped the APF project, chosen an approach (Component 1), and analysed both the country's adaptive capacity and the situation that gives rise to a need for adaptation to climate change (Components 2-3). The analysis may have taken an approach that emphasises vulnerabilities, climate risks, adaptive capacity, and/or policies and measures.

The broad objectives of the strategy will be tailored to the country and its decision-making process, stakeholder considerations, the particular vulnerabilities and climate risks being addressed, and the resources available (adaptive capacity). These efforts provide the basis for developing and implementing an adaptation strategy for subsequent monitoring and evaluation (Component 5).

## 8.3.    Key concepts

A list of brief definitions of terms used throughout the APF can be found in the Glossary. For important concepts specific to this TP, more complete definitions are described in this section. These key concepts are the overall adaptation strategy, policies, measures, and the time horizons. The no-regret and low-regret options are also briefly defined, as well as the top-down and bottom-up approaches.

The *adaptation strategy* for a country refers to a general plan of action for addressing the impacts of climate change, including climate variability and extremes. It will include a mix of policies and measures with the overarching objective of reducing the country's vulnerability. Depending on the circumstances, the strategy can be comprehensive at a national level,

addressing adaptation across sectors, regions and vulnerable populations, or it can be more limited, focusing on just one or two sectors or regions. In the Least Developed Countries, a National Adaptation Programme of Action (NAPA)[2] could well be developed into an adaptation strategy using the APF.

Generally speaking, *policies* refer to objectives, together with the means of implementation. In an adaptation context, a policy objective might be drawn from the overall policy goals of the country – for instance, the maintenance or strengthening of food security. Ways to achieve this objective may include: farmer advice and information services, agricultural research and development, seasonal climate forecasting, and subsidies or incentives for development of irrigation systems.

*Measures* are focused actions aimed at specific issues. Measures can be individual interventions or they can consist of packages of related measures. Specific measures might include actions which promote the chosen policy direction such as implementing an irrigation project, setting up a farmer information, advice and early warning programme, developing a new scheme for crop insurance, establishing a system of grain storage to protect against drought or crop failures, or providing financial incentive(s) to grow a specific crop, etc. Each of these measures would contribute to the national goal of food security.

When defining these concepts, their distinctions are not always clear. An example of the relationship among these terms follows. In Holland, the strategy to avoid rivers overflowing dikes is referred to as "giving water more room". A change in building codes forbidding building or constructing any obstacle in the riverbed is one of the policies intended to carry out the strategy. Deepening and widening riverbeds and appointing overflow areas are specific measures. The following are examples of either policies or measures: water conservation, investments in agricultural infrastructure, including roads to markets, drought and flood control or alleviation measures, crop diversification, and alternative off-farm employment in rural areas. Many policies and measures relevant to adaptation may already exist in countries.

Setting of *time horizons* is needed when defining a strategy, policy, or measure, and also for monitoring the implementation of an adaptation strategy (TP9). Generally, strategies would be long-term in nature, and policies targeted at the medium- to long-term. Measures may have an implementation time of any length, but are expected to have sustained effects. Prioritisation, mostly of measures, but in some cases also of (alternative) policies, will take the whole period into account. A NAPA, which is meant to communicate the most urgent needs of Least Developed Countries, is likely to contain measures with a short implementation time, but with immediate and, preferably, long-term effects. A further discussion on planning and policy horizons can be found in TP5.

---

[1] Autonomous adaptation refers to adaptation action taken by individuals or systems without involvement by government. While not considered in this TP, autonomous adaptation may lead to ideas for government-assisted adaptation and is an important determinant of the "adaptation environment".

[2] NAPAs are a quick participatory vulnerability and adaptation process through which Least Developed Countries will be able to develop project proposals relative to their immediate and urgent adaptation needs addressing the most vulnerable areas, systems (UNFCCC Decision 28/CP7).

*No-regret options* are measures or activities that will prove worthwhile even if no (further) climate change would occur. *Low-regret options* are no-regret options that require small additional outlays to cater to the negative effects of climate change. The notion of "no regret" or "low regret" may be useful when trying to obtain (outside) finance (Box 8-1).

The notions of *top-down* and *bottom-up* generally apply to the context of planning. Top-down planning would typically emanate from higher levels in government such as planning ministries. A bottom-up planning process would start at the local level and, from there, progress "up" to the decision-makers. The APF places equal importance on the bottom-up approach in order to take account of existing adaptation and coping mechanisms at the local level. Effective planning processes may be those that successfully integrate the two approaches. For both approaches, stakeholder involvement is equally important.

## 8.4. Guidance on formulation of an adaptation strategy

Adaptation teams can choose to use a top-down or bottom-up approach for formulating an adaptation strategy, policies and measures. While the bottom-up approach plays an important role throughout the APF, this TP emphasises the top-down approach and assumes the highest level of government buy-in. In practice, the selection of the approach means choosing the path of least resistance and taking into account the centralised/decentralised nature of decision-making in that country.

With the top-down approach, the overall policy direction of the country will guide the design of the adaptation strategy. In turn, the objectives of the strategy will guide the selection, design, and implementation of new policies and measures – a process that requires a high level of political will. At the same time, an adaptation strategy can assist a country in meeting existing obligations to international agreements – e.g., in biodiversity and desertification. Development or revision of national plans, such as those for drought management, coastal management, biodiversity conservation, and forest management, can present added opportunities for integrating climate change concerns into related planning processes. Given the opportunities for synergies and policy coherence, all of these imply a top-down approach. But a top-down approach requires significant capacity for policy design (TP1) and can be more ambitious than its alternative. In circumstances where this capacity is limited, a bottom-up approach may be more pragmatic.

A project team may prefer to begin the APF process with a set of articulated measures and/or policies that have already been analysed in terms of vulnerability, climate risks, future socio-economic conditions, and adaptive capacity. In this case, a bottom-up approach to developing the adaptation strategy – i.e., formulating the strategy that coheres with the measures – is obviously more appropriate. An example of a bottom-up approach can also be found in Klein et al. (1999), where policies and measures are first designed, then embedded in an adaptation strategy. Generally, these policies and measures

have been developed for reasons other than climate change, and climate change is included as an additional consideration.

This TP presents the five different activities involved in formulating an adaptation strategy (Figure 8-2). Depending on national circumstances, some of these tasks may already have been performed, mainly through Initial National Communications to the UNFCCC and NAPAs:

1. Synthesise outputs of previous APF Components and other studies
2. Design the adaptation strategy
3. Formulate adaptation options for policies and measures
4. Prioritise and select adaptation policies and measures
5. Formulate an adaptation strategy

While this paper emphasises the use of prioritisation methodologies, the APF recognises the value of expert judgement in policy-making processes when applied rigorously. Such alternative approaches will be presented in related APF publications.

### 8.4.1. Synthesise outputs of previous Adaptation Policy Framework Components and other studies

The first activity is to synthesise what is known about the country's vulnerability and adaptive capacity with regard to the potential impacts of climate change. Within the APF framework, this will involve assessing the adaptation options identified by previous Components – namely, the outputs of the project scoping process, as well as the current and future vulnerability assessments. If sufficient information exists, some users may choose to begin with Component 4, by applying this paper.

Most likely, this information will include responses around the following questions:

- What ecosystems, sectors, regions, and populations are particularly vulnerable to climate change? (APF Components 2 and 3; TPs 3- 6).
- What is the current level of adaptation and adaptive capacity that constitutes the adaptation baseline? (APF Component 2; TPs 3 and 7). The determinants of adaptive capacity may include technological advances, institutional arrangements, new and existing policies, availability of financing, level of information exchange, etc.
- What sets of indicators were chosen for vulnerability, socio-economic and adaptive capacity analyses? (TPs 3, 6, and 7). These should lay the basis for evaluating and prioritising alternative adaptations.
- Which set of stakeholders represents the various sectors, regions, and populations being considered? (TP2).

Teams using the APF will want to revisit their initial analysis of the policy process within their country (Component 1; TP1), particularly in light of the current socio-economic context and the scenarios of potential future conditions they will have developed (TP6).

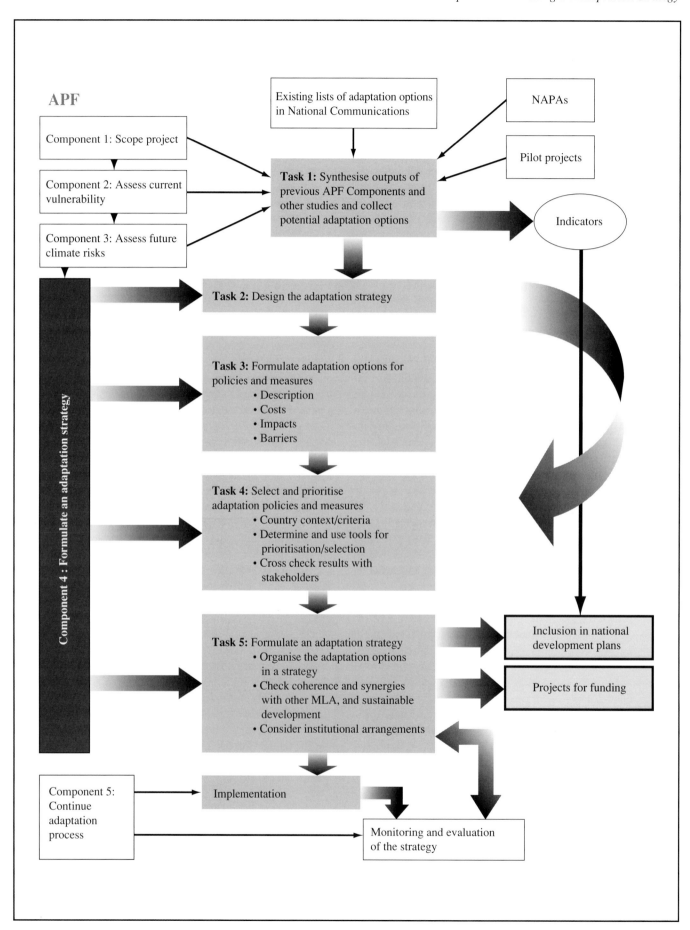

**Figure 8-2:** Component four in the Adaptation Policy Framework context

As stressed throughout the TPs, the APF process is expected to be embedded within the context of current policies and to build upon earlier national assessments. These assessments include, for example, the National Communications to the UNFCCC, NAPAs, previous vulnerability and adaptation studies, and pilot projects or programmes. NAPAs will likely only address a few priority sectors and systems, and the analysis may need to be extended to other systems. Guidance from the Intergovernmental Panel on Climate Change (IPCC)[3] may provide additional ideas.

Upon completing this task, the team should have a clear idea of the vulnerability of the country with regard to climate change, the capacity of the country to adapt, and the overall policy context to be considered while developing the adaptation strategy. A detailed plan to engage stakeholders in strategy development should be also prepared, as well as indicators that could be used to assess alternative adaptation strategies.

### 8.4.2.  Design the adaptation strategy: broad principles and considerations

The adaptation strategy consists of a broad plan of action to be implemented through policies and measures over the short-, medium- and long-term. The objectives of the adaptation strategy can be very specific (e.g., reduce the vulnerability of a sector), or quite broad (e.g., reduce poverty, achieve the Millennium Development Goals, etc.). From overall objectives, more specific goals can be derived. In any case, stakeholders, including government, will define these objectives (TP2) under Component 1 of the APF process.

An adaptation strategy is best supported by a set of instruments designed collectively. Without regulatory and economic instruments, adaptation to climate change will remain at the level of education and awareness raising. Many economic and especially regulatory instruments, however, may not work effectively without enforcement and compliance. A package of policies and measures should be designed to complement and reinforce each other. Policy instruments can be selected using both formal and informal methodologies in decision making, including decision-support tools (described later in the text).

A core set of policy instruments for implementing strategic decisions can be incentive-based or "control and command" interventions. The selection of instruments is closely linked to the socio-economic analysis, in which the barriers that impede adaptation have been identified. TP9 discusses common barriers to implementation in the context of environmental governance. Common types of policy instruments are outlined in Box 8-2.

As mentioned above, an adaptation strategy should also achieve synergy with other environmental strategies. Climate change issues are closely linked to the Convention on Biological Diversity as well as to the Convention to Combat Desertification. For example, drought early warning systems and contingency plans, food security systems, the development of alternative livelihood projects or sustainable irrigation programmes for both crops and livestock each could be considered an adaptation option in arid and semi-arid areas. At the same time, each of these could serve as a Component of a National Action Plan to combat desertification and drought in these same areas. To ensure greater efficiency and enhance the impact of all strategies, the APF team should ensure synergy among the responses to these different Rio Conventions.[4]

---

**Box 8-2: Common types of policy instruments**

- *Legislative, regulatory, and juridical instruments.* Legal instruments set limits and provide sanctions, but can be difficult to enforce. Examples are: laws, by-laws, regulations, standards, constitutional guarantees, and national agreements based on international conventions.

- *Financial and market instruments.* Fiscal instruments can influence behaviour by sending price signals. They are a powerful set of instruments for raising revenue for environmental management, but tend to be difficult to implement politically. Examples of market-based approaches are: property-rights based approaches (concessions, licences, permits), price-based approaches (taxes, payments for amenities, user fees, tax credits for investment funds, performance bonds), perverse subsidy removal, and market-based measures (labelling, procurement policies, product certification, information disclosure requirements).

- *Education and informational instruments.* Education instruments raise awareness, and over time, they change societal values. Examples are: consumer information, public awareness campaigns, and professional development.

- *Institutional instruments.* Private companies, corporations, and communities often adopt such policy instruments. Examples are: environmental management systems, management policies and procedures for service contracts.

---

[3]  Especially the 1995 and 2001 Working Group II Reports (Watson et al., 1996; McCarthy et al., 2001).

[4]  As it is expressed under point 39c of the Plan of Implementation adopted by the World Summit on Sustainable Development.

The strategic planning process will require "cross-sectoral co-operation, an interdisciplinary approach and considerable political will" (Least Developed Countries Expert Group, 2001; Annex A, OECD, 2002). It will need, among other things, engagement of each of the ministries responsible for development planning in the country. An adaptation strategy may contain several objectives (Box 8-3).

Based on the key priority systems identified earlier in the APF process, indicators will have been defined that can be used to assess the success, difficulties, failure of the adaptation strategy once implemented. Some of these indicators (vulnerability and socio-economic) are discussed respectively in TPs 3 and 6.

### 8.4.3.  Formulate options for adaptation policies and measures

Once the broad objectives of the adaptation strategy have been determined, it is possible to formulate policies and measures to achieve these objectives. If included at this stage of the process, several factors will facilitate integration of adaptation policies and measures later on.

- An important step in the process of formulating options is the integration of adaptation policies and measures between different sectors – and with existing policies and measures. This step builds on the synergies identified early on in strategy design by ensuring that the overlaps and intersections between adaptations and existing policies and measures are co-ordinated to the benefit of both. With integration, potential conflicts between adaptations in different sectors, and between proposed adaptations and existing policies and measures, can be avoided or limited. It is well known, for example, that developing an adaptation strategy in the agriculture sector without considering the water sector is not really feasible due to the relationship between the two. The same can be said for human health and water. Even purely structural adaptations (e.g., seawall construction, changes in

agricultural practices, establishment of early warning systems), therefore, will require integration. Of course, integrating adaptations leads to the issue of common (shared) costs and benefits, and the problem of how to attribute those costs and benefits to the different sectors/projects. An example of integration is given in Box 8-4.

- An evaluation of the relevant sectoral policies is therefore recommended. For example, a country's agricultural policy may have development objectives that are threatened by climate change. (TPs 1 and 6 discuss the adaptation baseline.) Such development objectives may include maintenance or strengthening of food security, the promotion of commercial crops for export or the production of crops that serve as industrial raw materials or substitute for imports. To achieve these objectives, adaptation policies may require additional effort for creation and/or improvement of all or some of the following: farmer advice and information services, agricultural research and development, seasonal climate forecasting, taxes and/or subsidies or incentives, irrigation, water conservation, investments in agricultural infrastructure including roads to markets, drought and flood control or alleviation measures, crop diversification, alternative rural off-farm employment and so on. Many different policy mixes are possible, and an actual policy usually tries to satisfy as many objectives as possible. Where the objectives are found to be incompatible or in conflict, an assessment of the trade-offs is necessary.

- Specific measures can be developed to support the chosen policy direction, e.g., an irrigation project, a farmer information, advice and warning programme, new scheme for crop insurance, a system of grain storage to be held as protection against drought or crop failures, financial incentives to grow a specific crop.*

- Adaptation options can be considered at different time scales (Parry and Carter, 1998), as some will have longer-term policy impacts than others. This factor may influence how urgently these policies and measures need to be

---

**Box 8-3: Five generic objectives of adaptation to climate variability and change**

1. Increasing robustness of infrastructure designs and long-term investments – e.g., by extending the range of temperature or precipitation a system can withstand without failure and changing the tolerance of loss or failure;
2. Increasing the flexibility of vulnerable managed systems – e.g., by allowing mid-term adjustments (including changes of activities or location) and/or reducing economic lifetimes (including increasing depreciation);
3. Enhancing the adaptability of vulnerable natural systems – e.g., by reducing other (non-climatic) stresses and removing barriers to migration (including establishing eco-corridors);
4. Reversing trends that increase vulnerability (also termed "maladaptation") – e.g., by introducing setback lines for development in vulnerable areas, such as floodplains and coastal zones;
5. Improving societal awareness and preparedness – e.g., by informing the public of the risks and possible consequences of climate change and setting up early-warning systems.

Source: Klein and Tol (1997)

---

* Many developing countries may need to consider low-technology options to overcome difficulties arising with the maintenance of "hard" adaptation measures. Experience with imported technologies indicates that it is not always wise to simply "paste" new technologies in the context of developing countries.

---

**Box 8-4: Considering sea-level rise in reconstruction of seawalls in Belize**

In 1998 Hurricane Mitch stalled offshore of Belize for several days. This powerful hurricane generated huge waves that slammed onto Belize's coast and generated unusually high tides. Over 90% of the country's piers were destroyed. There was tremendous erosion and mechanical damage to the reef and coastal infrastructure, including the seawall in Belize City. The following year the government embarked on a project to rebuild parts of the seawall. Belize was in the process of preparing its Initial National Communication, and the project co-ordinator wrote to the Ministry of Works, which had commissioned the construction of the seawall, explaining the projected increase in sea level expected in the coming decades. The project co-ordinator advised the Ministry to either consider making the seawall high enough to retain a higher sea level, or to build it in such a way that it could be raised in the future. The chief engineer agreed to factor in climate change and the seawall is higher than originally planned.

---

implemented over the whole planning horizon of the adaptation strategy. Examples are:

- long-term adaptations that are responding to mean changes in climate (river basin planning, institutional changes for water allocation, education and research);
- tactical adaptations concerned with mid-term considerations of climate variability (flood-proofing, water conservation measures);
- contingency adaptation related to short-term extremes associated with climate variability (emergency drought management, flood forecasting);
- analytical adaptations considering climate effects at all scales (data acquisition, water management modelling).

- To ensure that the adaptations identified are suitable to the challenge, it is important to engage stakeholders that can provide perspective on the feasibility of proposed options.

Box 8-5 gives examples of adaptations. Other categories of adaptation are organised in Smit et al. (2001) by functional outcome, type of policy instrument, and level of application. Sectoral measures are also available in the programmes of international organisations, such as the Food and Agricultural Organisation, government ministries and technical departments, research centres and non-governmental organisations, etc. Different adaptation measures have also been described in existing adaptation planning guidebooks and reports.[5]

Many adaptations will have been identified in the previous APF Components (especially Component 2; TPs 3, 4 and 6). Such adaptation may be currently in place to address climate variability (e.g., interannual variability of precipitation, ENSO) and extremes, such as droughts, floods and cyclones. Many of these practices are developed by local communities and especially by highly vulnerable people, existing on the margins of society. Such adaptations need to be considered when the strategy is developed, not only because these measures have been tested in the field, but also because they are more likely to be accepted by the communities. The team should develop not only lists of adaptations but also include

assessments of their experiences (what has worked and what has not?) in order to develop new and revised adaptation policies and measures (Figure 8-3). Other adaptations may be deduced from the analysis of future climate risks (APF Component 3 and TP5), and obtained from the literature, research centres, and clearinghouses (e.g., transferring available technologies at an international level). Experiences from adaptation policies and measures implemented in other countries could also serve to illuminate.

Figure 8-3 outlines the generic process involved in identifying and assessing adaptations. The first step involves the identification of existing and potential adaptations. The second step reviews these options in light of their actual or potential effectiveness in addressing current climate vulnerability and risk. The next step involves an assessment of the effectiveness of these options in light of potential climate futures. The fourth and final step involves prioritising certain adaptations over others, based on agreed criteria.

Once identified, adaptations have to be formulated in such a manner that their selection and prioritisation is possible using various methods. Since options will vary widely, it is only possible to outline typical information requirements, rather than give a prescribed format. Typical requirements are:

- ***Description of the measure***, indicating objective(s), location (e.g., international, regional, national, subnational, or local), timing of and responsibilities for implementation, and financing. This description would address the technical feasibility of measures, barriers to their implementation (e.g., cultural, social), the capacity to implement and sustain the measure, the cultural acceptability of the technology involved, etc.

- ***Estimated costs of the measure.*** The cost is a prerequisite for ranking a measure and including it in the (national, provincial, etc.) budget, or in a wider adaptation programme. Costs could be a one-time expenditure for capital investments, and recurrent costs (e.g., in the case of certain public health campaigns), includ-

---

[5] Such as the UNEP handbook for V&A studies (Feenstra et al., 1998), the US Country Studies Programme guidebook (Benioff et al., 1996), and the IPCC reports on impacts and adaptation (Watson et al., 1996; McCarthy et al., 2001).

**Box 8-5: Types of adaptation measures**

Adaptation measures may be grouped according to whether they are sectoral (e.g., introduction of improved agricultural varieties), multi-sectoral (e.g., use of improved watershed and coastal zone management methods), or cross-sectoral (e.g., promotion of public awareness, climate research, and data collection).

- *Sectoral measures* relate to specific adaptations for sectors that could be affected by climate change. In agriculture, for example, reduced rainfall and higher evaporation may call for the extension of irrigation. For infrastructure, sea level rise may necessitate improved coastal protection or relocation of population and economic activities. In most cases, measures will mean a strengthening of existing policies, emphasising the importance of basing climate change policies on existing coping mechanisms and the necessity of integrating them into national development plans.

- *Multi-sectoral measures* relate to the management of natural resources that span sectors – e.g., water management or river basin management. Integrated coastal zone management is also considered an appropriate framework to consider technical adaptation measures such as dike building, beach nourishment, etc. (Bernthal et al., 1990). The ecosystem approach to climate change adaptation involves the integrated management of land, water and other resources that promotes their conservation and sustainable use in an equitable way (Orlando and Klein, 2000).

- *Cross-sectoral measures* can span several sectors and include the following:

  *Education and training:* Introduction of climate change issues at different levels of the educational system is an ongoing process that can help to build capacity among stakeholders to support adaptation in the future, and can help to develop appropriate research activities and a greater awareness among citizens.

  *Public awareness campaigns:* Such campaigns can raise awareness and disseminate information in order to increase the concern and involvement of the broad array of stakeholders. These campaigns can also be an opportunity for adaptation decision makers to better understand the perception and views of the public on climate change and adaptation.

  *Strengthening/changes in the fiscal sector:* Public policies may encourage and support adaptation of individuals and the private sector, particularly through the establishment of fiscal incentives or subsidies.

  *Risk/disaster management measures:* These measures include the development of early warning systems, in particular for extreme events like cyclones (that can be predicted only a few hours before), and for droughts, floods, El Niño-Southern Oscillation (ENSO) (that can be predicted several months before). Emergency plans, extreme events relief and recovery measures also belong to this type of measure. Generally, the success of these measures depends upon good communication systems and a certain level of trust among users.

  *Science, research and development (R&D) and technological innovations:* R&D and innovation are needed to enable responses to climate change in general, and to enable specific responses to climate change vulnerability, including economic valuation of adaptations, technological adaptations (development of drought- or salt-resistant crop varieties), and investigations of new sources of groundwater and better resource management. It may also be necessary to adapt existing technologies to fit with the adaptation demands – e.g., the development of more energy-efficient air conditioning systems, low-cost desalination plants, and new technologies to combat saltwater intrusion.

  *Monitoring, observation and communication systems:* These systems may have to be created or strengthened, particularly for climate-related parameters, but also for other indicators of climate change and impacts (e.g., sea-level rise, changes in species composition of ecosystems, modification of piezometric levels, etc.). This monitoring will allow policy-makers to adjust the adaptation strategy based on confirmed changes in the climate (TP9).

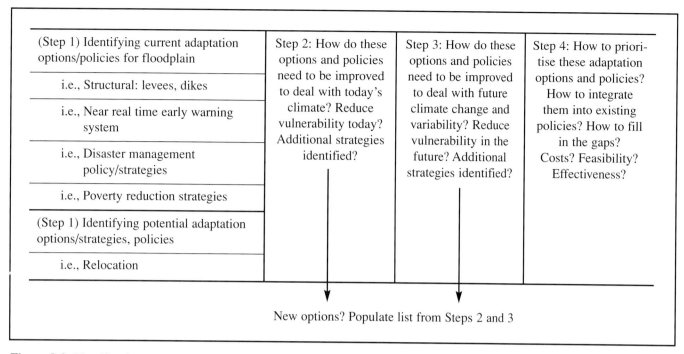

| (Step 1) Identifying current adaptation options/policies for floodplain | Step 2: How do these options and policies need to be improved to deal with today's climate? Reduce vulnerability today? Additional strategies identified? | Step 3: How do these options and policies need to be improved to deal with future climate change and variability? Reduce vulnerability in the future? Additional strategies identified? | Step 4: How to prioritise these adaptation options and policies? How to integrate them into existing policies? How to fill in the gaps? Costs? Feasibility? Effectiveness? |
|---|---|---|---|
| i.e., Structural: levees, dikes | | | |
| i.e., Near real time early warning system | | | |
| i.e., Disaster management policy/strategies | | | |
| i.e., Poverty reduction strategies | | | |
| (Step 1) Identifying potential adaptation options/strategies, policies | | | |
| i.e., Relocation | | | |

New options? Populate list from Steps 2 and 3

**Figure 8-3:** Identification, analysis and prioritisation of adaptation options

ing operational costs for project-type measures. Apart from direct costs, there are often indirect costs (e.g., in the form of an additional burden to the administrative system of the country) and external costs (linked e.g., to negative impacts in another sector). Costs should – to the extent possible – be expressed in monetary form. When this is not possible – as may be the case, for instance, in relation to changes in ecosystems – these factors have to be incorporated qualitatively. Methods have been developed to successfully quantify and value the use of resources for which there is no market price (Annex A.8.1); such methods can be used in the formulation process.

- **Estimated benefits of the measure.** The impacts of the measures on the environment and on society can be determined by comparing the "with" and "without" case.[6] These impacts need to be described in terms of their contribution to the objectives or criteria, again preferably in monetary terms. As is the case with costs, impacts may be system specific (e.g., human health, agriculture, environment, biodiversity, infrastructure, etc.), and be multi- or cross-sectoral. Costs and benefits are mirror images and often benefits result in reduction of the (social) costs. Examples are the reduction of typhoon damages resulting from installing an early warning system and a reduction of flood damages with increased heights of dikes. The evaluation of options should include equity consider-

ations, and an assessment of to whom benefits accrue is therefore needed.

The main output of this task is a portfolio of adaptation measures and policies. The next task will be to select and prioritise these options.[7]

### 8.4.4. Prioritise and select adaptation policies and measures

After adaptation policies and measures have been formulated, they can be prioritised with various methods and, subsequently, rejected, postponed, or selected for implementation. Given the range of climate change impacts and the measures to avoid or mitigate these impacts, it is unlikely that one single method can handle all possible cases. From a methodological point of view, the threats caused by climate change are not essentially different from what people have been experiencing in the past. Therefore, evaluation methods used in the selection and prioritisation exercise need not differ either. However, the increase in frequency and intensity of extreme events puts more emphasis on the treatment of uncertainty and risk. Sensitivity and risk analysis are therefore valuable elements in the decision-making process (Annex A.8.1).

Formal methods for prioritisation can most easily be applied to project-type (sectoral, and multi-sectoral) adaptation measures. In the case of cross-sectoral measures, such as institutional reform and legislation, it is often difficult or impossible to quantify the benefits or impacts of a measure. For these measures, it may then

---

[6] The method of comparing "with" and "without" is not confined to cases where costs and/or benefits can be quantified and/or expressed in monetary values. In qualitative reasoning the concept helps to avoid including impacts and costs resulting from autonomous development. An often observed fallacy is that of comparing the "before" and "after" circumstances.

[7] Many Initial National Communications to the UNFCCC included lists of adaptation options; however, these were not necessarily described and analysed (task 3) and prioritised (task 4) in a manner that could facilitate informed adaptation planning.

---

**Box 8-6: Four main methods for prioritising and selecting adaptation options**

The four major methods used for prioritising and selecting adaptation options – cost benefit analysis, multi-criteria analysis, cost effectiveness analysis, and expert judgement – vary in a number of ways. Some of these are outlined here:

- CBA can handle optimisation and prioritisation; it also provides an absolute measure of desirability, albeit judged by only one criterion, i.e., economic efficiency. CBA has comparatively heavy data requirements.
- MCA is suitable when more criteria are thought to be relevant, and when quantification and valuation in monetary terms is not possible. MCA is normally used for the ranking of options. But if the "do-nothing" case is included as an alternative, it can also help to clarify whether the measure is better than simply "bearing with the situation". Subjective judgement plays an important role in this method, making outcomes more arbitrary than that of CBA.[8]
- CEA is a method that falls somewhere between CBA and MCA. As is the case with MCA, CEA only produces a ranking.
- Expert judgement is a discipline in its own right and has its own place in the domain of policy making (Section 8.4).

Given that CBA is the more objective method and can handle optimisation, it may be the most desirable option. However, this depends on the purpose and stage of the analysis. In cases where important criteria cannot be accommodated in CBA (such as sociological or cultural barriers), or when benefits cannot be quantified and valued (such as the benefits of preserving biodiversity), MCA is preferred. If desired, the outcomes of CBA can be incorporated into MCA, making the overall analysis a hybrid one.

---

be necessary to employ informal, qualitative and subjective ways to determine their attractiveness.

Four main methods are likely to be particularly useful to the prioritisation process. These are:

- Cost Benefit Analysis (CBA)
- Cost Effectiveness Analysis (CEA)
- Multi-Criteria Analysis (MCA)
- Expert judgement

Box 8-6 presents the pros and cons of each method. *The Compendium of Decision Tools* lists a number of additional methods, including sector-specific tools (UNFCCC, 1999). The *Handbook on Methods for Climate Change Impact Assessment and Adaptation Strategies* (Feenstra et al., 1998) discusses the selection issue in great detail.

The selection of a method to evaluate policies and measures should be based on the real-life situation of the country, including available data and resources, and on the requirements of the prospective financier of the measures (government, outside financing). Formal methods such as CBA (discussed further in Annex A.8.1) can best be applied if outside financing is required or if planning authorities in the country so demand. It is important at this stage that planners are involved, either directly or in writing terms of reference for the studies. In many cases the policy process of a country will involve expert and political analysis and judgement. If a plan is to be entered into the budget of the government, an estimate of the cost will normally be required, with non-monetary elements such as institutional/organisational costs and cultural realities, and types of adaptive capacity needed for implementation quantified to the extent possible.

The flowchart in Figure 8-4 explains the reasoning presented in Box 8-6 and – because it attaches great importance to the accuracy of results – this chart applies especially in the *later stages* of the assessment of adaptation policies and measures, i.e., just before they are ready to enter into the adaptation strategy, the national development plan, or the national or sectoral budgets.

For evaluation and ranking of measures, it will be necessary to choose *criteria* to weigh the different concerns. These criteria can also act as indicators of the success or failure to realise the objectives, and can be used by a monitoring-evaluation programme for the adaptation strategies, policies and measures (TP9). The adaptation measures and policies should be evaluated within the same policy context as measures and policies introduced to alleviate poverty, or to foster economic development. The country's policy context should be taken into account when choosing criteria for the evaluation of adaptation measures. A sample set of criteria is offered in the NAPA guidelines (GEF, 2002), as outlined in Box 8-7. The NAPA Guidelines stress that the selection of criteria should be a country-driven process and that the list of criteria is not meant to be prescriptive. Other criteria that may be applicable are gender, sustainable development, equity, etc. Annex A.8.2 suggests a way to explore the impact of various measures on sustainable development; the team may find it helpful to organise data in that respect. In general, a useful way to organise data on adaptation policies and measures is to express the effects of measures in an "impact matrix" in which the measures to be compared are explored against the relevant criteria (see Table A-8-2-1 in Annex A.8.2 for an example).

---

[8] For more detail on CBA and MCA (or decision analysis in general), and risk analysis, refer to Annex A.8.1 and to relevant textbooks (see the references section).

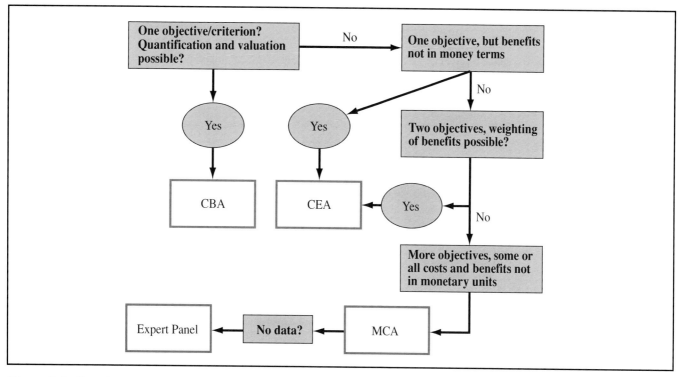

**Figure 8-4:** Choosing a tool for the prioritisation and selection of adaptation options

The process of prioritisation and selection of adaptations must involve a wide array of stakeholders. Multi-criteria analysis may be the best-known selection method involving stakeholders at the grass-root level. It is important to stress that during all steps of the process (choice of the method, choice of criteria, use of the method), stakeholders should be involved so that the process of implementation can be facilitated.

Given the uncertainties and the long time frame of climate change impacts (TP5; Willows and Connell, 2003), the so-called no-regret and low-regret adaptation options may be among the most attractive. The merits of an adaptation can also be compared for different climate scenarios, including the assumption that no (further) climate change will occur (TP6).

---

**Box 8-7: Sample criteria for selecting adaptation options**

As a NAPA may well precede a full-blown adaptation project, these criteria may be of use in both, and are therefore briefly discussed below:

- The **expected level of damage** is an indication of the benefits to be gained by preventing or mitigating this damage. Damage levels may need to be disaggregated into sector- or system-specific impacts, depending on the measure and the situation in the country with regard to adaptive capacity, the health situation, food security, and so on. If this implies that this criterion has to be broken up into sector- or system-specific criteria, the appropriate evaluation method will often become multi-criteria analysis (Annexes A.8.1 and A.8.3).
- **Poverty reduction** will enhance adaptive capacity and could be a goal in itself (and then a criterion) or a "by-product" of a measure (e.g., adaptation in agriculture). The two criteria can easily overlap and lead to double counting. PRSPs provide information on ("autonomous"[9]) development. Important synergies may be found between efforts under the PRSP and adaptation plans.
- **Synergies with multilateral environmental agreements** can have the form of cost savings or of additional benefit (e.g., the introduction of drought resistant crops which reduces desertification).
- **Cost effectiveness** (or just "costs") is a main criterion at the same level as "impacts/benefits".

Source: NAPA Guidelines (GEF, 2002)

---

[9] Autonomous in the vulnerability and adaptation analysis.

---

**Box 8-8: No-regrets and low-regrets options**

Given uncertainties and the long time frame of climate change impacts (TP5; Willows and Connell, 2003), two general types of adaptation options discussed here may often be most appropriate and most readily funded:

- *No-regrets:* These are options that are justified by current climate conditions, and are further justified when climate change is considered. For example, reducing water pollution could improve potable water supplies. The pollution reductions may be more valuable should climate change reduce water supplies or degrade water quality. The same can be said for introducing market reforms. However, an irrigation scheme for a drought-prone area may become more attractive when periods of drought, as a result of climate change, occur more often or become more severe.

- *Low-regrets:* Low regrets changes are those made because of climate change, but at a minimal cost. Thus, there is "low regret" if the investment proves not to be needed under future climate conditions. For example, incorporating risks of climate change in design of infrastructure may offer improved protection against current extreme climate events, as well as potential future events under climate change, while increasing costs only marginally (hence the "low" regret).

---

At the end of this task, the team would have designed a set of reasonable policy alternatives and have evaluated them in terms of criteria and objectives. Usually no one option will be superior in all respects. Option A may be more efficient, but less acceptable to some vested interests with political power; Option B may be less efficient, but less vulnerable to climate change in the medium- and longer-term. It is the task of the prioritisation and selection analysis to give a ranking based on explicit criteria and their weights. It is the up to the policy process to make the final choice. In other words, the policy analysis is subsidiary to the policy choice, which in the final analysis is made at the political level. A good policy process constrains the political choice to a limited set of viable alternatives.

### 8.4.5.  *Formulate the adaptation strategy*

Once the prioritisation process is completed, an adaptation strategy can be prepared with a combination of different measures and policies. The adaptation strategy consists of a plan containing the collection of measures selected for implementation, a time frame and other operational modalities for implementation[10]. This document should describe the scoping of the issues, identification of options, approaches taken to examine and evaluate the options, and transparency of the assessment process.

An implementation plan can be developed in which policies and measures are categorised according to:

1)  How they are to be incorporated into existing sectoral strategies, national development plans, poverty reduction strategies, etc. (e.g., management plans, education and research programmes, laws to be developed or enforced)

2)  Additional plans, policies, measures and/or projects that specifically address climate change that may be needed if gaps have been identified in the current policy framework. Some measures will likely require financing, either government or external, while others could be taken aboard within the regular national budget.

3)  A further distinction could be made between urgent policies, measures and projects, and those that are somewhat less urgent. Some of the measures may be implemented right away, while others may require detailed feasibility studies.

During this formulation and adoption process, it is important to include stakeholders at all levels (national to local) to gain public acceptance of the strategy (TP2). The resulting strategy document must be formally recommended for adoption, whether through a government decision, or through stakeholder consultation. TP9 suggests how adaptation strategies can be implemented, monitored and evaluated.

### 8.5.  **Conclusions**

A key objective of the APF is to facilitate the development and implementation of an adaptation strategy. However, formulating a plan that is only motivated by climate change may be unrealistic, not only because adaptation involves different sectors, regions and populations that are vulnerable to climate change, but because climate change is often far from being the first concern of most of decision-makers. Instead, decision-makers tend naturally to be more concerned with urgent goals such as poverty reduction and national development. It is thus of uppermost importance that, over the course of an adaptation project, efforts are made to build an understanding among key stakeholders that adaptation to climate change may become a necessary undertaking to achieve these same objectives.

---

[10] Sources of adaptation funding are expected to be provided under the UNFCCC through the GEF, such as the GEF Trust, Special Climate Change, and Least Developed Country Funds, and bilateral funds and national budgets.

Some adaptation measures may already be in place in some countries. This paper provides guidelines for developing an adaptation strategy by building on existing mechanisms, and should not be seen as prescriptive. The main output of the Component 4 is an adaptation strategy with an implementation plan for formal adoption. TP9 deals with both implementation and evaluation of adaptation. Since adaptation is a continuous process that needs to be informed regularly by evaluation of the adaptation strategy, implementation and monitoring are treated as integral parts of the strategy development process.

# References

**Abramovitz**, J., Banuri, T., Girot, T, P.O., Orlando, B., Schneider, N., Spanger-Siegfried, E., Switzer, J. and Hammill, A. (2001). *Adapting to Climate Change: Natural Resource Management and Vulnerability Reduction.* Gland, Switzerland, IUCN, World Watch Institute, IISD, Stockholm Environment Institute, Boston.

**Adger**, W.N. and Kelly, P.M. (1999). Social Vulnerability to Climate Change and the Architecture of Entitlements. *Mitigation and Adaptation Strategies for Global Change,* **4 (3-4)**, 253-266.

**Arrow**, K.J., Cline, W., Maler, K G., Munasinghe, M., Squitieri, R. and Stiglitz, J. (1996). *Intertemporal Equity, Discounting and Economic Efficiency.*

**Bell**, M., Hobbs, B., Elliott, E., Ellis, H. and Robinson, Z. (2001). An Evaluation of Multi-Criteria Methods in Integrated Assessment of Climate Policy. *Journal of Multi-Criteria Decision Analysis,* **10**, 229-256.

**Belli**, P., Anderson, J.R., Barnum, H.N., Dixon, J.A. and Tan, J-P. (2001). *Economic Analysis of Investment Operations: Analytical Tools and Practical Applications.* World Bank, Washington, pp. 264.

**Benioff**, R., Guiff, S. and Lee, J. (1996). *Vulnerability and Adaptation Assessments. An International Handbook,* eds., Dordrecht: Kluwer Academic Publishers.

**Bernthal**, F., Downdeswell, E., Luo, J., Attard, D., Vellinga, P., Karimanzira, R. et al. (1990). Eds. *Climate Change. The IPCC Response Strategies.* WMO/UNEP, pp. 270.

**Burton**, I. (2000). Adaptation to Climate Change and Variability in the Context of Sustainable Development. In: Gomez-Echeverri, L. ed. *Climate Change and Development,* Yale School of Forestry and Environmental Studies, pp. 153-173.

**Burton**, I., Kates, R.W. and White, G.F. (1993). *The Environment as Hazard.* New York: Guilford Press.

**Fankhauser**, S. (1998). *The Costs of Adapting to Climate Change.* GEF Working Paper 16.

**Feenstra**, J.F., Burton, I., Smith, J.B. and Tol, R.S.J. (1998). *Handbook on Methods for Climate Change Impact Assessment and Adaptation Strategies.* UNEP/IVM, Nairobi/Amsterdam.

**GEF** (2002). *Operational Guidelines for Expedited Funding for the Preparation of National Adaptation Programs of Action.* GEF, Washington.

**Jepma**, C.J., Asaduzzaman, M., Mintzer, I., Maya, R.S., Al-Moneff, M., Byrne, J., Geller, H., Hendriks, C.A., Jefferson, M., Leach, G., Qureshi, A., Sashin, W., Sedjo, R.A. and Van Der Veen, A. (1996). A generic assessment of response options. In: Bruce, J.P., Lee, H. and Haites, E.F. eds., *Climate Change 1995. Economic and Social Dimensions of Climate Change,* Cambridge: Cambridge University Press, pp. 225-262.

**Kelly**, P. and Adger, W.N. (1999). *Assessing Vulnerability to Climate Change and Facilitating Adaptation.* Working Paper GEC 99-07, Centre for Social and Economic Research on the Global Environment, University of East Anglia, Norwich.

**Klein**, R.J.T., Tol, R.S.J. (1997). Adaptation to Climate Change: Options and Technologies. An Overview Paper. Institute for Environmental Studies, Amsterdam, pp. 33.

**Klein**, R.J.T., Nicholls, R.J. and Mimura, N. (1999). Coastal Adaptation to Climate Change: Can the IPCC Technical Guidelines Be Applied? In: *Mitigation and Adaptation Strategies for Global Change,* Kluwer Academic Publishers, 4 (3-4), 239-252.

**Kuyvenhoven**, A. and Mennes, L.B.M. (1985). *Guidelines for Project Appraisal.* Erasmus University. The Hague: Government Publishing Office.

**Least Developed Countries Expert Group** (2001). *Guidelines for the Establishment of National Adaptation Programmes of Actions.* UNFCC-CC, Bonn.

**McCarthy**, J.J., Canziani, O.F., Leary, N.A., Dokken, D.J. and White, K.S. (2001). *Climate Change 2001: Impacts, Adaptation, and Vulnerability.* Cambridge: Cambridge University Press.

**Munasinghe**, M. (2002). *Framework for Analyzing the Nexus of Sustainable Development and Climate Change Using the Sustainomics Approach,* Draft, Munasinghe Institute for Development (MIND).

**Munasinghe**, M. and Swart, R. eds. (2000). *Climate Change and its Linkages with Development, Equity, and Sustainability.* Proceedings of the IPCC Expert Meeting held in Colombo, Sri Lanka. LIFE/RIVM/World Bank.

**Mustafa**, D. (1998). Structural Causes of Vulnerability to Flood Hazard in Pakistan. *Economic Geography,* **74(3)**, 289-305.

**OECD** (2002). *The DAC Guidelines. Integrating the Rio Conventions into Development Co-operation.* OECD, Paris. pp. 104.

**Orlando**, B.M. and Klein, R.J.T. (2000). *Taking an Ecosystem Approach to Climate Change Adaptation in Small Island States.* 2nd Alliance of Small Island States Workshop on Climate Change Negotiations, Strategy and Management, Samoa.

**Parry**, M. and Carter, T. (1998). *Climate Impact and Adaptation Assessment. A Guide to the IPCC Approach,* London: Earth Scan.

**Ramakrishnan**, P.S. (1998). Sustainable Development, Climate Change ad the Tropical Rain Forest Landscape. *Climatic Change,* **39**, 583-600.

**Smit**, B., Pilifosova, O., Burton, I., Challenger, B., Huq, S., Klein, R.J.T., Yohe, G. et al. (2001). Adaptation to Climate Change in the Context of Sustainable Development and Equity. In: McCarthy, J.J. et al. eds. *Climate Change 2001: Impacts, Adaptation and Vulnerability,* Cambridge: Cambridge University Press, 875-912.

**UNFCCC** (1999). Compendium of Decision Tools to Evaluate Strategies for Adaptation to Climate Change. UNFCCC, Bonn.

**Van Pelt**, M.J.F. 1992. Sustainability-oriented Project Appraisal for Developing Countries. PhD Thesis, Rotterdam.

**Wang'ati**, F.J. (1996). The Impact of Climate Variation and Sustainable Development in the Sudano-Sahelian Region. In: Ribot, J.C., Magalhaes, A.R. and Panagides, S.S. eds. *Climate Variability, Climate Change and Social Vulnerability in the Semi-Arid Tropics,* Cambridge: Cambridge University Press, 71-91.

**Watson**, R.T., Zinyowera, M.C. and Moss, R.H. eds. (1996). *Climate Change 1995. Impacts, Adaptations and Mitigation of Climate Change: Scientific-Technical Analyses.* Cambridge: Cambridge University Press, 879.

**Willows**, R.I. and Connell, R.K. eds. (2003). *Climate Adaptation: Risks, Uncertainty and Decision-Making.* UKCIP Technical Report, Oxford.

**Winpenny**, J.T. (1992). *Values for the Environment. A Guide to Economic Appraisal.* London: Overseas Development Institute.

# ANNEXES

## Annex A.8.1. Methods for the prioritisation and selection of adaptation policies and measures

This annex discusses methods for selecting and prioritising adaptation policies and measures, as well as using experts:

- Cost Benefit Analysis (CBA)
- Cost Effectiveness Analysis (CEA)
- Multi-criteria Analysis (MCA)
- Expert judgement

Numerous textbooks exist on cost-benefit analysis, including cost-effectiveness analysis. Sensitivity and risk analysis is of special importance when dealing with the many uncertainties of climate change and extreme events. Reference is made to the literature for full treatment of the methods.[11]

### A.8.1.1. Cost-benefit analysis

CBA involves comparing costs and benefits of a measure with a view to deciding whether it is attractive to undertake an activity (a project or a project-type adaptation measure). It is normally applied at the country level and for estimating the contribution of the measure to the national economy or society. However, the method can also be applied at the international or provincial level, as well as for private enterprise. Application requires making estimates of costs and benefits, involving three steps: (1) identify which costs and benefits are relevant, (2) quantify them, and (3) give them monetary value. Although benefits are not always quantifiable and/or cannot always be expressed in monetary values, the costing of measures is possible as long as priced resources are used. The non-monetary use of scarce resources, such as utilising existing capacity in government, should also be estimated and be taken into account.

The method is data intensive and needs specialists. However, before a proposed measure can be entered into any plan, its financial costs have to be known. Data on costs and all kinds of parameters needed in economic analysis are normally available at planning bureaus, ministries of planning, line ministries, etc.

Typically, professional economists perform CBA using spreadsheet software. The technical nature of the method precludes doing more here than indicating some specific issues of impor-

tance when performing CBA. IPCC (Jepma et al., 1996) and others (e.g., Belli et al., 2001) provide guidance on this method. Techniques for introducing equity considerations into CBA are provided in Kuyvenhoven and Mennes (1985) and Van Pelt (1992). The valuation of environmental benefits is discussed in Winpenny (1992), among others.

CBA allows analysts to optimise both the extent and the timing of a measure. When the growth of benefits decreases with increased intensity or extent of area of a measure, there comes a situation when marginal or incremental costs (MC) exceed incremental or marginal returns (MR). At the optimum the well known condition MC=MR applies, where the NPV (net present value) is at its maximum.[12] Postponement of a measure may result in a higher NPV.

*Costing,* or more generally, the valuation of costs and benefits is an important Component of CBA, CEA and often also in MCA. It is treated in a separate section below.

### A.8.1.2. Cost-effectiveness analysis

If benefits cannot be measured in a reliable manner, as is the case often with environmental goods and services, for instance, CEA is the appropriate method. It principally involves the costing of different options, which achieve *the same objective,* and compares those in order to find out how a well-defined objective can be reached in a least-cost way. If there are multiple objectives, CEA can only be applied if one objective can, quantitatively, be expressed in the other by assigning importance (weight) to the objectives to arrive at a single yardstick. This is called "weighted CEA".

### A.8.1.3. Multi-criteria analysis

MCA has become increasingly popular, not least in relation to environmental issues, including climate change (Arrow et al., 1996; Belli et al., 2001). Methods and software have proliferated. Some authors have attempted to compare different methods.[13] These reviews have provided useful insights. Provisional conclusions on MCA are:

- Method uncertainty: Different methods produce different results and it therefore appears preferable to apply several MCA methods (Belli et al., abstract, p.229);

---

[11] Software programmes that may be useful for the prioritisation and selection process include the following: Multicriteria Analysis: Manual of DTLR 2001 http://www.dtlr.gov.uk/about/multicriteria; HIVIEW for Windows http://www.enterprise-lsa.co.uk; DEFINITE Institute for Environmental Studies, Vrije Universiteit, Amsterdam, The Netherlands, http://www.vu.nl/ivm@RISK http://www.palisade.com/html/risk

[12] Samuel Fankhauser (1998) expresses the optimisation slightly differently. He minimises the sum of adaptation costs and residual damages. "Residual damages", when avoided by a measure, become "benefits" in CBA language. Note that, also in that reasoning, it is normally not economically efficient to avoid *all* (residual) damages.

[13] Belli et al., (2001) distinguish three groups of MCDM methods: weighting methods, deterministic ranking methods and uncertainty ranking methods.

- Ease of manipulation coupled with subjectivity and lack of transparency has contributed to lack of confidence in MCA methods. Some recommend simpler methods, preferably without use of computer software;
- MCA is very useful for structuring problems and decisions, not necessarily for solving problems (holistic assessments are preferred for ultimate decisions).

The ingredients of MCA are objectives, alternative measures/interventions, criteria (or attributes), scores that measure or value the performance of an option against the criteria, and weights (applied to criteria). Defining objectives and formulating different options is not different from CBA or CEA. The difference lies in the selection of criteria and their weights. As indicated above, these are judgmental elements. For the determination of weights, procedures exist that more or less guarantee that the set of weights is consistent. For example, the computer MCA model "DEFINITE" contains a separate routine (pair-wise comparison) to arrive at a consistent set of criteria. That model further allows using different ways of MCA, from simple to quite sophisticated, and includes a routine for CBA (Annex A.8.3).

A major task is determining the scores (or effects), i.e., assessing the impact of alternative measures on the different criteria. Assessing causal relationships between measures and effects is a matter of research. This may in practice be the most important task, since there is no method that can make up for unreliability of input data.

The selection of a set of criteria is subject to a number of pitfalls. Probably the most serious danger is overlap (double counting) or interdependency. Another danger is that only those criteria are selected to which effects can easily be attributed. Health and biodiversity are criteria that may fall victim to the difficulties of estimating and attributing effects. Too many criteria may be taken into account, leading to a "splitting bias". According to Van Pelt (1992), MCA is most reliable if the number of alternative options lies between three and eight, the number of criteria does not exceed seven, the impact can be quantified, and if different MCA techniques give comparable outcomes.

On the positive side are:

- Apart from forcing the user to frame the problem (see above), MCA provides a checklist of data required and the sensitivity of inputs and the result can easily be analysed. In a way, MCA *guides* the data collection process;
- MCA is particularly suitable for use in a "participative setting" (especially in determining the relevance and the weights of criteria) and so allows stakeholder participation in a systematic way through the various stages of the APF, such as scoping of adaptations, problem definition, determining relevance of input data and feeding back results to stakeholders.

In Annex A.8.3, a hypothetical example is given to further explain the selection/prioritisation procedure using MCA.

### A.8.1.4. Costing

Costing is of importance for all methods, and for each, the same principles apply. Basically, three steps are involved:

- Identification: determining which costs and benefits are relevant.
- Quantification: measuring inputs and effects in terms of, say, labour days, tons of produce, number of casualties.
- Valuation: pricing the in- and outputs.

There is, in this respect, no difference between costs and benefits, defined respectively as the decrease and increase of scarce resources. Often benefits are a decrease of costs that would be incurred in the absence of the project.

*Financial costs*

An estimate of financial costs is the starting point for costing economic or social costs. Financial costs are the outlays for the project to be made by the agency implementing the project. Economic or social costs are losses of scarce resources from the point of view of the whole society. The two notions rarely coincide and corrections on financial cost will have to be made.

Prices of production factors (labour, capital, expertise) and of goods/services are often distorted. Main sources of distortion are indirect taxes/subsidies and other deliberate government policies and mal-functioning of markets. Corrections are usually necessary for:

- taxation/subsidisation;
- wages;
- discount rate (interest, the price of capital);
- foreign currency (exchange rate).

A cost-benefit exercise presumes that, with regard to benefits, the changes in vulnerability can be measured. Issues of estimation and uncertainty make this difficult. On both fronts, however, progress is made (references, especially Belli et al., and Winpenny). Here the focus is on costing as this is a complex exercise in itself.

*Social costs*

While financial costs are fairly straightforward, social costs, as mentioned earlier, need to account for market distortions, transfer payments, and external effects.

**Market distortions**: A resource or activity may be zero-priced in its present use. For instance, land leased by the government for agriculture is now being used by a project to plant forests in watersheds to prevent soil erosion, and mangroves to act as storm breakers. Or, the land may be priced at the lease value. That value, however, may be too low relative to the value of the crops produced. Revaluation is often needed to reflect the real value of the resource or related activity to the community. A guiding concept here is that of "opportunity costs". When applied to labour in the project, the reasoning is that the true cost of labour is the added value forgone (e.g., in terms of rice produced). Applying the opportunity costs reasoning to material inputs leads to the use of "border prices" or long-run world market prices. The "shadow" exchange rate is the rate that would prevail in the absence of undue protection (more than of trading partners) and when the rate is left to fluctuate freely. The estimation is specialised work. There is a debate on the discount rate to use for investments with a very long time horizon and/or dealing with irreversible effects. Some argue for a lower rate in those cases in order to foster acceptance of such projects, often environmental projects (refer also to Adger and Kelly, 1999).

**Transfer payments**: Taxes and subsidies are costs/income for the project, but for the community as a whole these are mere money transfers and should therefore be eliminated in economic costing. Note, however, that a user charge (such as a levy for irrigation water or a road user charge) represent a use of a resource (preferably, at least from the point of economic efficiency) equal to the actual cost made for the provision of the water or the road system.

**External effects**: The use of the resource may entail additional costs or benefits outside the measure or project being considered. These may or may not be quantifiable. For instance, afforestation may generate additional recreational or biodiversity benefits. Similarly, mangrove plantation may improve spawning conditions for aquatic species, improve biodiversity and provide timber, fuelwood and fodder. These are external benefits. $CO_2$ produced during construction of (protective) infrastructure would be an external cost[14]. Also additional monetary and non-monetary benefits may be associated with an adaptation project/activity. For instance, in coastal zones, water conservation measures could increase soil productivity by raising the water table and reducing salinity. Negative costs (external benefits) occur when such secondary benefits/co-benefits/joint benefits more than offset the additional investment in adaptation. Win-win or no-regrets is another term applied to such projects/activities. When faced with competing priorities and scarce financial resources, identifying such projects becomes imperative. Conserving natural resources is a classic case, offering a range of social and environmental co-benefits, such as biodiversity conservation, enhanced sink capacity, poverty alleviation and reduced demand for international assistance (Abramovitz et al., 2001).

*Incremental cost*

Incremental costs refer to marginal costs increasing in concrete steps. In the climate change community, the term incremental cost as used, *inter alia*, by the Global Environment Facility (GEF) is defined as the additional cost a country incurs when undertaking a climate mitigation project, compared with the social cost of the activity the project substitutes and that has no provisions for mitigation of green-house gases.

While the IPCC TAR applies the incremental cost criteria across mitigation and adaptation projects, its relevance for adaptation projects is less obvious. Such projects have no global rationale, except when they intersect with mitigation projects. On the other hand, a case for applying such criteria to funding adaptation projects can be made. For instance, a dam wall is raised to address the threat of increased flooding but also stores more water for later irrigation releases. Applying these criteria means the latter benefit would need to be netted out of the project cost. This is an important consideration, especially in view of the limited available funding for adaptation.

*Time discounting*

The present value of a future cost stream is defined as:

Present cost = (future costs)/(1+discount rate)$^t$

Where the exponent "t" refers to the time stream of costs. The discount rate in the economic analysis can range from a low "ethical" rate, based on social considerations, to a rate which reflects the opportunity cost of capital. Discount rates vary between developed and developing countries and, basically, the scarcity of capital should determine the choice of the rate applied.

As the discount rate is very important for the outcome and often difficult to estimate, it is often made subject to sensitivity analyses to determine how sensitive the results are to the choice of the discount rate. Also, when estimating the time stream of cost it is necessary to spell out clearly the assumptions underlying the forecasts and how these assumptions are used to generate the forecasts, including linkages and feedback effects.

*Implementation costs*

Estimating direct project costs is not sufficient. It is also important to assess the institutional, economic and technical barriers to implementing the project because additional costs (financial or in kind) are involved in removing them. The required changes may be institutional (e.g., improving adaptive research capacity) or economic (e.g., establishing markets and incentives for new products). The cost of these changes ought to be added to the project.

---

[14] $CO_2$ emission is an external cost at global level (leads to climate change), and may – in a national-economic analysis – be negligible, especially for small developing countries.

Arguably, barriers to implementation may be less of a concern in the case of adaptation projects, as such projects tend to be mainstreamed into the policy system – for instance, early warning and mitigation of damages in relation to floods/droughts/cyclones, afforestation, water conservation and health interventions. Admittedly, there are different degrees of mainstreaming.

### Combining methods

As discussed above, a single method will not normally suffice. Often a combination of methods may be called for and also results of different methods may be checked against each other. The MCA model DEFINITE provides four MCA methods and CBA/CEA.

### A.8.1.5. Using an expert panel

It may also be that data unavailability or the complexity of the problem suggest the use of expert judgment. Employing a panel of experts may aim at taking a decision or at producing suitable information for decision making. A structured way to engage experts is the DELPHI method. It involves sending questionnaires to experts (rather than getting them in one meeting), collating the answers, and feeding those back to the experts and/or send a second questionnaire. An important ingredient of DELPHI is that experts give their opinions independently and anonymously. The results are given in the form of a statistical analysis of answers. DELPHI is mostly applied to forecasting (*e.g.,* of technological development), but may be of some use here, for instance for formulating adaptation options.

### A.8.1.6. Handling uncertainty and risk

Climate change is a process that is characterised by a number of uncertainties relative in particular to the magnitude, timing and nature of the changes. Decision makers are more familiar with processes/problems that are not subject to this degree of uncertainty. To take into account this situation, various methods can be used.

A common method in project appraisal is sensitivity analysis. Main inputs in the analysis (such as certain cost and/or benefits, the discount rate, etc.) are varied to see how sensitive the outcome is to these changes. A practical approach is to determine "switching values", i.e., those values of major inputs, either alone or in combination with others, that render an activity uneconomic. A similar procedure can be used in (computerised) MCA analysis.

Risk analysis uses Monte Carlo simulation on key inputs in the analysis. The analyst has to determine the probability distribution (normal, skewed, etc.) of the occurrence (say of an increase in cost, or number or intensity of extreme events) and the possi-

ble co-variance between these inputs.[15] The computer model, using a random number generator, makes a large number of runs to determine the (average) outcome. If a probability (or combinations) is an input into the analysis, the output, naturally, is also a probability distribution (e.g., of the NPV or the rate of return). There are commercial computer packages that perform the work, (Burton, 2000). Also in MCA models, routine risk analysis can be built in (Burton et al., 1993).

Another possibility to deal with uncertainty is scenario development. This method corresponds to the analysis proposed in TP6 and to climate change scenarios developed by IPCC. For a systematic analysis the likelihood of a situation occurring and its probability distribution could be used as input in risk analysis.

### Annex A.8.2. Linking climate change and sustainable development policies

Munasinghe (2002) developed a method to link climate change and sustainable development.

### Integrated assessments – The Action Impact Matrix (AIM)

There is a two-way linkage between climate change and sustainable development. Future development paths (hence the need for charting socio-economic scenarios) will determine not only projected greenhouse gas (GHG) emissions and the severity of climate change, but also the adaptive and mitigative capacity available to mount an effective response strategy. Conversely, climate change will have significant impacts on the three main elements of sustainable development (economic, social and environmental).

This dynamic interaction should become a consideration in development cooperation. Integrated sustainable development and climate change policies should account for the powerful economy-wide reforms in common use – including both sectoral and macroeconomic adjustment policies, which have widespread effects throughout the economy.

The highest priority needs to be given to policies that promote all three elements of sustainable development (economic, social, environmental). This is especially in recognition of the fact that there are sustainable development issues, which affect human welfare more immediately – such as hunger, malnutrition, poverty, health, and pressing local environmental issues. With other policies, trade-offs among different objectives need to be analysed. Economy-wide policies that successfully induce growth, could also lead to environmental and social harm, unless the macro-reforms are complemented by additional environmental and social measures.

The AIM provides a way to link integrated sustainable development and climate change policies explicitly. It can help find "win-win" policies, which not only achieve conventional

---

[15] See Belli *et al.* for a concise treatment of risk analysis.

*Table A-8-2-1:* A simple version of the Action Impact Matrix

| Activity/Policy (PRSPs, NSSDs) | Economic objectives | Impacts on key sustainable development issues | | | | | |
|---|---|---|---|---|---|---|---|
| | | *Land degradation. Biodiversity loss* | *Water scarcity and pollution. Adverse health impacts* | *Air emissions. Adverse health impacts* | *Other social effects* | *Institutional impacts* | *Vulnerability (socio-economic and biophysical)* |
| *Macroeconomic and sector policies* | Macroeconomic and sector improvements | Positive impacts Negative impacts Indeterminate impacts (I) | | | | | |
| *Exchange rate* | Improve trade balance and economic growth | Deforest open access areas | Water pollution (I) Adverse health impacts in low income areas | Air pollution (I) Adverse health impacts in low income areas | Forced migration to other areas | | Increase vulnerability |
| *Water pricing and management* | More efficient water use and economic efficiency | Reduced waterlogging/ salinity | Water use efficiency | | Improved access to water for poor farmers | Integrated water and drainage management | Reduce vulnerability |
| *Energy pricing and management* | Increase energy use efficiency | Reduced biomass use | | Reduced air pollution. Lower health risks | | | Reduce vulnerability |
| **Comple- mentary measures** | Socio-econom- ic and market gains | Enhance positive impacts Mitigate negative impacts | | | | | |
| *Market based* | | | Pollution charge | Emission charge | | | Reduce vulnerability |
| *Non-market based* | | Institute property rights | Voluntary compliance | Voluntary compliance | | Amending environmental laws/regulations | Reduce vulnerability |
| **Adaptation projects** | Reduce vulnerability | Investment decisions guided by broader policy and institutional framework Positive impacts Negative impacts Indeterminate impacts (I) | | | | | |
| *Re-forestation/ aforestation* | | Increase sinks capacity. Reduce soil erosion and downstream sedimentation | Reduce flooding | | Provide fuel, timber and fodder to poor communities. Reduce damage caused by flooding | | Reduce vulnerability |
| *Raising dam walls* | | Seepage, inundate forests | Flooding effects (I) | Potential hydropower. Reduced air emissions | Socio-econom- ic effects of flooding (eco- nomic losses, displacement, mortality) (I). | | Vulnerability effects (I) |
| *Drought miti- gation (relief, early warning, infrastructure and services)* | | Reduce pressure on land (crop and grazing) | | | Improved socio-econom- ic conditions | | Reduce vulnerability |

Note:     PRSP = Poverty reduction strategy paper
             NSSD = National strategy for sustainable development

Source: Munasinghe and Swart (2000)

macroeconomic objectives (like growth), but also make local and national development efforts more sustainable, and address climate change issues. AIM demonstrates in practical and qualitative terms that economic growth, social justice and environmental sustainability can co-exist.

The rows list the main development interventions (both policies and projects), while the columns indicate key sustainable development issues and impacts (including climate change vulnerability). Thus the elements or cells in the matrix help to:

- Identify explicitly the key linkages.
- Focus attention on methods of analysing the most important impacts.
- Suggest action priorities and remedies.

At the same time, the organisation of the overall matrix facilitates the tracing of impacts, as well as the coherent articulation of the links among a range of development actions – both policies and projects.

A simple version of the AIM is presented in Table A-8-2-1.

### Annex A.8.3. A hypothetical example of the use of multi-criteria analysis

This illustration of MCA uses a hypothetical example a situation, which may typically exist in the wet season in Bangladesh. The steps normally taken in a MCA analysis are as follows:

1) **Problem definition:** Because of rising sea level, higher intensity of precipitation and increased run-off in upstream areas, rain and smelt water reaches Bangladesh in a shorter period than before and also drains less easily. Floods, as Bangladesh has been experiencing for a long time, are thus getting worse.

2) The **objective** of the intervention is to get rid of superfluous water in order to safeguard agricultural production, to avoid the spread of waterborne diseases, and to avoid damages to buildings, nature, infrastructure, etc. (called "environment" in the example).

3) The **criteria** used to measure effects are: (a) agricultural production, (b) health, (c) expected damage of the environment, and (d) the cost of the intervention.

4) The following are considered as **alternative interventions**: (a) installing pumps at strategic sites, (b) improving the existing drainage infrastructure, (c) organising manual labour at a big scale (not unusual in Bangladesh). An alternative option is always to do nothing (bear the losses).

5) Estimating **effects** in a reliable manner naturally is of paramount importance. This is the area where risk analysis is especially valuable. It is here assumed that there is insufficient data at this stage to perform either CBA or CEA. For the study area, however, there are rough estimates of the extent and duration of floods

*Table A-8-3-1: Scores on Criteria*

|  | Cost (million $) | Effect (million HA days) | Health (million DALYs) | Environment cost (million $) |
|---|---|---|---|---|
| Pump | -700 | 1000 | 10 | -70 |
| Infrastructure | -800 | 800 | 8 | -10 |
| Labour | -900 | 300 | 3 | -10 |
| Bear losses | 0 | 0 | 0 | -50 |

*Table A-8-3-2: Scores standardised (0-1 scale), weighted summation and ranking*

|  | Cost | Effect | Health | Environment | Weighted summation | Ranking |
|---|---|---|---|---|---|---|
| Pump | 0.22 | 1.00 | 1.00 | 0.00 | 0.56 | 2 |
| Infrastructure | 0.11 | 0.80 | 0.80 | 1.00 | 0.68 | 1 |
| Labour | 0.00 | 0.30 | 0.30 | 1.00 | 0.40 | 3 |
| Bear losses | 1.00 | 0.00 | 0.00 | 0.33 | 0.33 | 4 |
| Weight | 0.25 | 0.25 | 0.25 | 0.25 | 1.00 |  |

that could be avoided, of the daily adjusted life years (DALYs) (Belli et al., 2001) that could so be gained, of the damage done to the environment (in money terms) and of the costs of the different interventions (also in money terms).

6)  The last step is to give **weights** to the different criteria.

All steps lend themselves to stakeholder participation, especially Steps 3, 4 and 6. Under Step 4, traditional coping mechanisms would be brought in, and under Step 6, the preferences of people affected by the floods and the measures to avoid them. Table A-8-3-1 gives the basic data and Table A-8-3-2 the results in terms of ranking after the effects (expressed in various units) have been standardised by scaling them (0 – 100 or 0-1 scale) and weights have been assigned to the criteria.

The data was input in a spreadsheet (Table A-8-3-1) and the calculations (standardisation and summation of contribution) are made in Table A-8-3-2. When the calculations are formulated in the spreadsheet, performing sensitivity is easy.[16] It would be logical to investigate the sensitivity of the effects (scores in Table A-8-3-1) and of the weighting (Table A-8-3-2) on the outcome (ranking). If pumping would be less expensive (for instance, only 400) and infrastructure more expensive (950) the two alternatives would get the same ranking. If cost would be given a weight of 0.45 and environment of 0.05, then pumping becomes the better alternative. Also risk analysis could be done on the scores (Burton, 2000), but first every

effort should be done to improve the estimates of effects. MCA can be done also using computerised models. Both HIVIEW and DEFINITE (Annex A.8.4) support weighted summation, as done above. Doing sensitivity using a computerised MCA model is relatively easy.

Both models are easy to apply. HIVIEW does sensitivity, allows relative and absolute scaling, and accepts inputs of scores in various forms (numbers, but also "yes" and "no"). HIVIEW can also present the structure of weighting graphically. DEFINITE is a full-fledged decision support programme. It includes four different MCA methods, CBA and graphical evaluation methods. It allows all formats for inputs, including +, ++, -, —. DEFINITE leads the analyst through rounds of interactive assessments of options, weights, scores, etc.; there is a routine to check internal consistency of the weight set using pair-wise comparison. Reporting is in text and numbers, but graphs can also be produced, as shown here in Figure A-8-3-1. Some training in the models is highly recommended before using them to propose important decisions.

The danger of the programme (as with all MCA) is that the emphasis comes on the method, rather than on the (hard) work to develop estimates of costs and benefits of options.

### Annex A.8.4.  Useful internet addresses

UNFCCC Secretariat: www.unfccc.int
UNITAR: www.geic.or.jp/cctrain
IPCC: www.ipcc.ch/pub/tar/wg2/069.htm
World Bank: www.worldbank.org
Stratus Consulting: www.stratusconsulting.com
Stockholm Environment Institute (SEI): www.tellus.com
Assessments of Impacts and Adaptations to Climate Change (AIACC): www.start.org/Projects/AIACC_Project/aiacc.html
United Kingdom Climate Impacts Programme (UKCIP): www.ukcip.org.uk

### Models

HIVIEW for Windows http://www.enterprise-lsa.co.uk; DEFINITE Institute for Environmental Studies, Vrije Universiteit, Amsterdam, The Netherlands, http://www.vu.nl/ivm@RISK http://www.palisade.com/html/risk

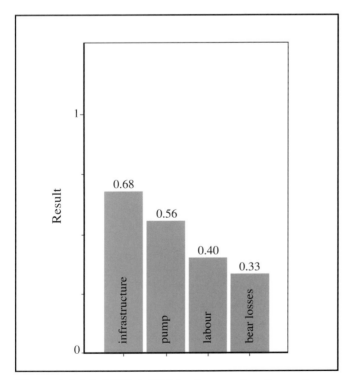

**Figure A-8-3-1:** MCA 1: Weighted summation {interval; direct (cost: 0.5)}

---

[16] Note that the scaling of scores is always done on the highest/lowest score in a column and that this order can change in the sensitivity analysis.

# 9

# Continuing the Adaptation Process

ROSA T. PEREZ[1] AND GARY YOHE[2]

Contributing Authors
*Bo Lim[3], Erika Spanger-Siegfried[4], David Howlett[3], and Kamal Kishore[3]*

Reviewers
*Mozaharul Alam[5], Suruchi Bhadwal[6], Henk Bosch[7], Nick Brooks[8], Moussa Cissé[9], Qin Dahe[10], Mohamed El Raey[11], Ulka Kelkar[6], Martin Krause[3], Maynard Lugenja[12], Hubert E. Meena[12], Mohan Munasinghe[13], Atiq Rahman[5], Roland P.A. Rodts[14], Samir Safi[15], Juan-Pedro Searle[16], Barry Smit[17], Juha Uitto[3], and Tom Wilbanks[18]*

[1] Philippines Atmospheric, Geophysical and Astronomical Services Administration, Manila, Philippines
[2] Wesleyan University, Middletown, United States
[3] United Nations Development Programme – Global Environment Facility, New York, United States
[4] Stockholm Environment Institute, Boston, United States
[5] Bangladesh Centre for Advanced Studies, Dhaka, Bangladesh
[6] The Energy and Resources Institute, New Delhi, India
[7] Government Support Group for Energy and Environment, The Hague, The Netherlands
[8] Tyndall Centre for Climate Change Research, School of Environmental Sciences, University of East Anglia, Norwich, United Kingdom
[9] Enda Tiers Monde, Dakar, Senegal
[10] China Meteorological Administration, China
[11] University of Alexandria, Alexandria, Egypt
[12] The Centre for Energy, Environment, Science & Technology, Dar Es Salaam, Tanzania
[13] Munasinghe Institute for Development, Colombo, Sri Lanka
[14] Independent consultant, Ouderkerk a/d Ijssel, The Netherlands
[15] Lebanese University, Faculty of Sciences II, Beirut, Lebanon
[16] Comisión Nacional Del Medio Ambiente, Santiago, Chile
[17] University of Guelph, Guelph, Canada
[18] Oak Ridge National Laboratory, Oak Ridge, United States

# CONTENTS

## 9.1. Introduction

The Adaptation Policy Framework (APF) advocates that any adaptation process should support a country's overarching development objective, and be integrated into its current plans, policies and programmes. In some cases, structural adjustments may be justified on the basis of cost-benefit analysis. However, given the uncertainty of climate change impact projections, the APF recommends a dynamic and process-orientated approach to adaptation.

As with any policy-making process, incorporating adaptation to climate change, including variability, into regular development planning is a challenge. But because climate change can potentially affect all sectors of the national economy, adaptation requires both an interdisciplinary approach and cross-sectoral policy analysis. Careful monitoring and evaluation (M&E) of implemented adaptation measures can enable the user to assess what is working, what is not working, and why.

If the original adaptation strategy anticipates the type of information required for a *post hoc* examination, this type of evaluation will be possible as well. A good M&E framework depends on at least two key ingredients: a framework with clearly formulated goals, objectives, and output measures; and the availability of quality data. Participatory M&E can sustain the impetus for continuous feedback-correction cycles.

This Technical Paper (TP) suggests how adaptation may be incorporated into national development processes. It focuses on M&E as a tool for establishing a learning process initiated by an adaptation project. The M&E process can reveal how social, economic, institutional and political factors support or impede adaptation. In this way, countries can incrementally adjust their adaptation strategies to ensure that they are increasingly effective.

TP9 is organised as follows: Section 9.2 highlights the relationship between TP9 and the larger APF. Section 9.3 introduces key concepts important to this TP. In Section 9.4, guidance is provided on constructing an M&E framework, conflicts and unintended consequences, and incorporating adaptation into the development process.

## 9.2. Relationship with the Adaptation Policy Framework as a whole

TP9 outlines the adaptation process as framed by the APF. This paper aims to tie each of the TPs together, but its relationship to TPs 1, 7 and 8 are the most direct. Adaptation project goals that are articulated in TP1 later become the focus of M&E activities in TP9. For example, a goal such as "to increase the adaptive capacity of vulnerable coastal communities" can translate into goals of the M&E process (i.e., "to monitor

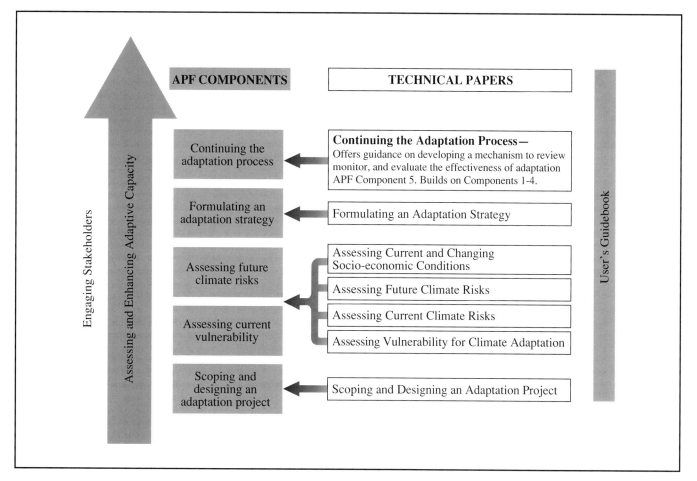

**Figure 9-1:** Technical Paper 9 supports Component 5 in the Adaptation Policy Framework

changes in the adaptive capacity of coastal communities"). Similarly, indicators of adaptive capacity developed in TP7 can be used for monitoring changes in adaptive capacity. Ideally the adaptation strategy developed in TP8, will have been developed with an M&E plan for each of the adaptation goals. Figure 9-1 shows the relationship of this TP to the overall APF and to Component 5 in the APF process.

### 9.3.    Key concepts

Adaptation to climate change, including variability, fits into a broader conceptual framework on responses, as outlined in Figure 9-2. This broad framework subsumes many of the Components described in the APF into its "Planning Design" and "Implementation" boxes. It also highlights the M&E function of Component 5 as an important part of any adaptation process. As a result, the design of any adaptation to a long-term climate hazard should include specific plans for careful *ex post* evaluation of performance with performance indicators. This point is made in Figure 9-2 (shaded area), where the feedback mechanisms can improve adaptation practices.[1] Well-constructed M&E mechanisms can do more than that. Properly conducted, these mechanisms can contribute to an evolutionary

"learning by doing" function that will provide insight into how the adaptation process can evolve most efficiently. For example, if a climate hazard manifests itself through repeated extreme events, then monitoring the frequency of these events, as well as evaluating the sensitivity of adaptation to the intervals between their occurrence, will suggest how an adaptation might best evolve (e.g., become more robust, find stable sources of funding, and so on).

### 9.3.1.    Monitoring

The purpose of monitoring is to keep track of progress in the implementation of an adaptation strategy and its various components in relation to the targets. This enables management to improve operational plans and to take timely corrective action in the case of shortfalls and constraints. As part of the management information system, monitoring is an integral part of the function of management, and should be conducted by those responsible for the project/programme implementation. The resulting data, in whatever form, must be archived so that they can be readily accessed for internal or external evaluation. Monitoring should be carried out during implementation, as well as during the lifetime of the project. Both the selection of

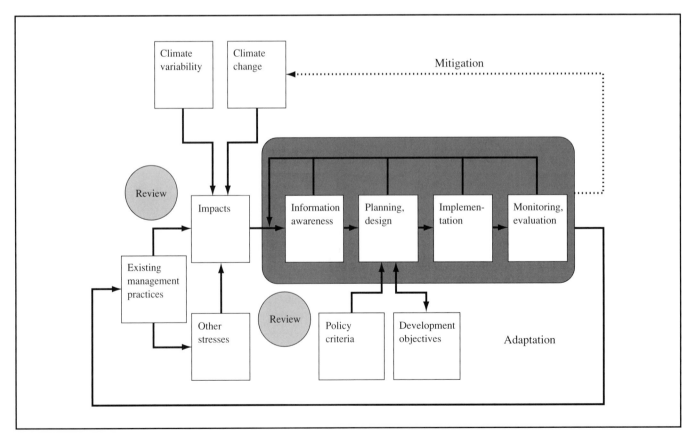

**Figure 9-2:** Conceptual framework on responses to climate change, including variability (adaptation/mitigation) (after R. Klein et al., 1999)

---

[1] The review portion of an adaptation strategy looks at which adaptations or suites of adaptations have been proposed or implemented. A comprehensive review of existing management practices and policies, policy criteria, development objectives and existing levels of adaptive capacity would have been undertaken prior to selection and prioritisation of adaptation strategies (TPs 3, 7 and 8), so earlier work will have already identified the targets for review.

indicators for monitoring and the frequency of monitoring can evolve over time as the adaptation process matures; this evolution may continue as the adaptation process is incorporated into a country's overall policy mix. The most important point is that monitoring continues.

### 9.3.2. Evaluation

M&E must go hand-in-hand. In the context of adaptation, evaluation is a process for systematically and objectively determining the relevance, efficiency, effectiveness and impact of an adaptation strategy in light of its objectives. Whereas monitoring is carried out only during implementation, evaluation is carried out during implementation (ongoing evaluation), at the completion of a project (final evaluation) or some years after completion (post evaluation). Much of the evaluation activity can be based on self-assessment of the responsible operational staff, but external evaluation is also a common practice.

Formal M&E processes should be practical. In principle, a network of concerned institutions and stakeholders (data suppliers and users) could be established. Increasingly, the trend in this field is towards participatory M&E, which includes the most vulnerable group(s) in decision-making. The concept of a central M&E unit to co-ordinate all of the functions could be established within, or under the jurisdiction of, a strategic government agency (e.g., a Ministry of Finance, Planning or Environment). While institutional barriers can impede M&E, these barriers can be assessed during project design and addressed during its implementation.

Comprehensive adaptation strategies consist of policies, measures and projects. Appropriate M&E processes may be quite different for each strategic level. Furthermore, gaps in the structure and design of the strategy can impede progress toward long-term goals of sustainability. Policies that exist without tangible measures are paper tigers; conversely, projects that exist outside of a clear policy context can be redundant or contradictory. Monitoring for gaps of this sort can pay enormous dividends.

### 9.3.3. Performance indicators

Monitoring alone is useless if the raw data and basic information it generates are not analysed in the evaluation process. M&E processes depend on carefully developed sets of indicators by which the performance of adaptation activities can be assessed. These indicators provide the basis for before-and-after analyses and describe the effects (positive and negative) of project interventions – anticipated and unanticipated, intended and unintended. Indicators are quantitative or qualitative measures that can be used to describe existing situations and measure changes or trends over time (Glossary and TP6). Performance indicators are criteria for success. In the context of the logical framework approach, at least one indicator should be defined as a performance standard to be reached in order to achieve an objective (GEF, 2002). Indicators should include both outputs

and outcomes (impacts), with explicit statements of how the indicator demonstrates that the project goal has been met, and what the functional relationship is between a change in the indicator and the outcome of a project.

### 9.3.4. Learning by doing

Exploring the success or failure of the adaptation process depends on more than just the success or failure of implemented projects. More critically, it depends upon the concept of learning by doing. This approach enables users to: a) undertake midcourse corrections in implemented adaptations, so that they meet their objectives more efficiently; and b) improve their understanding of the determinants of adaptive capacity so that capacity development activities can be more successful from the start.

To accomplish these tasks, two earlier insights can be revisited. First, Component 4 of the APF and TP8 both establish the necessary criteria for evaluation. Second, the M&E process will eventually have historical evidence of what actually happened over a period of time; this can be compared to the conjectural characterisation of future conditions. To learn from mistakes and successes, it is important to combine these insights to:

- compare actual experience with the initial characterisation, and with the criteria; and
- construct a revised adaptation baseline that describes how the system would have performed in the absence of the implemented adaptation.

This revised adaptation baseline will differ from the adaptation baseline described as a part of the APF Component 2; it will be more accurate, based on actual experience and on the evolution of the structural, economic, and political context. This can be critical, since it will suggest whether an adaptation to climate is "swimming uphill" against some non-climatic impediment or "being carried along" by other reforms. Thus, an evaluation could improve the team's forecasting capability. A review of the criteria used for making the original implementation decision will yield insights about needed changes, and will improve the next adaptation decision.

Annex A.9.1 suggests what could be learned under various decision criteria by simply asking a series of questions as the future unfolds; the questions themselves will identify what to monitor. They are offered merely as suggestions of insights that might be uncovered by returning to past decisions using the same criteria, informed by new information.

### 9.3.5. Participatory monitoring and evaluation

Participatory processes in support of adaptation can add value and enhance feasibility. Engaging as many stakeholders as possible can democratise the overall process of adapting to climate

**Box 9-1: Sample case study from Tlaxcala, Mexico**

This is an example of an ongoing project in Tlaxcala, Mexico, initiated by the National Autonomous University of Mexico in 1997, and now in the hands of the Autonomous University of Tlaxcala, Mexico.

*1.  Scoping and designing the project*

Project objective: To develop climate forecasts for use by public agricultural agencies and farmers to improve production strategies in face of climatic variability.

Information review: The project incorporated the findings of Mexico's first national study on vulnerability to climate change (Country Study Mexico), as well as an extensive review of the history and recent changes in agricultural policy, crop production variability and trends, the agro-ecology of maize production in central Mexico and existing studies on El Niño Southern Oscillation (ENSO) linkages to crop yields.

Project development: An interdisciplinary research team was formed, consisting of climate specialists, agro-meteorologists and agro-biologists, a specialist in socio-economic aspects of vulnerability and representatives of the state agricultural research agency. A farmer's group was contacted to provide feedback on the project's focus, content and objectives. Regular contact was maintained with this farmer's group during the project's implementation. Additional interviews and surveys were initiated with other farmers to diversify and expand the scope of stakeholder consultation.

*2.  Assessing current vulnerability*

Climate Risks: Early frosts in the fall and late frosts in the spring were found to constrain crop choice and affect harvest yields and quality. Rainfall was extremely variable both in distribution and in quantity. The timing of the onset of the rainy season and the duration and intensity of the mid-summer drought were particularly important for farmers.

Socio-economic conditions: Most farm households produce maize largely for subsistence purposes, together with beans and barley (depending on landholding size). Yields are uniformly low and variable depending on climatic conditions and access to inputs. Recent neoliberal reforms are related to rising fertiliser prices and the loss of guaranteed producer prices for most crops. Credit and crop insurance are largely unavailable to smallholders. Extension advice is quite limited and irrigation not possible in much of the state.

Vulnerability: Anomalous frost events associated with ENSO and irregular rainfall distribution were the sources of crop losses in 1997, 1998 and 1999. Maize was particularly sensitive; shorter-cycle crops such as oats and barley were less affected. Despite extensive experience with managing risk, farmer's coping strategies were limited by the economic insecurity they faced from declining producer prices, rising input costs and lack of institutional support for alternative crops.

Adaptations: Households have developed a range of risk-adverse strategies to address anticipated climatic variability including planting rapid-maturing (typically lower-yielding) maize varieties, changing planting dates, altering the timing of crop tasks to conserve moisture, and altering their crop mix and land use choice.

*3.  Assessing future climate risks*

Climate trends: The distribution of precipitation in the state has become more variable in the 1990s with an accentuated mid-summer drought. Anomalous frost events have become more frequent, although climate change scenarios illustrate declining frost risk and a prolonged growing season. If ENSO conditions are viewed as representative of future climate conditions, the state may experience increased frost risk during the rainy season.

Socio-economic trends: Farmers will face increased pressures to participate in commercial markets, increased competition in the market for basic grains, and continued reductions in public investment and support for agriculture. Without significant investment in alternative economic activities, rates of rural-urban and international migration are likely to continue to rise.

## 4. Formulating adaptation strategies

The state research agency collaborating in the project invested in the preparation of "crop suitability maps" in 1998 to guide crop recommendations and land use in the state, taking into account the probability of increasing impacts from ENSO events. Although maize is the dominant crop in Tlaxcala, it is not considered by governmental agencies appropriate for much of the state's area given its sensitivity to frost and drought. Barley and oats are thought to be more appropriate because of the shorter growing seasons of these crops, which makes them less affected by early frost events or a prolonged mid-summer drought. Conversion from maize to oats and barley was encouraged by Tlaxcala's agricultural ministry via the distribution of free packages of oats seed to help farmers recuperate from losses to maize in from 1998 to 2000. The state research agency also disseminated the experimental forecasts prepared by the research team in 1998 with recommendations to farmers on planting strategies. After attending a stakeholder workshop in which the forecasts were introduced, a group of commercial farmers used the university's experimental forecast of drought conditions in 1998 to buy seeds for hardy varieties of oat in advance of the rainy season. They reported that their strategy successfully mitigated some of the worst impacts of the 1998 season. This adaptation strategy appeared to depend, however, on the farmers' organisational and financial capacities. The success of this strategy also depends on the existence of a viable commercial market for oats, or a demand for oats as an input into livestock production.

## 5. Continuing the adaptation process

From interviews and households surveys, the research team learned that: a) farmers identified many resource and institutional obstacles to being able to act on possible forecast information, b) some farmers felt the information could be used to plan investments and the timing of farm activities, c) others thought that the forecasts might need to be far more spatially explicit than what climatologists were proposing in order to be useful, and d) because farmers' own methods of forecasting were no longer reliable because of the changes they perceived in climate patterns, they also assumed that any new climate forecasting method would be equally unreliable for the same reasons. The farmers were sceptical that the forecasts would be reliable, however other studies have illustrated that, with time and personal experience, such scepticism can be overcome.

The project is ongoing, although still on an experimental basis. Current efforts involve working with a select few farmers to conduct experiments on changing cropping patterns and choices on the basis of forecast information. This effort is designed to address problems of scepticism, as well as to facilitate the technical aspects of crop switching on the basis of climate information.

Incorporate into development plans: The project contributed to raising awareness of ENSO impacts and the potential utility of forecasts in the state's agricultural research agency. The project did not, however, account for the political structure of the state's agricultural institutions, and thus when a new political party came to power mid-way through the project, the project lost much of its linkages to the state's formal institutions. The project has also illustrated that co-ordination with other sector policies and programmes (e.g., extension, research, input supports) is required to improve the general flexibility of farmers' strategies. Work is still needed to improve the geographic specificity and temporal accuracy in the forecasts in order for the information to be widely disseminated.

Source: Conde et al. (1998), Ferrer (1999), Eakin (2000), Conde and Eakin (2003)

change, including variability. It follows that participatory M&E can be productive, but care must be taken to note the potential pitfalls. In the example provided in Box 9-1, stakeholder engagement uncovered obstacles, including a healthy degree of initial scepticism on the part of farmers about the information provided by the government. A similar result emerged from the earlier "MINK" study that focused attention on the vulnerability of agriculture in four midwestern United States (US) states to a return to the dust-bowl climate of the 1930s (Easterling, 1996). Farmers in the US found information about the coming season offered by seed salesmen to be far more credible than anything provided by state or federal agencies and/or their extension services.

### 9.3.6. Mainstreaming

In the context of adaptation, "mainstreaming" refers to the integration of adaptation objectives, strategies, policies, measures or operations such that they become part of the national and regional development policies, processes and budgets at all levels and stages. The idea is to make the adaptation process a critical component of existing national development plans. Likely entry points for mainstreaming climate adaptation include: environmental management plans (particularly when they incorporate environmental impact assessments), national conservation strategies, disaster preparedness and/or management plans and sustainable development plans for specific sectors (e.g., agriculture, forestry, transportation, fisheries, etc.) Moreover, working through the determinants of adaptive capacity makes it clear that promoting capacity can complement or even advance the broader objectives of poverty reduction and sustainable development. The issue is to recognise an opportunity for mainstreaming and to use it.

The ability of adaptation to ameliorate climate impacts is fundamentally path dependent and site specific (Box 9-2). As a result, an adaptation that works well in one place and time may or may not work in a different place or time. Whether it does or does not is essentially an empirical question, and M&E can inform the framing of such a question. This diversity should not, however, discourage mainstreaming. Indeed, the crafters of development plans already cope with the "sometimes the magic works, and sometimes it doesn't" character of the real world when they ponder, e.g., the effect of increased trade or market reform on productivity growth, equity, the incidence of poverty, and so on.

## 9.4.    Guidance on continuing the adaptation process

This section is divided into three major elements – the first on M&E, the second on synergies and conflicts, and the third on mainstreaming. While the bulk of guidance in TP9 focuses on M&E, all elements are essential to continuing the adaptation process. Mainstreaming adaptation into existing policy processes and priorities is a central goal of the entire APF process. Annex A.9.2 offers greater detail on M&E.

### 9.4.1.    *Establishing the monitoring and evaluation framework*

This section provides guidance on how to develop a mechanism for reviewing, monitoring and evaluating the effectiveness of adaptation. The design of a monitoring and review system can support a learning-by-doing empirical approach. In the project design stage, initial M&E plans should describe systems by which results can be incorporated into the management process, and indicate how the proposed activities will con-

tribute to the establishment of long-term M&E capability in the country (TP1, Section 1.4.4).

With an understanding of the generic aspects of M&E outlined above, APF users may next choose to consider a few specific actions for continuing the adaptation process. All of these actions presume that the teams – comprised of responsible individuals and institutions – will have met some organisational prerequisites in identifying goals and objectives. These actions fall into two main groups.

The first set of actions falls under the rubric of *devising a framework for monitoring the progress of an adaptation strategy*, and can include the following steps:

1.  Define issues, goals, and targets – determination of what should be monitored and evaluated and why.
2.  Define tools to collect and process data/information – selection of methods may depend on which one will allow minimal data collection and yet produce accurate information. Sampling methods, tracking points for inputs and outputs, and data sources should all be considered.
3.  Assess and interpret the results of implementing strategies – quantitative or qualitative assessment. (See TP6 Section 6.4.2 for information on indicators and TP8 for evaluation of adaptation strategies such as multi-criteria analysis, cost effectiveness analysis, and expert judgment.)
4.  Ensure that the options and strategies to achieve the goals and targets are feasible.
5.  Involve stakeholders – M&E planning and identification of relevant indicators should, as much as possible, involve those communities or institutions likely to be affected by the adaptation activities, whether positively or negatively. (See TP2 for description of level of participation, e.g., participatory or interactive.)
6.  Monitor and evaluate what can be accomplished using alternative approaches (e.g., Table 9-1) to support "learning by doing", self-assessment through workshops and/or periodic reporting, external review by consultants, etc.
7.  Institutionalise M&E.

---

**Box 9-2: Vaccination as an example of the site dependency of potential adaptations**

Vaccination is one possible adaptation to climate change in the health sector. However, as this example shows, the side effects of vaccinations can vary across countries. Kremer (2002) reports that rotavirus kills almost one million children per year in developing countries. The virus, though, is only a minor nuisance in developed countries; it causes diarrhoea, but few deaths are reported (CVI, 1999, Murphy, 2001a). An oral vaccine received regulatory approval in the US in 1998, but it was quickly withdrawn across the developed world when it became evident that intussusceptions (a form of intestinal blockage) produced a direct side effect. Sadly, pharmaceutical companies stopped production even though children in developing countries are not susceptible to intussusceptions, and a casual cost-benefit analysis for developing countries overwhelmingly supported using the vaccine in the developing world (Murphy, 2001b). Experience shows that the effectiveness of this oral vaccination is very site-specific, and that the pros and cons of its application should be evaluated in an appropriate context.

**Table 9-1:** *Evaluation approaches*

| | **Conventional Evaluation** | **Participatory Evaluation** |
|---|---|---|
| Why | Accountability, usually summary judgments about the project to determine if funding continues | Empowerment of people to initiate, control and take corrective action |
| Who | External experts, community members, project staff, facilitator | Community members, project staff, facilitator |
| What | Predetermined indicators of success, principally cost, are used to assess project impact | People identify their own indicators of success |
| How | Focus on scientific objectivity: distancing of evaluators from other participants, uniform complex procedures, delayed limited access to results | Self-evaluation; simple methods adapted to local culture; open, immediate sharing of results through involvement in the local evaluation process |
| When | Midterm and completion; sometimes *ex-post* | Any time; any assessment for programme improvement through the merging of M&E functions, hence more frequent small evaluations |

Source: Department for International Development (DFID), 2002

This monitoring framework could involve a national coordinating structure. The goal is to include the adaptation strategy (policies/measures/operations) in an institutional planning process and to coordinate the process by which priorities are established. Fundamentally, the framework can only succeed if inputs are well established (e.g., locating sources of information and data, devising performance or process indicators, recognising and covering additional costs over the long term, establishing control mechanisms such as period of review or evaluation, etc.). Annex A.9.3 provides a sample planning matrices for M&E in the adaptation process.

The second set of actions for continuing the adaptation process can be conceptualised as *devising an inquiry-based framework for evaluating the adaptation strategies* and can involve responses to the following questions:

1. Was the endorsement of the adaptation strategies from project level to the national level reached? Why or why not? And how could stakeholders tell (i.e., what indicators might demonstrate this endorsement)?
2. Was an M&E process institutionalised properly? Were the institutional arrangements established? Who did the monitoring? Was a national consultative structure established with a central coordinating body identified? Did project teams identify inputs for M&E such as sources of information and data, performance or process indicators, additional costs, and/or control mechanisms such as period of review or evaluation?
3. Were the adaptation strategy, policies, measures and operations incorporated into institutional planning processes? How did they fit in overall processes of prioritisation?

A satisfactory analysis need not address each action listed above. Perhaps more to the point, no project team should abandon an M&E plan because it cannot progress past any particular point on these lists. Indeed, determining what to monitor and making arrangements for the careful archiving of the information in accessible locations is the only absolute necessity.

Once the framework has been chosen, the project teams should decide what to monitor. In the APF context, a particular methodological approach will have already been selected under Component 1. The four approaches are: hazards-based, vulnerability-based, adaptive-capacity, and policy-based. These approaches, together with the prioritising process carried out under Component 4, should have identified which adaptations and associated indicators (Components 2 and 3) will be monitored. Clearly, the monitoring function extends well beyond keeping track of the climate risk. Socio-economic drivers of exposure and sensitivity that frame the adaptation baseline must also be monitored (TP6).

The evaluation of an adaptation process can begin once the information is generated. Solid evaluations can be carried out with simple, careful examinations of success, relative to what was expected. The following list provides examples of questions that can contribute to this evaluation:

- If, e.g., an adaptation involved investing in a protection project in response to a climate hazard, then the evaluation should determine if losses have continued, grown, or been abated.
- If the protection project simply tried to reduce sensitivity to extreme events, has it worked and how?

- Have episodes of intolerable exposure become more or less frequent?
- Has the definition of "intolerable" in terms of physical effects changed?
- Has the investment expanded the coping range, reduced exposure to intolerable outcomes that exceed the range,or both?
- Have things stayed the same or grown worse because the adaptation was ineffective, or because unanticipated stresses have aggravated the situation?
  - Is there a causal relationship?

The purpose of this exercise is to determine whether or not the objectives of an adaptation project have been satisfied. More complete evaluations of specific adaptations should identify the root causes of both successes and failures. As an aid, a questionnaire specific to the particular adaptation can be constructed to understand the reasons why an adaptation succeeded or failed to meet its objectives. In the example provided earlier in Box 9-1, the five Component APF process was applied to traditional maize agriculture in Mexico for an evaluation of two on-going adaptations.

While evaluation can occur at any stage in the adaptation process, the final evaluation may require additional funding following the project's completion. To enable the lessons learned to feed back into and inform subsequent actions, it is essential that the necessary resources (e.g., human, financial, technical) be factored in during the project design phase. This step is recommended, but is often neglected.

### 9.4.2. *Working with synergies, conflicts and unintended consequences*

For successful continuation of the adaptation process, isolated evaluations are not sufficient. The notion of opportunity cost, expressed as monetary units, is really an observation that any action occurs at the expense of another. These costs are diminished if adaptations complement one another either directly or by promoting synergies across the underlying determinants of adaptive capacity; they are exaggerated when adaptations contradict and/or create obstacles for each other or with other developmental objectives (maladaptation). Careful evaluation of any adaptation will therefore contemplate the interaction of a suite of adaptations in the context of a more general pursuit of social and economic objectives. A review and evaluation should repeat the analysis – following all the Components in the APF – incorporating new and/or updated information from the intervening years. Care must be taken, though, not to apply insights derived from one location to another location, without careful review of the underlying analysis. Adaptation is, by its nature, site-specific and path-dependent.

### 9.4.3. *Mainstreaming adaptation into the development process*

Current thinking assumes that stand-alone adaptations are neither desirable nor cost-effective. In developing countries, one

group of stakeholders responsible for facilitating adaptation include the international development agencies and donor governments. Like other environmental issues, this group has collectively agreed that climate change adaptation would be cost-effective if mainstreamed into the development processes. As the term "mainstreaming" implies, the approach places environment squarely in the centre of development poverty reduction. This approach is warranted because global environmental issues remain marginalised in all but a few countries – even ten years after Rio – leading to conclusion that rather than introducing additional environmental plans at this stage, governments should renew effort on implementing those plans. Note that mainstreaming is not unique to adaptation; it is a policy principle for introducing all multilateral environmental issues onto the policy agenda.

Environmental mainstreaming is seen as both a popular and elusive goal. In reality, the process is poorly documented, and the gap between theory and practice is acute. Recognising these constraints, this section provides suggestions for tackling mainstreaming and draws upon experience learned from other domains. This section addresses the APF's core issue: "How can societies best adapt to changing climate?" It considers the boundaries of, and entry points to, the priority system, its socio-economic context, criteria for mainstreaming, and roles and responsibilities of stakeholders. While adaptation to climate change is new, the practices used for coping with climate variability are not. Box 9-3 provides an example of mainstreaming.

*Defining system boundaries and identifying entry points*

Both mainstreaming and adaptation are extremely broad concepts. To develop an approach to mainstreaming, it is absolutely essential to define the system boundary and to be as specific as possible about the scale and type of intervention. In other words, what is being mainstreamed into what, and how?

First, the entry point for the adaptation should be identified. For example, the approach to mainstreaming climate change adaptation into a national water policy and sectoral programmes would be very different from mainstreaming at the community level. While the two strategic levels are interrelated, the entry points for the intervention would vary.

- A "top-down" approach could involve changes in policies and procedures at the strategic, programming and operational levels. For example, at the country level, critical entry points for programming lie within the different development agencies. Recognising that greater harmonisation is required among development efforts, the United Nations (UN) has recently launched a programming tool that is a common country framework used in all UN programmes. Examples of programming tools are the World Bank's Poverty Reduction Strategy (PRS) (Box 9-3) and the Asian Development Bank's "climate proofing" approach (Box 9-4). These guide-

---

### Box 9-3: Mainstreaming environment in Tanzania

Environment is one of the priority crosscutting issues in the development of Tanzania's second generation Poverty Reduction Strategy (PRS) Paper. Development partners have been working with the government on environment and the PRS for over three years. As result, a programme has been developed to integrate environment into the PRS process. The programme is under the Poverty Eradication Division of the Vice President's Office, and the Ministry of Finance is fully involved. Environment is now becoming seen an essential element for sustainable growth and the achievement of poverty reduction targets. Included in this is the reduction in vulnerability of the poor from environmental risk, and the need to address issues of drought and floods, and in the longer term how these risks may increase from climate change. In this context the development of the new PRS is looking at how to integrate commitments under multilateral agreements and include actions on adaptation for climate change and desertification. (For more information, see www.povertymonitoring.go.tz.)

One of the central principals of the PRS is national ownership. Consequently the PRS is becoming harmonised with the budgetary and other planning processes. Key steps to mainstreaming environment in Tanzania are:

- A strong national group of "champions" from government and non-government organisations on environment has been active since mid-1990s.
- Provision of catalytic support by development partners.
- Focus on identifying the practical links between poverty and environment.
- Development of a cross-sectoral working group on the environment.
- Public expenditure review on environment to assess the contributions environment to growth and poverty reduction and levels of expenditure on the environment.
- Development of poverty-environment indicators for local and national monitoring systems.
- Development of environment issues and appraisal are integrated into planning processes, particularly at the local level.
- Focus on how multi-lateral commitments on environment (e.g., Climate Change convention) can be integrated into national policies and strategies.

Source: David Howlett, UNDP Tanzania, Energy and Environment Practice Network E-discussion on "Mainstreaming Environment into the PRS".

---

lines and strategies can provide opportunities for introducing climate change adaptation and other issues into national sectoral policies and programmes.

- For community-based actions, the entry points could be at the household level. In certain interventions, experience in development initiatives has shown that gender considerations are important. For example, one project showed that women in Indonesia, Cambodia, and Vietnam placed a high value on household latrines (water) (Mukherjee, 2001). Therefore, any awareness-raising efforts designed for adapting through water conversion should target women at the household level using a "bottom-up" approach.

Second, a sectoral or multi-sectoral approach to mainstreaming should be chosen since climate change will affect all sectors to some degree. In general, the fewer the sectors, the easier the mainstreaming process. However, a cross-sectoral approach is recommended where possible, because of the large potential impact of the measure, and the synergies and conflicts among sectors, for a given adaptation measure. For the same reasons, it is preferable to aim to influence the policy process at as high a level as possible. Project teams should decide what adaptations are politically feasible, if the capacity exists to implement the measures, and to tailor the mainstreaming strategy accordingly.

*Describing the socio-economic context and identifying opportunities*

The socio-economic context for a given system will govern how decisions are made, and will largely determine whether adaptation will actually take place (TPs 6 and 7). In order to select adaptations that have potential for mainstreaming, the policy analysis carried out in TP6 should distinguish between elements that are "inside" and "outside" the locale or sector. For "outside" elements, this simply means that adaptation mainstreaming needs to consider the predominant policy drivers for a country in its region. In Eastern Europe, no socio-economic analysis is complete without consideration of the European Union accession process. In some countries, such as those in Latin American and Asia, decentralisation and privatisation may be the predominant policy processes. In China, the five-year economic plan lies at the heart of the country's economic development, and provides opportunities for adaptation (e.g., water use) to be mainstreamed into the country's targets through industrial structural adjustments, agricultural and rural economic development. Box 9-5 provides an example of India's approach to managing climate risks. The country's policy directions will help to identify where opportunities for mainstreaming exist.

TP8 discusses procedures, and the National Adaptation Programmes of Action (NAPA) process offers guidance on incorporating adaptation strategies into sustainable development.

---

**Box 9-4: "Climate proofing" development in the Pacific:**
**The Asian Development Bank's efforts to mainstream adaptation to climate change**

The Asian Development Bank (ADB) provides a significant source of revenue to the Pacific Islands for their development. Currently, the ADB funds a variety of investment projects, ranging from road infrastructure to coastal development. However, under climate change, the climatology of the Pacific will be perturbed, with expected changes in the frequency and severity of extreme climatic events. If climate change is left unabated, the damages associated with future climate will impose social and economic costs on the Pacific Islands, and may even severely limit their governments' capacity to pay back its loans. Surprisingly, the ADB's investment process had not factored in the risks of climate change until very recently.

For the first time, the ADB is now piloting a climate risk reduction strategy to reduce the impacts of climate change. As part of a strategy for "climate proofing" its investments, the ADB is mainstreaming adaptation to climate change through its country programming and project preparation processes. The objective is to reduce the exposure of climate risk of the ADB's investments.

Operationally, the ADB's risk reduction strategy involves a suite of procedures. The steps taken include:

- Developing guidelines for mainstreaming adaptation ("climate proofing");
- Using climate information and analysis of climate sensitivities, providing recommendations on country strategies and programming;
- Identifying projects that are sensitive to climate change impacts, and assessing further the current and future climate risks of these projects;
- Incorporating risk reduction into project preparation processes, including recommendations on how adaptation measures and policies can be used to reduce climate risks; and generally
- Increasing the ADB staff's awareness on climate risk reduction.

Following these procedures, adaptation is being piloted through the application of risk reduction strategies in six case studies. These case studies have implications for, e.g., coastal communities, road infrastructure projects, harbour expansion and the human health and environment components of National Strategic Development Plans.

In each case study, the project follows the procedures laid out in the ADB's mainstreaming guidelines. As a result, adaptation policies and measures are being mainstreamed into national development planning, land-use planning, and through legislative instruments, such as modified building codes, environmental impact assessments, and health regulations. Various initiatives are being "climate proofed" in this way.

Over the long term, such adaptation efforts should reduce the exposure of the ADB's investments to climate hazards and associated risks. However, the impacts of climate change are uncertain, and it will be several decades before the effectiveness of adaptations can be truly evaluated. The ADB's mainstreaming guidelines, nonetheless, provide a planning tool for making mid-course changes to reduce damages of climate change, as the impacts of climate change manifest themselves.

Source: Brotoisworo, E, Perez, R. T. and King, Wayne, 2004: Climate Change Adaptation Program for the Pacific (CLIMAP), a presentation during the UNFCCC Consultation for the preparation of the Second National Communications for non-Annex I Parties, Manila, April 26 – 30, 2004).

---

*Analysing socio-economic barriers*

Another dimension of mainstreaming is the analysis of barriers (also in TP7, Section 7.4). Barrier analysis will help to identify appropriate policy instruments for adaptation including: legislative/regulatory/juridical, institutional, financial/market, and education/information mechanisms. (TP8 describes these policy instruments in further detail.)

At the national level, common barriers may be:

- *Institutional framework:* It would be equally important to identify the institution responsible for adaptation, define its mandate, and assess its human capacity,

financial resources, and organisational effectiveness in convening different sectoral bodies. For example, the institution may need to bring together the climate change and the disaster risk management communities.
- *Legal framework:* Successful mainstreaming may require new laws and regulations or the improvement and enforcement of existing regulations (e.g., a building code to limit the elevation of construction to above the 50- or 100-year flood level).

At the local level, it could be useful to consider additional factors:

- *Social institutions* and arrangements that discourage concentration of power and prevent marginalisation of

---

**Box 9-5: Managing climate risks in India: A historical perspective**

Over the last 125 years, India's approach to dealing with climate variability has evolved. As much of the country's population is dependent on rainfed agriculture, any fluctuations in monsoon patterns can have serious implications for food security and the economy. India's experience in dealing with this threat represents a paradigm shift from the reactive approach of drought management towards the proactive management of climate risks. It also suggests how current approaches for managing climate risks may be used by societies to better adapt to future climate risks.

Until the middle of the 20th century when India was under British rule, the approach to dealing with the consequences of a failed monsoon season was to provide relief. This response mechanism was activated only when food shortages became severe and food famines were already well advanced. After independence from the British, India's policies to cope with climate variability and famine changed significantly.

From the mid-1950s onward, after democracy took hold, the Indian government became progressively more active in addressing food scarcity at an early stage. By the early 1970s, the government placed great emphasis on domestic food production as a way to cope with recurring food scarcity. The practices introduced ranged from continuous monitoring of the climate system to enable timely adjustments in agricultural systems, to regulation of food grain markets, to public distribution systems. These practices marked the transition from drought relief to drought management, one that is cross-sectoral in nature and provides an institutional response during, rather than after the monsoon.

Since the early 1980s, there has been an increased emphasis on the use of seasonal forecasting. In April or May of each year, a statistical model is used to produce a seasonal forecast of the southwest monsoon. As the model has improved over the years, seasonal forecasting now provides a basis for national decision-making. However, recent experience indicates that vulnerability to drought is related to both rainfall patterns and the role of government. In India, the federal government plays a significant role in relief operations and fiscal incentives may be required to complete the paradigm shift from drought relief to management.

Here the 2002 monsoon is relevant. In this year, the southwest monsoon was forecast as "normal". Indeed, it was normal until the first week of July, when a break of monsoon impacted agriculture severely. Although temporary, this two-week break triggered large-scale drought relief operations, which continued well into August, even after the monsoon was revived. The operations continued because states are entitled to central "relief" in times of disasters and there is little incentive to offset losses, which occur during the late monsoon season.

Given recent experiences with drought management, UNDP promotes an integrated approach to risk management that uses historical climate information in conjunction with climate forecasts over short, intra-seasonal and seasonal time scales. Integrated risk management allows for continuous adjustments of management decisions with the potential of minimising negative impacts as well as maximising potential benefits of climate variability. However, it must be recognised that climate information is only *one* of the several factors that needs to be taken into account in decision-making.

At the same time, in order for climate information to translate into both region and sector-specific information products and beneficial action for risk management, adaptive capacity needs to be developed. This will require much greater dialogue among the various stakeholders that constitute the end-to-end system – from the understanding of regional-scale climate variability, to its local manifestations, to its specific local and sectoral consequences to sectoral action planning for risk reduction. Integrated risk management is being piloted by UNDP as a policy tool to adapt to future climate risks.

---

sections of the local population (Mustafa, 1998), arrangements to ensure the representativeness of decision-making bodies and maintenance of flexibility in the functioning of local institutions (Ramakrishnan, 1998).

- Diversification of income resources, particularly for poorer sectors of the society (Wang'ati, 1996; Adger and Kelly, 1999).
- Formal and informal arrangements for collective security (Kelly and Adger, 1999) (Box 9-6).

Barrier analysis is based on a governance framework. Practically every organisation has its own definition of governance, but one of the simplest definitions is "a process or method by which society is governed" (Rhodes, 1997). Elements for effective governance are provided in Table 9-2. This table maps a governance structure onto a capacity assessment framework at three different levels. Project teams may use this framework both to identify barriers that impede adaptation and to identify capacity needs for mainstreaming adaptation (discussed below).

---

**Box 9-6: Case study: The Lakkenahally micro-watershed**

Lakkenahally micro-watershed in India covers 210 acres divided by ravines into three micro-catchments. This land was owned by 62 families and was vulnerable to floods. Crops were often washed away, and many fields had not been cultivated for several years.

In 1991, three Credit Management Groups (CMGs) were formed with a total membership of 54. A Watershed Development Association (WDA) was also established in 1992. The CMGs were small, homogenous, voluntary, and autonomous groups that mobilised savings. One was a women's group with 14 members. The CMGs developed their own rules and regulations governing the purpose and size of loans, interest rates, schedules of recovery, and sanctions. They provided credit and group support to help their members meet their livelihood needs, e.g., by providing loans for various forms of consumption, small business and cottage industries. The main problems faced by farmers were erratic rainfall, low moisture-holding capacity of soils, and declining productivity. As part of a collective exercise, 75 farmers (including 35 women) outlined a plan of action, and agreed to contribute towards the costs. The following activities were taken up on a priority basis:

1.  Construction of silt traps to build up adequate soil in areas with high water storage potential. The improved water holding capacity would reduce the risk of crop failure in these areas. Farmers who cultivate lands in the tank bed downstream would also benefit in terms of reduced damage by floods rushing through the ravine.
2.  Excavation of small open wells near reclaimed areas. This provided farmers with protective irrigation, and they were able to introduce a paddy crop.
3.  Wasteland development was taken up on 16 acres of land. This included regeneration of a hillock, with farmers working at lower wages to construct protection walls and plant saplings around the hillock. The grasses were harvested and sold locally by the credit group, indicating the sustainability of the exercise.

Hence, this approach combines different categories of adaptation options viz. prevention and modification of impacts and events, and changes in land use.

Source: Fernandez, 1993

---

*Identifying partners and change agents*

No mainstreaming strategy would be complete without an analysis of partners, change agents, their roles and responsibilities, and their capacity development needs (TP7, Sections 7.4.4 and 7.4.5). It is important to ask, "Who does the mainstreaming? What is the role of agencies, and governments and other stakeholders?" To promote adaptation through mainstreaming, partners are needed in sectors such as health, water, agriculture, and risk management.

To summarise, a checklist of questions for mainstreaming might include:
- Has the system boundary been clearly delineated? (TP1)
- Are the entry points for mainstreaming clear?
- Has a sectoral or multi-sectoral approach been selected for mainstreaming?
- Are the socio-economic context and policy processes well understood? (TP6)
- Have the political opportunities for mainstreaming been identified? (TP6)
- What are the socio-economic barriers to implementation? (TP7)
- Does the adaptation project target barriers and, by doing so, create favourable conditions for implementing the proposed adaptation?

- Have the partners who are responsible for adapting been identified? (TP2)
- Do they have the adaptive capacity required? If not, does the adaptation project aim to enhance their capacity? (TP7)

## 9.5.   Conclusions

This TP presented a mechanism for monitoring and evaluating the effectiveness of adaptation. M&E supports opportunistic review of adaptation processes, particularly if a learning-by-doing approach is adopted, and if significantly informed by engaged stakeholders. More to the point, stakeholders can be important players in an assessment of the effectiveness of any adaptation strategy or suite of strategies. These stakeholders can provide valuable information about whether the proposed interventions have been successful in achieving the strategic objectives; they can also provide insight into how existing social, economic, institutional and political factors have supported or impeded implementation. More importantly, substantial findings from the M&E process will point to corrective action for the adaptation strategies, measures or policies.

Countries already have policies and plans with distinct sets of priorities. The message here is that countries need to add cli-

***Table 9-2:*** *Capacities at three levels required to perform the key functions of the three Rio Conventions*[1]

| Key functions to be performed to comply with Conventions [2] | Capacity required to perform key functions | | |
|---|---|---|---|
| | **System level**[3] | **Institution level**[4] | **Individual level** |
| **Conceptualise and formulate policies, legislation, strategies and programmes**<br>• Analyse global, regional and national socio-economic conditions<br>• Visualise and develop long-term strategies<br>• Conceptualise sectoral and cross-sectoral policies<br>• Prioritise, plan and formulate programmes | **Institutions and laws**<br>• Rules for using natural resources formulated and enforced at the appropriate level (national/regional/local)<br>• Rules and penalties for violating the rules in place<br>• Appropriate mechanism to resolve disputes established<br><br>**Participation rights and presentation**<br>• Public can influence and contest the rules over national resources<br>• People who use or depend on natural resources appropriately represented when decisions on using these resources are made | **Corporate governance**<br>• Consistent strategic direction established<br>• Corporate risk managed appropriately<br>• Management structure acts on performance results<br><br>**Corporate strategy**<br>• Corporate strategy based on mandate<br>• Corporate plan linked to management plans<br>• Appropriate corporate goals and targets established with clear indicators to measure progress | **Job requirements**<br>• Job requirements clearly defined<br><br>**Monitoring performance**<br>• Clear reporting and accountability system in place<br>• Reliable and transparent performance measurement system in place<br><br>**Incentives**<br>• Appropriate salaries and incentives provided<br>• Possibility of career advancement provided<br><br>**Skill development**<br>• Adequate training provided to gain skills necessary to conduct tasks effectively |
| **Implement policies, legislations, strategies and programmes**<br>• Mobilise and manage human, material, and financial resources<br>• Execute and manage programmes and projects effectively<br>• Select effective technologies and infrastructure | **Authority level**<br>• Authority over resources reside at the appropriate level (local/regional/national/international) | **Resource management**<br>• Resource allocation in line with management plan<br>• Adequate financial control mechanism established | |
| **Engage and build consensus among all stakeholders**<br>• Identify and mobilise stakeholders<br>• Create partnerships<br>• Raise awareness<br>• Find "win-win" approaches<br>• Appropriately involve all stakeholder groups in decision-making and implementation<br>• Accept sharing arrangements and resolve conflicts | **Accountability and transparency**<br>• Appropriate mechanism established for the public to question authority on decisions on natural resources<br><br>**Property rights and tenure**<br>• Property rights and tenure allocated to the users appropriately | **Operational management**<br>• Efficient operational procedures established<br>• Clear operational targets set<br><br>**Quality assurance**<br>• Adequate internal guidance and review in place<br>• Adequate monitoring and supervision mechanism established<br>• Well-functioning internal audit process in place<br>• Well-functioning evaluation office in place | |
| **Mobilise information and knowledge**<br>• Gather, analyse and synthesise information<br>• Identify problems and potential solutions | **Markets and financial flows**<br>• Financial practices, economic policies, and market behaviour influence authority over natural resources | **Staff quality**<br>• Transparent recruitment exercised<br>• Transparent promotion mechanism established<br>• Appropriate staff performance management system in place | |
| **Monitor, evaluate, report, and learn**<br>• Monitor and measure progress<br>• Identify and distribute lessons learned<br>• Use lessons learned for policy dialogues and planning<br>• Report to donors and global conventions | **Science and risk**<br>• Ecological and social science incorporated into decisions on natural resource use to reduce risks and identify new opportunities | | |

---

[1] This table provides a sample structure to analyse national capacities needed to respond to requirements under the 3 Rio Conventions. The table is not exhaustive or definitive.

[2] Capacity Development Indicators – UNDP/GEF Resource Kit (No.4), Page 4. http://www.undp.org/gef/undp-gef_monitoring_evaluation/sub_undp-gef_monitoring_evaluation_documents/CapDevIndicator%20Resource%20Kit_Nov03_Final.doc.

[3] World Resources 2002-2004 by World Resources Institute, Page 7, Box 1.3 "Seven Elements of Environmental Governance".

[4] Presentation on "UNDP's evolving approach to managing for results".

mate variability and climate change, including variability into the portfolio of risks to which they are applying their adaptation planning processes. More importantly, the inclusion of adaptation into the development mainstream must focus not only on the pre-decision stages of the process (i.e., project design stage, climate risk assessment), but also on M&E in the implementation and post-implementation stages. Neglecting these important steps can prevent the adaptation process from being an effective management tool. On a larger scale, it can cause countries to miss important opportunities to correct past mistakes and improve current practices.

# References

**Adger**, W.N. and Kelly, P.M. (1999). Social Vulnerability to Climate Change and the Architecture of Entitlements. *Mitigation and Adaptation Strategies for Global Change*, **4 (3-4)**, 253-266.

**Brotoisworo**, E., Perez, R.T., and King, W. (2004). Climate change adaptation program for the Pacific (CLIMPP). ADB paper presented at the UNFCCC consultation for the preparation of Second National Communications for non-Annex I Parties. 26-30 April, 2004, Manila, Philippines.

**Conde**, C., Ferrer, R., Gay, C. (1998). Variabilidad Climática y Agricultura. *GEO UNAM*. **5(1)**: 26-32.

**Conde**, C. and Eakin, H. (2003). Adaptation to Climatic Variability and Change in Mexico. In: *Climate Change, Adaptive Capacity and Development*. J. Smith, R.J.T. Klein, and S. Hug, eds, London: Imperial College Press. 241-261.

**CVI** (Children's Vaccine Initiative), 1999: *CVI Forum* Number **16**, Geneva.

**Department for International Development** (DFID). (2002). *Tools for Development: A Handbook for Development Activity* /Version **15** / September 2002.

**Eakin**, H. (2000). Smallholder maize production and climatic risk: A case study from Mexico. *Climatic Change*, **45**, 19-36.

**Easterling**, W.E. (1996) Adapting North American agriculture to climate change in review. *Agricultural and Forest Meteorology* **80**, 1-53.

**Fernandez**, A.P. (1993). The MYRADA experience: The interventions of a voluntary agency in the emergence and growth of people's organisations for sustained and equitable management of micro-watersheds. Unpublished report. MYRADA, Bangalore, India.

**Ferrer**, R.M. (1999). *Impactos del cambio climático en la agricultura tradicional en el municipio de Apizaco, Tlaxcala*. Facultad de Ciencias, Universidad Nacional Autónoma de México. México.

**Global Environment Facility** (GEF). (2002). *Monitoring and Evaluation-Policies and Procedures* http://gefweb.org/ResultsandImpact/ Monitoring___Evaluation/M_E_Procedures/m_e_procedures.html

**Kelly**, P. and Adger, W.N. (1999). *Assessing Vulnerability to Climate Change and Facilitating Adaptation*. Working Paper GEC 99-07, Centre for Social and Economic Research on the Global Environment, University of East Anglia, Norwich.

**Kremer**, M. (2002). Pharmaceuticals and the developing world. *Journal of Economic Perspectives*, **16**, 67-90.

**Klein**, R.J.T., Nicholls, R.J., and Mimura, N. (1999). Coastal adaptation to climate change: can the IPCC Technical Guidelines be applied? *Mitigation and Adaptation Strategies for Global Change*, **4**, 51-64.

**Mukherjee**, N. (2001) Achieving sustained sanitation for the poor: policy lessons from participatory assessments in Cambodia, Indonesia and Vietnam *Jakarta, Indonesia, Water and Sanitation Program for East Asia and the Pacific*. Is this a report? Has it been published?

**Murphy**, T.V., (2001a). "Intussusceptions Among Infants Given an Oral Rotavirus Vaccine", *New England Journal of Medicine*, **344**, 564-572.

**Murphy**, T.V., (2001b). "Intussusceptions and Oral Rotavirus Vaccine", *New England Journal of Medicine* 344, 1866-1867.

**Mustafa**, D. (1998). Structural Causes of Vulnerability to Flood Hazard in Pakistan. *Economic Geography*, **74(3)**, 289-305.

**Ramakrishnan**, P.S. (1998). Sustainable Development, Climate Change ad the Tropical Rain Forest Landscape. *Climatic Change*, **39**, 583-600.

**Rhodes**, R.A.W. (1997). *Understanding Governance, Politics, and the State*, Basingstoke, UK: MacMillan,

**Wang'ati**, F.J. (1996). The Impact of Climate Variation and Sustainable Development in the Sudano-Sahelian Region. In: Ribot, J.C., Magalhaes, A.R. and Panagides, S.S. eds. *Climate Variability, Climate Change and Social Vulnerability in the Semi-Arid Tropics*, Cambridge: Cambridge University Press, 71-91.

# ANNEXES

## Annex A.9.1. Cases of learning by doing

### *Learning from applications of a cost-effectiveness criterion*

An evaluation of an adaptation process may appear to be easy if we confine our considerations to projects that were selected on the basis of cost-effectiveness. However, the analysis is more complicated than simply computing whether an adaptation intervention cost more or less than anticipated:

- Were there unanticipated inefficiencies that exaggerated costs?
- Were these inefficiencies part of the adaptation strategy or part of the underlying socio-economic context within which the adaptation project was implemented (i.e., would eliminating distortions and/or improving governance in the economy reduce the cost)?
- How were the costs distributed?
- Did the beneficiaries of the adaptation bear the costs, or were they borne elsewhere?
- If the adaptation was a policy (like retreat from the sea) rather than a specific construction, did it create another set of distortions with ancillary costs (or benefits)?
- Would benefits still exceed costs by enough to sustain the high priority ranking that was assigned originally (in Component 4)? Or is the internal rate of return, when recalculated *ex post*, similar to those projected during planning?

Exploring these and other questions could provide insight into when and where projects may not be as effective as possible, or at least as anticipated.

### *Learning from applications of a precautionary criterion*

Application of the precautionary principle (or more elaborate risk analysis) could easily produce adaptations designed explicitly to reduce exposure. Success or failure might then simply be measured in terms of the frequency with which actual experience exceeded a coping range that was established when the implementation decision was made, but such a measure could be very misleading. An examination for purposes of learning and adjusting would compare the actual experience against the revised adaptation baseline described above, because a higher or lower than anticipated frequency of threshold crossings could simply be a manifestation of variability that was higher or lower than anticipated.

Moreover, cost-effectiveness should also be evaluated against that baseline in large measure because implementing one adaptation over one or more alternatives (whether or not they might have addressed a climate-related stress) imposes an opportunity cost that grows with the degree to which actual costs exceeded anticipated expenses.

### *Learning from applications of a cost-benefit criterion*

Application of the cost-benefit approach to the implementation decision means that much could be learned about the process by simply repeating the calculation using the revised adaptation baseline informed by monitoring the future as it unfolded. The questions are some of the same ones asked above regarding cost effectiveness:

- If you knew then what you know now, would benefits still exceed costs?
- Would benefits still exceed costs by enough to sustain the high-priority ranking that was assigned originally (in Component 4)?

Remember, as well, that the benefits for any new adaptation are costs avoided by its implementation assuming that existing adaptations continue to function. Applying the same procedure to new data would accommodate a more accurate portrayal of the future, but it need not be done blind. One might expect, e.g., that:

- Lower than anticipated economic growth could reduce sensitivity and lower benefits.
- Higher than anticipated economic growth could expand sensitivity and increase benefits for anticipated climate change while it increased exposure (e.g., development in a flood-prone area protected by the adaptation project) to surprises.
- Higher than anticipated exposure could reduce benefits unless the adaptation could be amplified.
- Delays in implementation could reduce benefits and increase costs.
- Cost-ineffectiveness would increase costs.

These are, of course, conjectures that need to be examined on a case-to-case basis; but observing patterns could provide new understanding.

### *Learning from applications of a multi-criteria analysis*

Repeating the analysis with new information measured against the revised adaptation baseline and based on the criteria variables identified in the analysis would again be the appropriate approach.

The point of this more detailed analysis, of course, is not to revisit old decisions for purposes of finding fault and attributing blame. It is, instead, to "learn by doing" and thereby glean some general insights into why some adaptations succeed while others fail. Each evaluation would be site-specific and path-dependent; but only careful evaluation of multiple implementing decisions across diverse contexts holds the promise of advancing our understanding of the process. It is appropriate to use the same multi-criteria analysis participative approach in the *ex post* evaluation.

### Annex A.9.2.  A closer look at adaptation monitoring and evaluation

*Monitoring*

As discussed in this TP, monitoring can serve a multitude of purposes as part of an iterative planning-implementation-evaluation process. These include:

- Meeting regulatory requirements;
- Discovering negative impacts of a particular strategy so that corrective actions might be applied;
- Providing insights for ongoing policy/decision-making processes; and
- Helping policy/decision-makers achieve a particular adaptation target or goal more effectively.

A successful monitoring process or activity can also provide documentation about whether or not regulatory requirements have been met in terms of:

- Quantitative indicators of performance;
- Qualitative indicators of performance; and
- A combination of quantitative and qualitative indicators.

The exact mix, of course, depends on the particular methodology chosen for the monitoring process – a decision that itself depends in large measure on the criteria chosen to evaluate adaptation alternatives.

The first question is what to monitor and what to measure. In Components 2 and 3 of the APF, the team has identified what to monitor as the future unfolds between implementation and evaluation. For M&E, the team revisits the goal of the project and asks, "How will we know if we've reached the goal?" The same indicators chosen to describe current vulnerability (Component 2) and characterise future risks (Component 3) will likely provide part of the monitoring process. However, Components 2 and 3 do not necessarily result in identifying who will do the monitoring. Nor will either Component necessarily fill gaps in knowledge that the project may be designed to fill; some decisions might have required making assumptions about information that was not available.

Incorporating an effective monitoring function into an adaptation plan must therefore: (a) assign the responsibility of collecting and maintaining appropriate data to a specific institutional location, and (b) devise procedures by which informational gaps and quality deficiencies might be overcome over time. For example, collection of new data may be assigned to a local organisation that would then be responsible for data quality and archiving.

*Evaluation*

Evaluation without quality data from effective monitoring processes will have no inputs with which to work and no basis for conclusion. Unsupported evaluations produce little more than hypotheses.

Evaluation interprets trends and changes relative to baselines. Do the indicators change as predicted? If not, what needs to be adjusted? An evaluation process can thereby lead to:

- Well-supported decisions (e.g., a careful determination of which strategy is most cost-effective);
- Well-documented responses to critical questions (e.g., a more complete understanding of how a particular adaptation strategy or suite of adaptation strategies can reduce vulnerability);
- Credible depictions of what is actually happening (e.g., a clear depiction of when and how a particular activity or collection of activities might be maladaptive); and
- Equally credible suggestions of how strategies might be improved or corrected (e.g., a systematic investigation of cross-sectoral issues).

*Performance indicators*

Indicators can be described as part of a causal chain. The interrelationships between natural and social processes have been demonstrated by many studies and summarised in the following way: human activities exert **pressures** on the environment, including climate, and change the **state** of the environment while society responds to these changes through environmental, economic, and sectoral policies (the social **responses**).

This Pressure-State-Response (PSR) framework, adopted by many international organisations for defining environmental indicators, can be used to monitor the implementation of adaptation strategies to address climate variability and change, e.g.:

- Indicators can describe pressures on the climate caused by human activities (e.g., greenhouse gas emissions).
- Indicators can describe the state of the environment in terms of environmental quality and aspects of quantity and/or the quality of natural resources.
- Response indicators can, in the context of the PSR framework, refer only to societal (not ecosystem) responses.

Indicators, however, can be described in at least four other dimensions:

- Indicators of *implementation* of the adaptation strategies in the various focal areas can enumerate the delivery of technical services, operating funds, and capital inputs with related disbursements and the resulting outputs generated (e.g., facilities created, activities and participatory processes organised).
- Indicators of *institutional change* can demonstrate capacity development, attitudinal and awareness shifts, and policy reorientations.

- Indicators of *impact in global and local terms* can reveal the environmental accomplishments of the adaptation strategies (e.g., disaster damages trend).
- Indicators of *socio-economic conditions* can be inter-related with the environmental results and impacts, including measures of the consequences of adaptation strategies interventions.

**Annex A.9.3.  Sample planning matrices for monitoring and evaluation in the adaptation process**

PLANNING WORKSHEET (a)

| Project Objectives (Goals, Purpose, Output) | Indicators | Data Collection | | | | | |
|---|---|---|---|---|---|---|---|
| | | Sources of information | Baseline data needed | Who is involved | Tools and methods | How often needed | Added information |

PLANNING WORKSHEET (b)

| Project Objectives (Goals, Purpose, Output) | Indicators | Data Analysis and Use | | | |
|---|---|---|---|---|---|
| | | How often | Who is involved | How are data to be used | Who gets information |

Source: Adapted from DFID (2000)

# Section III

# Case Studies

Section Co-ordinators:
BO LIM (UNDP) AND ELIZABETH MALONE (USA)

# Preface

## Why were these case studies assembled?

At the time of writing of this document, the first adaptation projects to use the Adaptation Policy Framework (APF) are in their early stages. Nonetheless, innovations of the APF have already been taken up in a number of on-going projects. In light of this, these case studies were assembled to further ground the discussion of the User's Guidebook and Technical Papers (TPs) in concrete experience. As experience with climate change adaptation grows, the array of case studies upon which to draw will increase. The preliminary collection here reflects the young and evolving body of adaptation experience.

## What are the objectives of these case studies?

These case studies illustrate the wide range of situations in which the APF may be applied. For purposes of demonstration, the range of situations in which the APF may be used is drawn from different sectors (e.g., agriculture, health, coastal zone management and water resources) and represent different stages of the adaptation cycle (e.g., issue identification, strategic planning and project implementation). They are not intended to be examples of APF "best-practice" – in some cases, their approach is quite different from that advised by the APF. But by indicating the variety of approaches that can be applied, and the steps that can be taken to suit different situations, these cases illustrate the flexibility of the APF.

## Who should read these case studies?

Both the adaptation community and adaptation practitioners will gain from reading these case studies. In addition, those in the policy-making community may find value in reviewing the concrete policy outcomes of real cases, framed in the context of the APF.

As discussed in the User's Guidebook, some project teams may want to emphasise only a subset of the APF Components and related methodologies in order to accomplish their principal objectives. These case studies represent projects with a range of different priorities and objectives, and show how each can be accommodated within the APF. The APF commentary sub-sections are intended to showcase this flexibility and indicate which APF Components and TPs are most relevant to the approach used, and steps taken within the case study.

It should be noted that the process outlined in TP1, *Scoping and Designing an Adaptation Project*, is the starting point for all APF projects, and that TPs 2, *Engaging Stakeholders in the Adaptation Process*, and 7, *Measuring and Enhancing Adaptive Capacity*, are cross–cutting papers with relevance to all components of the process.

# Case Studies

MAARTEN VAN AALST (THE NETHERLANDS), KRISTIE L. EBI (UNITED STATES), ANDREW GITHEKO (KENYA), GARY YOHE (UNITED STATES), ROGER JONES (AUSTRALIA)

# CONTENTS

## I.    Small island state case study: Kiribati[1]
**Author:** Maarten van Aalst, The Netherlands

This case study describes an ongoing adaptation project in Kiribati. It illustrates how the key elements of the Adaptation Policy Framework (APF) apply to planning for adaptation in a small island state, and across all relevant sectors and layers of government down to the community level.

### *Commentary*

This project could use the APF to support its overall objective of placing Kiribati on a more adaptation-friendly development path, i.e., to build adaptive capacity in parallel with its national development goals. The adaptation effort places a heavy emphasis on stakeholder involvement. To assess current and future vulnerability (APF Components 2 and 3), the project team is relying primarily on studies and assessments that have already been completed. Most of the project resources will be devoted to APF Components 4 and 5, developing an adaptation strategy that will be part of the National Development Strategy and continuing the adaptation process.

The following APF Technical Papers (TPs) may be especially useful for this project: TP2, *Engaging Stakeholders in the Adaptation Process*, TP7, *Assessing and Enhancing Adaptive Capacity*, TP8, *Formulating an Adaptation Strategy*, and TP9, *Continuing the Adaptation Process*.

### *Component 1: Scoping and designing an adaptation project*

Kiribati is among the most vulnerable countries to climate change, including variability, and sea level rise. With the current project, the government of Kiribati aims to place the country on a more adaptation-friendly development path. The project includes an extensive process of public awareness and consultation, leading to the formulation of a national adaptation vision and adaptation benchmarks that will be mainstreamed into the National Development Strategy, budget, sectoral plans and policies. The consultations are a key element, not only to develop suitable government policies, but also to facilitate adaptation by non-government stakeholders. This is particularly crucial in the remote outer islands, where traditional systems of governance remain of paramount importance. In these places, adaptation may involve difficult issues such as relocation, which will require a long-term dialogue among customary land owners. In order to have sufficient political leverage, the project is placed directly under the leadership of the Secretary to the Cabinet and the Director of Economic Planning – in cooperation with a working group of senior officers from all relevant government agencies, as well as representatives of NGOs, women's groups, and the private sector.

### *Component 2: Assessing current vulnerability*

Kiribati has a population of 93,000, spread over 33 low islands in the central Pacific, covering a landmass of only 730 sq km. The country is extremely isolated, with the nearest large markets 4,000 km away. About one-third of the population lives in the capital, South Tarawa, a very densely populated area with a population growth of 3% a year; this population density and rate of growth places great challenges on the fragile atoll environment. Most of the land in Tarawa is less than 3 meters above sea level, with an average width of only 450 meters. Kiribati's arid climate and poor atoll soil offer little potential for agricultural development. On the other hand, the immense area of ocean (an exclusive economic zone of 3.6 million sq km) harbours some of the richest fishing grounds in the world, and provides the country's most important source of revenue.

The islands are exposed to periodic storm surges and droughts, particularly during La Niña years. Kiribati is becoming increasingly vulnerable to climate events due to its high population concentration, accelerated coastal development, shoreline erosion, and increasing environmental degradation, including problems with solid and human waste disposal in South Tarawa. On the other hand, many strong traditional coping mechanisms remain in place, including strong community and family support structures and traditional construction methods.

Under the current project, further analysis of current vulnerability will focus on the key risks experienced by communities on all islands of Kiribati. It will complement local experiences and participatory risk mapping with technical analyses of, for instance, water resources and inundation risks.

### *Component 3: Characterising future climate risks*

Climate change impacts in Kiribati have been analysed for the 2000 World Bank Regional Economic Report (World Bank, 2000) on the basis of an integrated assessment model of climate change in the Pacific Islands region (Kenny et al. 1999, updated for World Bank, 2000). Kiribati is likely to experience higher temperatures, sea level rise, and a more El Niño-like mean state, while there is considerable uncertainty with respect to rainfall (Table I-1).

Among the most dramatic impacts is the increased risk of inundation, particularly during storm surges. For instance, up to 25-54% of South Tarawa areas and 55-80% of North Tarawa areas could be inundated by 2050. The combined effect of sea level rise, changes in rainfall, and changes in evapotranspiration due to higher temperatures could result in a 19-38% decline in the thickness of the main groundwater lens. Agriculture productivity – particularly for taro and pandanus – could decline due to storm-induced saltwater intrusion into groundwater lenses.

[1] The Kiribati Adaptation Project reflects many elements of the APF, but was planned independently. Its first phase started in 2002, to be followed by a second phase around 2005. Full project details can be found in the project information available at http://www.worldbank.org. This case study highlights only the elements relevant to the APF. The actual project is continuously evolving and may change substantially over time.

***Table I-1:*** *Climate change and variability scenario for Kiribati (World Bank, 2000)*

| Impact | 2025 | 2050 | 2100 | Level of Certainty |
|---|---|---|---|---|
| Sea level rise (cm) | 11–21 | 23–43 | 50–103 | Moderate |
| Air temperature increase (degrees Centigrade) | 0.5–0.6 | 0.9–1.3 | 1.6–3.4 | High |
| Change in rainfall (%) | -4.8-+3.2 | -10.7-+7.1 | -26.9-+17.7 | Low |
| El Niño Southern Oscillation (ENSO) | A more El Niño-like mean state | | | Moderate |

Note: Ranges reflect a best-guess scenario (lower value) and a worst-case scenario (higher value). For details, see Annex A of World Bank (2000).

***Table I-2.*** *Estimated annual economic impact of climate change, 2050 (millions of 1998 US$)*

| Impact | Average Annual damage[a] | Level of Certainty | Likely Cost of an Extreme Event[b] |
|---|---|---|---|
| **Impact on coastal areas** | | | |
| Loss of land to erosion | 0.1–0.3 | Low | ? |
| Loss of coastal land and infrastructure to inundation | 7–12 | Low | 210-430 (storm surge) |
| Loss of coral reefs and related services | 0.2–0.5 | Very low | – |
| **Impact on water resources** | | | |
| Replacement of potable water supply due to change in precipitation, sea level rise, and inundation | 1–3 | Low | ? |
| **Impact on agriculture** | | | |
| Agriculture output loss | + | Low | ? |
| **Impact on public health** | | | |
| Increased incidence of diarrheal disease | ++ | Low | ? |
| Increased incidence of dengue fever | + | Low | ? |
| Increased incidence of ciguatera | + | Low | ? |
| Impacts on public safety and on the poor | + | Very Low | ? |
| Potential increase in fatalities due to inundation, and water-borne and vector-borne diseases | + | Low | ? |
| Total | >8–16+ | | ? |

[a] Reflects incremental average annual costs due to climate change, equivalent here to the capital recovery cost factor of land and infrastructure damaged by inundation, using a discount rate of 10% and a 10-year period

[b] Reflects financial damages to land and infrastructure caused by sea level rise and storm surge during a 1 in 14 year storm event.

For assumptions, see World Bank (2000) Annex A.

Higher temperatures could also increase the epidemic potential for dengue fever by 22-33%, increase the incidence of ciguatera poisoning and degradation of coral reefs, and divert critical tuna resources away from Kiribati waters. In the absence of adaptation, these impacts were estimated to result in economic damages averaging US$8-$16 million a year, equivalent to 17 to 34% of the 1998 GDP (Table I-2).

These threats will be exacerbated by high population growth and the associated pressure on water resources and the environment, particularly on Tarawa. The atolls' extreme isolation will remain a critical factor in their vulnerability, particularly in the outer islands. To some extent, autonomous adaptation may include migration from outer islands to Tarawa, further exacerbating its problems of overcrowding.

The current project will analyse specific impacts based upon the concerns reflected by communities, partly on the basis their perception of trends in the local climate and/or their vulnerability. This may include, for instance, additional inundation and storm surge mappings and analysis of water resources.

### Component 4: Formulating an adaptation strategy

The initial consultations will yield a better sense of the vulnerability of the country and its communities. Subsequent technical and economic analyses together with further consultations will assess potential adaptation options, including costs and benefits (in the short- and long-terms), urgency, and barriers to implementation. Particular attention will be given to Kiribati's traditional community adaptation strategies and the extent to which these strategies will be sufficient to cope with future conditions, and where government support could help to bolster this local capacity. Finally, communities and the government will adopt a national adaptation vision and adaptation benchmarks that will be included in the National Development Strategy, the budget, sectoral plans and policies, and the regulatory framework. Priorities could include mangroves and coral reef protection; management of human and solid waste in lagoons; water conservation; changes in fisheries management; land use planning; protection of coastal infrastructure (e.g., by elevation or set-backs); promotion of traditional adaptation practices for agriculture (such as dry/wet season crop rotations and breeding for drought/salinity tolerance); control of mosquito vectors; and multi-year license fees for tuna fisheries to smoothen out inter-annual variations.

### Component 5: Continuing the adaptation process

Once the adaptation vision and benchmarks are mainstreamed into the National Development Plan, sectoral plans, the budget and the regulatory framework, they should become part of the regular monitoring of the government's strategies and expenditures, including through traditional local governance systems and local consultation.

In addition to solid analysis, this mainstreaming process will require intensive consultation and strong political will. However, several priority adaptations may also require upfront financial investments. Hence, the current Preparation Phase will also generate a project proposal for an Investment Phase to be submitted to donors. Rather than funding stand-alone measures, the Investment Phase could institutionalise adaptation across sectoral programs and policies by providing financial support for broad sectoral adaptation goals, measured against agreed adaptation benchmarks.[2] In addition, communities may identify adaptation strategies that could be executed with limited government support. A *social adaptation fund* could target their adaptation priorities that may include local efforts to replant mangroves, conserve water, control pollution, or experiment with drought-resistant crops.

### References

**Kenny**, G., Warrick, R., Ye, W. and de Wet, N.Z. (1999). *The PACCLIM Scenario Generator – System Description and User's Guide* (A report to: South Pacific Regional Environment Programme, Apia, Samoa), International Global Change Institute, University of Waikato, Hamilton, New Zealand.

**World Bank**. (2003). Kiribati Adaptation Project, Project Information Document, *http://www.worldbank.org* (downloaded April 2003).

**World Bank**. (2000). *Cities, Seas, and Storms, Managing Change in Pacific Island Economies. Volume IV: Adapting to Climate Change*. The World Bank, Washington DC.

---

[2] Such benchmarks could include a reduction in water leakage or an increase in rainwater catchment capacity by a certain percentage, allocating a minimum percentage of the fisheries budget to coastal management, or starting a national campaign to control dengue fever.

## II.     Highland malaria case study: Kenya[3]

**Authors**: Kristie L. Ebi[4] and Andrew Githeko[5]

In evaluating the health impacts of global warming, scientists have discovered that climate change may work against efforts to bring malaria (*Plasmodium falciparum*) under control. Rising temperatures could create environmental conditions that would actually increase the area that is climatically hospitable for malaria vectors worldwide. Clearly, understanding the range of possible impacts of climate change is imperative, particularly on national and sub-national scales in sub-Saharan Africa, where malaria is already prevalent.

### Commentary

The goal of this project is to develop an early warning system to increase Kenya's preparedness for malaria epidemics – in accordance with its existing national policy for control of this disease. Stakeholder input occurred during project design, formulation of the adaptation strategy, and continuing the adaptation process (Adaptation Policy Framework (APF) Components 1, 4 and 5). The team used their own experience and their research findings in assessing current vulnerability (APF Component 2), and the established IPCC scenarios to characterise future climate risks (APF Component 3). Given their familiarity with the research, the adaptation team considered that formulating a strategy would be straightforward (Component 4); they have identified barriers to continuing the adaptation process (Component 5) and are actively seeking solutions.

Among the most valuable APF Technical Papers (TPs) for this project are TP2, *Engaging Stakeholders in the Adaptation Process*, TP3, *Assessing Vulnerability for Climate Adaptation*, TP7, *Assessing and Enhancing Adaptive Capacity*, and TP9, *Continuing the Adaptation Process*.

### Component 1: Scoping and Designing an Adaptation Project

The objective was to develop an early warning tool using meteorological data to predict when and where malaria epidemics are likely to occur in western Kenya in order to reduce uncertainty in decision-making and to facilitate better resource and disease management.

The Kenyan government's policy on malaria control is based on quick diagnosis and effective treatment. The policy assumes the availability of sufficient manpower, drugs, and other resources, as well as prompt interventions to prevent epidemics. However, the number of people affected in recent epidemics was so high that the demand for drugs outstripped sup-

plies. Development of an early warning system would increase preparedness for malaria epidemics, thereby decreasing the burden of malaria – and the social and economic costs associated with outbreaks of the disease.

This case study is based on a research project. In other words, it was driven by researchers recognising a policy need and leading the policy-makers in the appropriate direction.

The project team consisted of Drs. Githeko and Ndegwa. They determined the scope of the project based on their experience with highland malaria in Kenya. Other scientists working on tools to predict malaria epidemics formed an informal stakeholder group. At the project's conclusion, policy makers were included; they were approached for funding to implement the early warning system they had developed.

### Component 2: Assessing current vulnerability

From a public health perspective, malaria significantly affects Kenyan health, society, and economy. The 1990 Global Burden of Disease study estimates that malaria accounted for approximately 10.8% of years of life lost across sub-Saharan Africa (Murray *et al.*, 1996). In sub-Saharan Africa, malaria remains the most common parasitic disease and is the main cause of morbidity and mortality among children under five and among pregnant women. Roughly 1 million deaths (0.74 to 1.3 million) from the direct effects of malaria occur annually in Africa, more than 75% of them in children. This estimate could double if the indirect effects of malaria (including malaria-related anaemia, hypoglycemia, respiratory distress and low birth weight) are included when defining the burden of malaria (Breman, 2001).

Although climatic factors can influence malaria transmission, the outcome of the clinical disease depends on the level of immunity of the infected person, how early the disease is treated, and the effectiveness of the anti-malarial drugs.

Epidemic malaria in the Kenya highlands generally occurs at altitudes of between 1500-2200 meters. Epidemics normally occur from May to August, following the long rains. Malaria epidemics usually require emergency measures that must be promptly implemented. Predicting when and where outbreaks will occur has been a matter of guesswork.

The project team determined the country's current vulnerability using data collected on malaria outbreaks from districts in the Kenyan highlands. They found that, during the past 13 years, malaria epidemics in western Kenya spread from 3 to 15 districts, frequently taking the population by surprise (Githeko and Ndegwa, 2001). The epidemics were associated with high morbidity and mortality in all age groups, with the prevalence

---

[3]  This example is based on research conducted in Kenya by Drs. Githeko and Ndegwa, malariologists working for many years on tools to better predict malaria epidemics in the Kenyan highlands.

[4]  Exponent, Alexandria, United States

[5]  Centre for Vector Biology and Control Research, Kisumu, Kenya

of the disease ranging from 20-60%. The case mortality rate was estimated at about 7.5%.

The project team determined the current climatic risks based on the numerous laboratory and field studies that document the influence of precipitation and temperature on the range and prevalence of malaria. In summary, climate and anomalous weather events directly influence malaria transmission by either hindering or enhancing vector and parasite development and survival, as follows:

- Climate suitability is a primary determinant of whether the conditions in a particular location are suitable for stable *Plasmodium falciparum* malaria transmission.
- A change in temperature may lengthen or shorten the season during which mosquitoes or parasites can survive.
- Changes in precipitation or temperature may result in conditions during the season of transmission that either increase or decrease the parasite and vector populations.

Changes in precipitation or temperature may cause previously inhospitable altitudes or ecosystems to become conducive to transmission. Higher altitudes that were formerly too cold or desert fringes that previously were too dry for mosquito populations to develop may be rendered hospitable by small changes in temperature or precipitation.

The project team obtained descriptions of Kenyan climatic conditions that could influence malaria transmission from the local meteorological service. In general, Kenya experiences significant interannual and decadal variability, whereas in particular, its total annual precipitation is characterised by strong variability, with global oceanic-atmospheric processes, such as the El Niño Southern Oscillation, as the dominant cause of precipitation variability (high rainfall is associated with El Niño years in the East African region). Analysis of climate data over the past 100 years showed that Kenya experienced a warming and drying trend. The instrumental surface air temperature suggested an increase of up to 0.8∞C in some regions. This warming trend was accompanied by a 10% reduction in rainy season precipitation. Analysis of data over the past decade suggested that there has been an increase in the frequency and intensity of anomalies in mean monthly maximum temperature.

### Component 3: Assessing Future Climate Risks

The project team used regional analyses conducted by the local meteorological service for the IPCC Third Assessment Report to determine future climatic conditions. Analyses of the regional impact of projected global climate change suggest an increase in temperature of 0.7-4.7°C by the 2050s (dependent on the emissions scenario, climate sensitivity, and general circulation model). The magnitude and direction of projected changes in precipitation is less consistent. Climate variability is expected to increase with climate change, resulting in an increased number of temperature anomalies.

### Component 4: Formulating an adaptation strategy

Githeko and Ndegwa analysed the available malaria and meteorological data and found an association between mean monthly rainfall and anomalies in mean monthly maximum temperatures and the number of inpatient malaria cases 3-4 months later. These meteorological variables were used to construct an epidemic prediction model for the Kenyan highlands. Because the model uses readily available temperature and rainfall data, health personnel can apply the early warning system with little training.

The early warning system is designed to enhance current malaria surveillance programs. As the model developed is weather-based, its use will increase adaptive capacity to current and future conditions, assuming that the weather/health relationships identified do not change with climate change. A formal assessment of the costs and benefits of the model was not conducted and was not necessary because the benefits of reducing the burden of malaria to Kenyan society are significantly higher than any costs associated with implementing and running the prediction model.

Stakeholders had two opportunities for input into model development: (1) feedback at professional society meetings and (2) peer review of manuscripts submitted for publication. The project team used this feedback to improve the model.

### Component 5: Continuing the adaptation process

Clearly this early warning system needs to be implemented as an adaptation measure to current and future climate. There have been a number of barriers to model implementation. One is the general unawareness by health personnel that weather is a factor in malaria transmission. Another is that the Ministry of Health established model validation as a criterion for implementation. However, funding was not available for the data collection and analysis required. A third is a lack of trained personnel to use the model.

The project team is working with the various stakeholder groups to increase awareness of the value of the model, and to address the concerns expressed by the Ministry of Health. Recently, after the initial scepticism about the application of the model, pressure from funding agencies and new field observations have cleared the way for collaboration between the Meteorological Department and health experts in model application.

### References

**Githeko**, A.K. and Ndegwa, W. (2001). Predicting malaria epidemics in the Kenyan highlands using climate data: a tool for decision makers, *Global Change & Human Health*, **2**, 54-63.

**Murray**, C. and Lopez, A. (1996). *The Global Burden of Disease*, Boston, MA: Harvard University Press.

**Breman**, J.G. (2001). The ears of the hippopotamus: manifestations, determinants and estimates of the malaria burden, *American Journal of Tropical Medicine and Hygiene* **64** (suppl 1), 1-11.

## III.    Agriculture case study: Mexico

**Author:** Gary Yohe, Wesleyan University, Middletown, United States

The vulnerability of smallholder maize production to climate risk provides a perfect context within which to illustrate how the Adaptation Policy Framework (APF) can be employed. This hypothetical example, based on methodological work by Yohe et al. (1999), will work through the five Components of the APF before applying the fundamental principles of monitoring and evaluation to one specific adaptation strategy – the introduction of a drought resistant hybrid.

### *Commentary*

This case study adopts a vulnerability-reduction strategy and focus. Hence, most of the resources in this hypothetical study are devoted to delineating the vulnerability and resilience of smallholders (APF Component 2), and to defining a range of future climate risks (APF Component 3). Also emphasised is the role of monitoring and evaluation (APF Component 5) in determining the success of the selected adaptation. Although stakeholder consultation is not discussed, this type of activity would clearly benefit from direct discussion with farmers.

The most helpful APF Technical Papers (TPs) would be TP3, *Vulnerability Assessment for Climate Adaptation*, TP4, *Assessing Current Climate Risks*, TP5, *Assessing Future Climate Risks*, and TP9, *Continuing the Adaptation Process*.

### *Component 1: Scoping and designing an adaptation project*

The objectives of this adaptation project were to reduce the vulnerability of small, traditional farming communities to climate vulnerability, and to increase their maize yields so that they can participate in commercial markets.

In a review of the literature, the local climate has never been particularly suited for agricultural production. Grazing and irrigated farming should be preferred over the more than 40% of the land that is classified as arid, but poor soils, limited water, and complex topography can frequently support only rain-fed agriculture.

Historically, the federal government has initiated a series of economic and land tenure reforms over the past several decades. They have promoted modest technological change, freer product markets, and general integration into international markets in addition to sporadic input subsidies and preferential structures for small loans.

### *Component 2: Assessing current vulnerability*

*Climate risks:* The major climate risks are early frosts in the fall and late frosts in the spring, so that the probability of a frost is

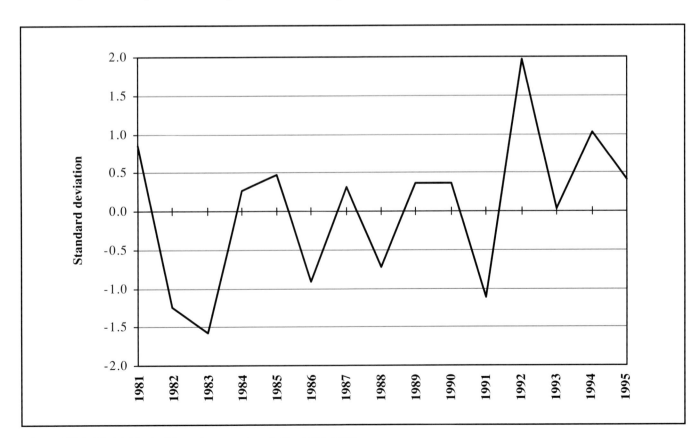

**Figure III-1:** Variability in Tlaxcalecan maize yields, 1981-1995.
Source: Figure 4 in Eakin (2000)

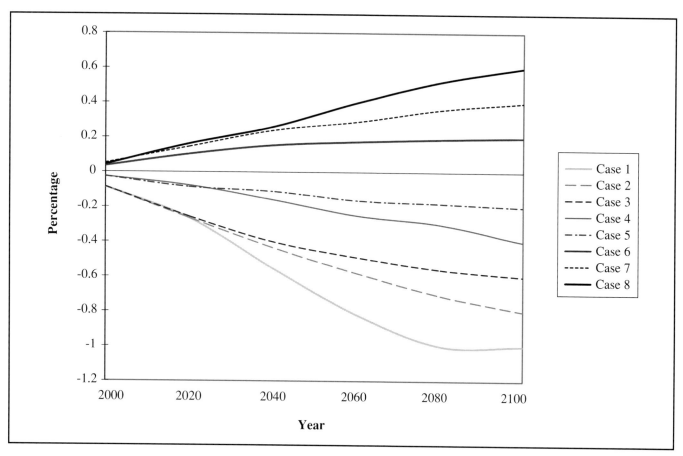

**Figure III-2:** Trajectories of July precipitation for eight representative climate scenarios.

Source: Figure 8 in Yohe, et al. (1999)

less than 50% for only 187 days per year (on average); extremely variable precipitation (400mm to 1200mm per year), particularly in July.

*Socio-economic conditions*: Maize agriculture dominates production for more than 50% of the households. Yields are extremely variable (Figure III-1). Recently, fertiliser costs have risen significantly, sources of financial credit have dwindled, and price guarantees have evaporated to the point where socio-economic uncertainty dwarfs climate uncertainty.

*Vulnerability*: Households suffer extreme hardships when yields fall below 2000kg/ha. This threshold defines a coping range whose boundary was crossed 30% of the time between 1967 and 1989. Precipitation in July is the critical climate variable, particularly when warming climate scenarios reduced the threat of early and/or late frost.

*Adaptations*: Households routinely adopt a range of risk-averse adjustments depending on their experience-based expectations of climate for the next growing season. These include planting shorter, fast-maturing maize varieties (with corresponding lower yields), changing planting dates, rescheduling labour-intensive tasks, building terraces and small scale irrigation projects, and diversifying crops across locations. Under extreme conditions, farmers must sell live-

stock and/or farm equipment for cash to support themselves and their families; and they rely on family and social community networks for assistance.

*Policy needs*: Interventions designed to reduce vulnerability to climate variability and uncertain economic conditions in the short-run and to reduce vulnerability to climate change and socio-economic trends over the longer term.

### Component 3: Characterising future climate risks

*Climate trends*: Figure III-2 displays a representative range of not-implausible scenarios of July precipitation in Mexico drawn by COSMIC (Schlesinger and Williams, 1999) from 14 different global circulation models and multiple climate sensitivities and emissions trajectories. Figure III-3 depicts the corresponding sustainability indices for each scenario; they reflect the likelihood (inferred from fitting a gamma distribution to historical monthly precipitation records) that rainfall in July will be high enough to sustain yields in excess of 2000 kg/ha along any given trajectory.

*Socio-economic trends*: Continued emphasis on commercialisation, globalisation, liberalisation of even domestic markets, and strong urbanisation.

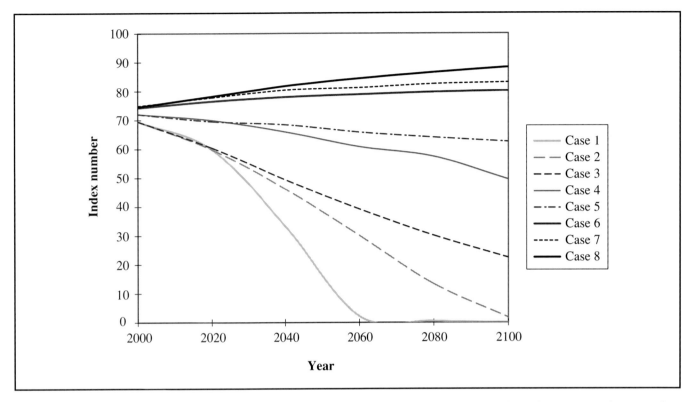

**Figure III-3:** Corresponding sustainability index trajectories for traditional maize agriculture along the representative scenarios. The sustainability index is the likelihood that July precipitation will be above a critical threshold in any given year.

Source: Figure 9 in Yohe, et al. (1999)

### *Component 4: Formulating an adaptation strategy*

The government developed drought-resistant hybrid varieties of maize and provided incentives for some farmers to participate in demonstration farms.

### *Component 5: Continuing the adaptation process*

*Incorporation*: The demonstration farm initiatives were implemented and were perfect subjects for applying monitoring and evaluation fundamentals.

*Monitor and evaluate hybrid varieties*: External analysis showed that the hybrids produced higher yields in good years but performed only slightly better in really bad years. Reliance on chemical inputs and irrigation increased economic vulnerability (through higher debt). Yields from the demonstration participants did not cross the critical 2000 kg/ha as frequently, so the sustainability indices are higher for any future climate scenario.

*Incorporate into development plans*: A switch to hybrid varieties would reduce the vulnerability of traditional agriculture to climate change for even the most dramatic changes reflected in the range of possible futures. This technological fix would not necessarily preserve traditional maize agriculture even though the sustainability index would stay above current levels for decades. Socio-economic threats would dwarf vulnerability

to climate even along the worst trajectories of Figures III-2 and III-3.

### References

**Eakin**, H. (2000). Smallholder maize production and climatic risk: a case study from Mexico, *Climatic Change* **45**, 19-36.

**Schlesinger**, M. and Williams, L. (1999). Country specific model for intertemporal climate, *Climatic Change* **41**, 55-67.

**Yohe**, G.W., Jacobsen, M. and Gapotchenko, T. (1999). Spanning 'not-implausible' futures to assess relative vulnerability to climate change and climate variability, *Global Environmental Change* **9**, 233-249.

## IV.    Coastal zone management case study

**Author:** Gary Yohe, Wesleyan University, Middletown, United States

Sea level rise is the best-identified impact of projected climate change. Thus, coastal zone management provides important opportunities for adaptation.

### Commentary

In this hypothetical project, the approach taken emphasises socio-economic development along the coast to evaluate two development strategies. The project team places most of its resources in assessing current vulnerability (Component 2), characterising future climate risks (Component 3), and developing an adaptation strategy (Component 4).

The Adaptation Policy Framework (APF) Technical Papers (TPs) that served as valuable resources for this case study are TP5, *Assessing Future Climate Risks*, TP6, *Assessing Current and Changing Socio-economic Conditions*, and TP8, *Formulating of an Adaptation Strategy*.

### Component 1: Scoping and designing an adaptation project

The team scoped and designed the APF process for coastal zone management. The team wanted to address the specific issue of superimposing coastal storms and long-term – global warming-induced – sea level rise can pose significant problems for coastal zone management, particularly along the country's developed coastlines. Some communities and states have responded to these dual threats by adopting policies designed to promote a systematic and long-term retreat from the sea. Restricting new development within vulnerable areas and, by extension, resettlement of displaced citizens from these at-risk locations, promotes retreat, but not without some cost. If other trends ameliorate those costs, however, then policies that restrict development might be more popular.

Mindful of this situation, the adaptation team designed a project focusing on three coastal areas currently experiencing pressure to develop and build new infrastructure along the coastline. The team members decided to use a primarily socio-economic analysis of the value of coastal property in these areas, together with a study of current and projected climate risks. They intended to examine the effect of a proposed policy to preclude rebuilding after the destruction of a building due to storm damage or sea level rise.

The design of the project included plans for stakeholder involvement from coastal property owners, community leaders and government officials, and representatives from the Ministries of Economic Development and Environment.

### Component 2: Assessing current vulnerability

The critical vulnerability of developed property to coastal storms works directly through flooding and wind damage and indirectly through beach erosion. The project team used West's characterisation of current vulnerability in terms of

- inundation from coastal storms with and without sea level rise,
- the probability of damage (for storms categorised in terms of the frequency of their occurrence),
- a net likelihood that damage will occur in any one year (given the current distribution of storm intensity on an annual basis), and
- a probability distribution of the degree of damage (contingent on the condition that some sort of damage has been observed).

The team members used data and distributions, calibrated by data on storm-damage claims from the National Flood Insurance Program in the United States.

Current adaptations – in communities where the mean value of property is established – include rebuilding damaged structures as long as the present value of future housing services (as indicated by current property values) exceeds the cost of renovation. Current policy allows relocation within the vulnerable area in the wake of complete destruction.

### Component 3: Characterising future climate risks

Using the same methodology, the project team represented future climate conditions in terms of warming-induced sea level rise (40 cm through 2100) and various rates of background erosion. They then used stochastic weather generators to produce a collection of trajectories characterising storm events over the next 100 years under the assumption that the distribution of future storms must be consistent with current climate variability. Combined with erosion and sea level rise futures, these stochastic futures generated economic cost trajectories defined by the relationships that characterised the current climate. The relationships characterised by the collection of trajectories served as anchors for measuring the degree of damage that could be expected as erosion and sea level rise moved the shoreline in towards the mainland.

### Component 4: Formulating an adaptation strategy

The project team investigated the effect of adding a new adaptation policy – restricting development to the point of prohibiting people who lose their homes to coastal storms from rebuilding in the same vulnerable community. They chose the economic cost of storm damage attributed to sea level rise as their indicator – a statistic measured by computing the discounted cost of storm damage with and without sea level rise of 40 cm through the year 2100. Two regimes were considered. In the first, displaced owners could relocate within the vulnerable locale, but they could not in the second. The precise numbers reported for this exercise by West and Dowlatabadi were hypothetical, but comparing the two policy regimes nonetheless produced some revealing results. The economic cost attributed to sea level rise was the same for the two regimes without background erosion, but the estimates diverged significantly as the rate of erosion grew. Indeed, high rates of erosion allowed for the possibility that closing the re-development option could actually eliminate storm damage attributed to sea level rise. This is not to say that storms would cause no damage in these cases; instead, storm damage eliminated vulnerable structures (because they could not be reconstructed in vulnerable areas) before they were in jeopardy from rising seas.

### *Component 5: Continuing the adaptation process*

The project team, including stakeholders, established monitoring and evaluation efforts that included tracking the number and value of structures that were not replaced after suffering significant storm damage in advance of rising seas. Another measure will be to track the success of court cases where owners challenged a policy that limited their options after suffering a loss.

Integrating into this policy change into development programs is involving construction of zoning rules, defence against legal challenges, and perhaps the definition of the degree of damage that would signify a total loss (from the perspective of the reconstruction prohibition). The Beachfront Management Act of South Carolina, enacted in 1988, sets this limit at two-thirds of the pre-storm value, and it has stood the test of time in the courts. In this APF process, the success of the policy will need to be evaluated over time.

## References

**Neumann**, J.E, Yohe, G.W. and Nichols, R. (2000). *Sea-level Rise and Global Climate Change*, Pew Center on Global Climate Change, Washington, D.C. *Human System Response Relevant to Global Environmental Change*, PhD. Dissertation, Carnegie Mellon University, Pittsburgh.

**West**, J. and Dowlatabadi, H. (1999), On assessing the economic impacts of sea level rise on developed coasts. in T. Downing, A. Olsthoorn, and R.S.J. Tol eds) *Climate, Change and Risk*, London: Routledge, pp. 205–220.

### V.    Water resources case study: The Murray-Darling Basin in Australia

**Author**: Roger Jones, Commonwealth Scientific & Industrial Research Organisation, Atmospheric Research, Aspendale, Australia

Water supply and quality is one of the primary areas of risk under climate change because of its importance in both natural and managed systems and its sensitivity to climate. Although a well-managed system can cope with a wide range of climate variability, climate change has the potential to threaten water resources in both developed and undeveloped systems. This case study uses a series of assessments, mainly carried out in the Macquarie River Catchment – part of the Murray-Darling Basin – in eastern Australia to demonstrate how uncertainty can be managed using risk assessment techniques.

#### *Commentary*

This project focuses on assessing current and future climate risks (Adaptation Policy Framework (APF) Components 2 and 3). Stakeholders are informed of the results at the end of Component 2 and are the principal factor in suggesting ways to continue the adaptation process; also, use of an existing model brought together water managers and users in the stakeholder group. The resulting adaptation options were not evaluated against others, perhaps because there was existing consensus on the policies and measures to be taken.

The most useful APF Technical Papers (TPs) for this type of project would be TP2, *Stakeholder Engagement in the Adaptation Process*, TP4, *Assessing Current Climate Risks,* and TP5, *Assessing Future Climate Risks.*

#### *Component 1: Scoping and designing an adaptation project*

The objective of this project was to determine whether climate change poses a sufficient risk to water resources in the Murray-Darling Basin to be incorporated into water policy and integrated catchment management plans.

The project utilised a catchment water management model operated by the state water management authority. This model was used because of its credibility with the managers and water users. Depending on availability, water management allocates irrigation supply from a dam with one year's streamflow each irrigation season. Some environmental flows and all domestic and industrial flows are high security. Excess supply may be sold as low security "off-allocation" water. The method of assessment was to take the existing model and perturb it by climate change scenarios for 2030 and 2070.

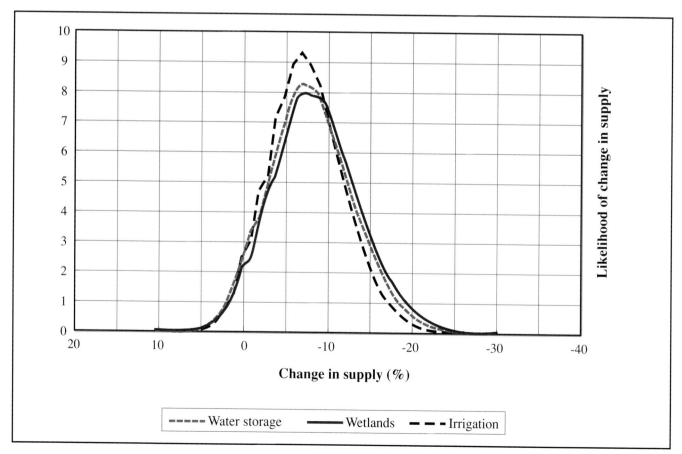

**Figure V-1:** Probability distribution for changes to mean annual water supply, Macquarie Marsh inflows and irrigation allocations for the Macquarie catchment in 2030

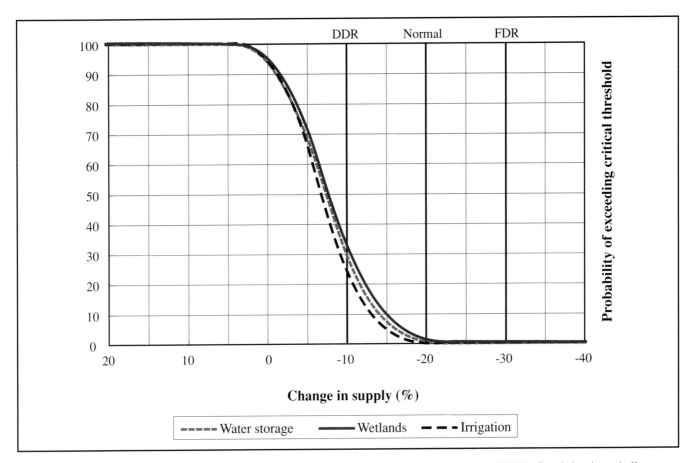

**Figure V-2:** Probability of exceeding critical thresholds under a drought-dominated climate (DDR), flood-dominated climate (FDR) and normal climate (Normal) for the Macquarie catchment in 2030.

*Component 2: Assessing current vulnerability*

A baseline climate record of rainfall and potential evaporation (A-Class Pan extended using temperature regression), allowed a 100+ year record of flows to be analysed, although only the past 50 years have good streamflow records. The widespread development of irrigation systems has occurred since the 1950s and irrigation development was largely uncontrolled until capped in the late 1990s. Extractions in the Macquarie catchment are above their sustainable limit (NLWRA, 2001).

The 20th century can be divided into a dry period for the first half and a wet period during the second half. These are described as drought-dominated and flood-dominated rainfall regimes, and denote a period of several decades where average rainfall decreases by more than about ±20% from the long-term mean. However, irrigation development has seen flows decrease throughout the twentieth century, seriously threatening the Macquarie Marshes, a Ramsar wetland of international significance. This has united local graziers and conservationists, both concerned over wetland degradation.

Irrigation has been economically successful, and cotton growing has expanded and moved south due to climate change and improved varieties. However, the catchment is threatened by both irrigation and dryland salinity, with elevated levels of

saline discharge threatening future water supply. Irrigators who learned their craft during the wetter second half of the 20th century may not yet have developed the adaptations to deal with reduced water supply if climate variability and/or climate change reduces supply below the current capped levels.

*Component 3: Characterising future climate risks*

Several levels of climate change information have contributed to the assessment of climate change risks.

*Climate projections:* Climate change projections for rainfall and potential evaporation from the set of climate models stored on the IPCC Data Distribution Centre were calculated for region in question. Rainfall was taken directly and potential evaporation was calculated from model output. These changes were converted into change per degree of global warming and scaled for the IPCC range of global warming in 2030 and 2070. They show the spread of models in terms of rainfall increase and decrease and which models are the driest and wettest for the region. Robust findings were 1) that late winter-spring rainfall usually decreased relative to summer-autumn rainfalls in most models and 2) potential evaporation change could be related to rainfall change across all models – where rainfall increased, the increase in potential evaporation was smaller than it was for rainfall decreases. This information was communicated to stakeholders.

*Sensitivity assessment:* A series of sensitivity experiments showed that this catchment was much more sensitive to cool season changes in rainfall than warm season changes in rainfall, and that low flows were much more sensitive to changes than median or high flows.

*Vulnerability assessment:* Critical thresholds defining vulnerability in terms of irrigation supply and environmental flows were set as five years of irrigation allocations below 50% of the water right and ten years of low flows into the Macquarie Marshes insufficient to trigger waterbird breeding. They were found to be breached when mean annual streamflow decreased by more than 10% in a drought-dominated rainfall regime, more than 20% in a normal regime and more than 30% in a flood-dominated regime. The assessment showed that both long-term rainfall variability and climate change acting together should be assessed as part of long-term climate risks. This is probably true for many regions in the world but is made difficult by the need for long-term baseline data.

*Natural hazard-based assessment:* A natural hazard-based assessment projected a number of scenarios through the model to estimate the most likely outcomes. Ranges of uncertainty for input variables describing changes in mean global warming, rainfall and potential evaporation were randomly sampled and used to perturb a simple algorithm relating rainfall and potential evaporation change to change in mean annual streamflow. The result was a probability distribution describing a wide range of possible changes that favoured the central tendencies at the expense of the extremes.

Figure V-1 shows the results for 2030. Although there is an increased flood risk with increased flows, the drier outcomes are considered worse in terms of lost agricultural and environmental productivity. The extremes of the range are about +10% to –30% but the most likely outcomes range from about 0% to –15%.

*Vulnerability-based assessment:* Probabilities of exceeding the critical thresholds described earlier were assessed in a vulnerability-based risk assessment. These showed that the likelihood of exceeding a critical threshold was subject to both the decadal rainfall regime and to the mean change in climate. The likelihood of exceeding the critical thresholds in 2030 in a drought-dominate rainfall regime is about 25% for irrigation and 35% for environmental flows, showing that environmental flows are subject to a higher risk. In a normal climate, these likelihoods are about 2% and 1%.

In 2070, likelihoods of critical threshold exceedances are much higher: 70% and 75% for irrigation and environmental flows respectively in a drought-dominated rainfall regime, 30% and 40% in a normal climate and 3% and 6% in a flood-dominated rainfall regime. Without adaptation, vulnerability becomes more likely as climate change progresses.

*Integrated assessment:* Limited integrated assessment was carried out looking at the combined impacts of climate change and re-forestation on streamflows. For a 10% increase in tree cover in the headwaters of the Macquarie, a 17% reduction in inflows to Burrendong Dam was estimated and for a 2% increase a reduction of 4% was estimated. Changes in flows shown in Figure V-1 would add directly onto these changes except for large reductions in flow. Therefore, if revegetation and carbon sequestration aims are both pursued in the upper catchment, they will combine with climate change to reduce flows. Revegetation targeted for mid-catchment areas to control dryland salinity have less of an effect on streamflow but are commercially sub-economic because of the lower rainfall in these areas.

### Component 4: Formulating an adaptation strategy

In this case, no distinct adaptation strategy has been developed. Instead, this case outlines the replication of successful activities as a means of advancing adaptation.

### Component 5: Continuing the adaptation process

A number of strategies to manage water resources in a more sustainable manner are currently underway or have been recently implemented. A cap on extractions has been imposed in the Murray-Darling Basin and is being extended to the rest of Australia as part of the National Water Reform Policy. The ultimate intention is that sustainable limits of extraction be set in each catchment. The Living Murray project is assessing environmental flows for the Murray River where over 70% of the available flow is currently being extracted. As part of the investigations Jones et al. (2001) extended the results from the Macquarie Study, to the Murray River to determine whether climate change could threaten the allocation of environmental flows. As a result, climate change was identified as a risk requiring further management by the catchment management authority, the Murray-Darling Basin Commission and further investigations are underway.

Actions consistent with adaptation to climate such as capped allocations, better environmental flows, improved irrigation management and moves to improve water quality are ongoing and climate change has been recognised as having an impact on the success of all these measures.

A recent stakeholder workshop identified the following items for further research:

- Predict inflows to dams and water allocations with three to six months lead times to help manage cropping risks
- Forecast temperatures and potential evaporation with three to six week lead times
- Predict flows in unregulated streams and link with environmental flow requirements

- Build climate variability and climate change research inputs into key national water reforms
- Program integrated research into recent rainfall reductions in eastern Australia, similar to that instituted to south-west Western Australia.

## References

**Herron**, N., Davis, R. and Jones, R.N. (2002). The effects of large-scale afforestation and climate change on water allocation in the Macquarie River Catchment, NSW, Australia, *Journal of Environmental Management*, **65**, pp. 369-381.

**Jones**, R.N. and Page, C.M. (2001). Assessing the risk of climate change on the water resources of the Macquarie River Catchment, in *Integrating Models for Natural Resources Management across Disciplines, issues and scales* (Part 2), Modelling and Simulation Society of Australia and New Zealand, Canberra, pp. 673-678.

**Jones**, R.N., Whetton, P.H., Walsh, K.J.E. and Page, C.M. (2001). Future impacts of climate variability, climate change and land use change on water resources in the Murray Darling Basin, [http://www.thelivingmurray.mdbc.gov.au/]. Murray-Darling Basin Commission, Canberra, ACT.

**NLWRA**, (2001). *Australian Water Resources Assessment 2000*. [http://www.nlwra.gov.au/] National Land & Water Resources Audit, Turner ACT, IPCC Data Distribution Centre http://ipcc-ddc.cru.uea.ac.uk.

# Adaptation Policy Frameworks
# for Climate Change:
# Developing Strategies,
# Policies and Measures: Annexes

# A

---

# Glossary of Terms

---

This section provides definitions for many of the concepts and terms used in the Adaptation Policy Framework (APF). In most definitions, references are given to the applicable Technical Papers (TPs) where additional details concerning the particular topic can be found. Citations to the technical literature on a given topic are, in turn, found in the TPs themselves.

For some terms, such as vulnerability and risk, definitions vary between disciplines and contexts. In these cases, a broad definition is provided, together with alternative definitions. The motivation here is to provide flexibility to users to adapt the APF to their own applications.

**Adaptation** – is a process by which strategies to moderate, cope with, and take advantage of the consequences of climatic events are enhanced, developed, and implemented.

**Adaptation baseline** – also referred to as an adaptation *policy* baseline, this includes a description of adaptations to current climate that are already in place (e.g., existing risk mitigation policies and programmes) (TP6). See also *project baseline*.

**Adaptation Policy Framework** – is a structured process for developing adaptation strategies, policies, and measures to enhance and ensure human development in the face of climate change, including climate variability. The APF is designed to link climate change adaptation to sustainable development and other global environmental issues. It consists of five basic Components: scoping and designing and adaptation project, assessing current vulnerability, characterising future climate risks, developing an adaptation strategy, and continuing the adaptation process (Executive Summary and User's Guidebook).

**Adaptive capacity** – is the property of a system to adjust its characteristics or behaviour, in order to expand its coping range under existing climate variability, or future climate conditions (TP7). The expression of adaptive capacity as actions that lead to adaptation can serve to enhance a system's coping capacity and increase its coping range (TPs 4 and 5) thereby reducing its vulnerability to climate hazards (TP3). The adaptive capacity inherent in a system represents the set of resources available for adaptation, as well as the ability or capacity of that system to use these resources effectively in the pursuit of adaptation. It is possible to differentiate between adaptive potential, a theoretical upper boundary of responses based on global expertise and anticipated developments within the planning horizon of the assessment, and adaptive capacity that is constrained by existing information, technology and resources of the system under consideration.

**Adaptive-capacity approach** – is one of several conceptual and analytical approaches that can be applied to adaptation projects. With this approach, a project can investigate a system with respect to its current adaptive capacity, and assess ways in which adaptive capacity can be increased (or ways in which it may be lessened) so that the system is better able cope with climate variability and change (TP7). See also *adaptation project approaches*.

**Adaptation capacity baseline** – includes a description of the current capacity within a priority system to cope with and adapt to climate variability (TP7). See also *project baseline*.

**Adaptation project approaches** – are conceptual and analytical approaches that can be selected to respond to the unique needs of adaptation projects (TP1). Four major approaches that can be applied to adaptation projects include the hazards-based approach, the vulnerability-based approach, the adaptive-capacity approach and the policy-based approach. See also the individual project approach definitions.

**Baselines** – used in two distinct ways in the APF, the term "baseline" can refer to either a *project baseline* (definitions) or a future baseline or *reference scenario* (definition). The project baseline describes where the project is starting from (for use in, e.g., subsequent monitoring and evaluation), while the reference scenario provides a plausible picture of a future in the priority system *without* adaptation, to allow for comparison of different adaptation strategies, policies and measures.

**Climate change** – refers to any change in climate over time, whether due to natural variability or because of human activity (IPCC, 2001). See also *climate variability*.

**Climate change vulnerability** – is the degree to which a system is susceptible to, or unable to cope with the adverse effects of climate change, including climate variability and extremes (IPCC, 2001) (TPs 4 and 5). See also *vulnerability*.

**Climate risk baseline** – includes a description of the current climate risk within the priority system (i.e., the probability of a climate hazard combined with the system's current vulnerability) (TPs 4 and 5). See also *project baseline*.

**Climate variability** – refers to variations in the mean state and other statistics (such as standard deviations, the occurrence of extremes, etc.) of the climate on all temporal and spatial scales beyond that of individual weather events. Variability may result from natural internal processes within the climate system (internal variability) or to variations in natural or anthropogenic external forcing (external variability) (IPCC, 2001). See also *climate change*.

**Coping range** – is the range of climate where the outcomes are beneficial or negative but tolerable; beyond the coping range, the damages or loss are no longer tolerable and a society (or system) is said to be vulnerable (TPs 4 and 5).

**Cost-benefit analysis** – is a quantitative method that makes a detailed comparison of the costs and benefits of a particular measure, or set of measures (TP8). A decision to fund a project, e.g., can depend on the ratio of benefits to costs – the higher the ratio, the more attractive the investment. Its major advantages are its verifiable bottom line and its familiarity to ministries and planning agencies. Disadvantages include limitations regarding the ability to directly address equity considerations and represent non-quantifiable benefits.

**Evaluation** – is a process for determining systematically and objectively the relevance, efficiency, effectiveness and impact of the adaptation strategies in the light of their objectives (TP9). See also *monitoring*.

**Exposure** – is the nature and degree to which a system is exposed to significant climatic variations (IPCC, 2001).

**Food insecurity** – a situation that exists when people lack secure access to sufficient amounts of safe and nutritious food for normal growth and development and an active and healthy life. It may be caused by the unavailability of food, insufficient purchasing power, inappropriate distribution, or inadequate use of food at the household level. Food insecurity may be chronic, seasonal, or transitory. More recent literature focuses on livelihood security – an expansion of food security to include multiple stresses and sectors to which livelihoods might be exposed (TP3).

**Hazard** – is used here to describe a physically defined climate event with the potential to cause harm, such as heavy rainfall, drought, flood, storm and long-term change in mean climatic variables such as temperature (TPs 4, 5, and 7).

**Hazards-based approach** – one of several conceptual and analytical approaches to adaptation projects, this approach places its starting emphasis on the biophysical aspects of climate-related risk – i.e., the climate hazard. With the hazards-based approach (also referred to as either the natural hazards-based (TPs 4 and 5) or climate risk-based approach), a project can assess current climate vulnerability or risk in the priority system (TP4), and use climate scenarios to estimate changes in vulnerability or risk over time and space (TP5). See also *adaptation project approaches*.

**Hybrid** – is used here to refer to approaches that apply uniform and site-specific methods in tandem and within an iterative process to develop and assess the range of adaptation strategies (TP8).

**Impacts** – are the detrimental and beneficial consequences of climate change on natural and human systems (IPCC, 2001).

**Indicators** – are quantitative or qualitative parameters that provide a simple and reliable basis for assessing change. In the context of the APF, a set of indicators is used to characterise an adaptation phenomenon, to construct a baseline, and to measure and assess changes in the priority system (TPs 1 and 6). See also *baseline, evaluation* and *monitoring*.

**Logical Framework ("Logframe") Analysis Approach** – is a project planning tool that includes project goals, objectives and activities, with specific outputs and measurable indicators of achievements.

**Measure** – see *policies and measures*.

**Monitoring** – is a mechanism or mechanisms used to track progress in the implementation of an adaptation strategy and its various components in relation to established targets (TP9). See also *evaluation* and *indicators*.

**Policies and measures** – usually addressed together, respond to the need for climate adaptation in distinct, but sometimes overlapping ways (TP8). *Policies*, generally speaking, refer to objectives, together with the means of implementation. In an adaptation context, a policy objective might be drawn from the overall policy goals of the country – for instance, the maintenance or strengthening of food security. Ways to achieve this objective might include, e.g., farmer advice and information services, seasonal climate forecasting and incentives for development of irrigation systems. *Measures* can be individual interventions or they can consist of packages of related measures. Specific measures might include actions that promote the chosen policy direction, such as implementing an irrigation project, or setting up a farmer information, advice and early warning programme. Both of these measures would contribute to the national goal of food security. See also *strategy*.

**Policy-based approach** – is one of several conceptual and analytical approaches that can be applied to adaptation projects. With this approach, a project can test a new policy being framed to see whether it is robust under climate change, or test an existing policy to see whether it manages anticipated risk under climate change (TP6). See also *adaptation project approaches*.

**Priority system** – is the focus of an adaptation project. The priority system (or systems) is generally characterised by high vulnerability to different climate hazards, as well as strategic importance at local and/or national levels. Socio-economic and biophysical criteria are often used to select priority systems by a given stakeholder group, and to set system parameters (indicators) for a given project (TPs 2 and 3). See also *system*.

**Probability** defines the likelihood of an event or outcome occurring. Probability can range from being qualitative, using word descriptions such as *likely* or *highly confident*, to quantified ranges and single estimates, depending on the level of understanding of the causes of events, historical time series and future conditions (TP4). See also *risk*.

**Project baseline** – is a description of where the project is starting from e.g., who is vulnerable to what, and what is currently being done to reduce that vulnerability (TP1). Project baselines are generally focused on the priority system, and are thus site-specific and limited to the duration of the project. Depending on the approach used in an adaptation project, a project baseline will be characterised by a set of quantitative and/or qualitative indicators, and may take the form, e.g., of one of the following:

- a *vulnerability baseline* (TP3)
- a *climate risk baseline* (TPs 4 and 5)
- an *adaptive capacity baseline* (TP7)
- or an *adaptation (policy) baseline* (TP6).

See also the individual baseline definitions. Project baselines can later be used in the monitoring and evaluation process to measure change (e.g., in vulnerability, adaptive capacity, climate risk) in the priority system, and the effectiveness of adaptation strategies, policies and measures.

**Reference scenario** – is an internally coherent description of a possible future without consideration of climate change; Depending on a project's needs and design, APF users may choose to develop reference scenarios, or future baselines, that represent future conditions in the priority system, in the absence of climate adaptation (TPs 1 and 6). Additional scenarios, in which various adaptations are applied, may also be developed and compared with reference scenarios to evaluate the implications of different adaptation strategies, policies and measures. Reference scenarios differ from project baselines in that they deal with the longer term and are used for informing policy decisions concerned with various development pathways at the strategic planning level.

**Resilience** – is the amount of change a system can undergo without changing state (IPCC, 2001).

**Risk (climate-related)** – is the result of the interaction of physically defined hazards with the properties of the exposed systems – i.e., their sensitivity or (social) vulnerability (TPs 3, 4, 5 and 7). Risk can also be considered as the combination of an event, its likelihood, and its consequences – i.e., risk equals the probability of climate hazard multiplied by a given system's vulnerability. See also *probability* and *vulnerability*.

**Scenario** – is a plausible and often simplified description of how the future may develop, based on a coherent and internally consistent set of assumptions about driving forces and key relationships. Scenarios may be derived from projections, but are often based on additional information from other sources, sometimes combined with a "narrative storyline" (IPCC, 2001) (TP6). See also *reference scenario*.

**Sector** – refers to a part or division, as of the economy (e.g., the manufacturing sector, the services sector) or the environment (e.g., water resources, forestry).

**Sensitivity (climate-related)** – is the degree to which a system is affected, either beneficially or adversely, by climate-related stimuli (IPCC, 2001). Sensitivity affects the magnitude and/or rate of a climate related perturbation or stress (while vulnerability is the degree to which a system is susceptible to harm from that perturbation or stress) (TPs 3 and 4). See also *climate change vulnerability*, *exposure* and *vulnerability*.

**Site-specific approaches** – seek to develop and assess detailed adaptation strategies on the basis of specific perceptions of vulnerability that have emerged from the full range of stakeholders at the site level (e.g., local communities, local project). See also *uniform approaches*.

**Socio-economic vulnerability** – is an aggregate measure of human welfare that integrates environmental, social, economic and political exposure to a range of harmful perturbations (TP6).

See also *climate change vulnerability* and *vulnerability*.

**Stakeholders** – are those who have interests in a particular decision, either as individuals or as representatives of a group. This includes people who influence a decision, or can influence it, as well as those affected by it (Hemmati, 2002) (TPs 1 and 2).

**Strategy** – refers to a broad plan of action that is implemented through policies and measures. A climate change adaptation strategy for a country refers to a general plan of action for addressing the impacts of climate change, including climate variability and extremes. It may include a mix of policies and measures, selected to meet the overarching objective of reducing the country's vulnerability. Depending on the circumstances, the strategy can be comprehensive at a national level, addressing adaptation across sectors, regions and vulnerable populations, or it can be more limited, focusing on just one or two sectors or regions (TP8). See also *policies and measures*.

**System** – may refer to a region, a community, a household, an economic sector, a business, a population group, etc., that is exposed to varying degrees to different climate hazards (TPs 1 and 3). See also *priority system*.

**Uncertainty** – is an expression of the degree to which a value (e.g., the future state of the climate system) is unknown (TP5).

**Uniform approaches** – seek to develop and assess broad adaptation strategies on the basis of a comprehensive perception of vulnerability that may exist – e.g., across sectors, across regions, across development challenges (TP8). See also *site-specific approaches*.

**Vulnerability** – The degree to which an *exposure unit* is susceptible to harm due to exposure to a perturbation or stress, and the ability (or lack thereof) of the exposure unit to cope, recover, or fundamentally adapt (become a new system or become extinct) (Kasperson et al., 2000). It can also be considered as the underlying exposure to damaging shocks, perturbations or stresses, rather than the probability or projected incidence of those shocks themselves (TPs 3, 4, and 5). See also *climate change vulnerability* and *socio-economic vulnerability*.

**Vulnerability-based approach** – one of several conceptual and analytical approaches to adaptation projects, this approach places its starting emphasis on the socio-economic aspects of climate-related risk. With the vulnerability-based approach (TP3), a project focuses on the characterisation of a priority system's vulnerability and assesses how likely critical thresholds of vulnerability are to be exceeded under climate change. Use of the vulnerability-based approach can feed into a larger climate risk assessment (TPs 3, 4 and 5). See also *adaptation project approaches*.

**Vulnerability baseline** – includes a description of current vulnerabilities to climate variability and events (TPs 3 and 4). See also *project baseline*.

# References

**Hemmati**, M. (2002). *Multi-stakeholder Processes for Governance and Sustainability*, Earthscan, London.

**IPCC**, (2001). *Climate Change 2001: Impacts, Adaptation and Vulnerability.* Contribution of Working Group II to the Third Assessment Report of the Intergovernmental Panel on Climate Change. IPCC/WMO/UNEP.

**Kasperson**, J.X., Kasperson, R.E., Turner, II, B.L., Hsieh, W. and Schiller, A. (2002). Vulnerability to global environmental change. In *The Human Dimensions of Global Environmental Change*, ed. Andreas Diekmann, Thomas Dietz, Carlo C. Jaeger, and Eugene A. Rosa. Cambridge, MA : MIT Press (forthcoming).

# B

## List of Additional Reviewers

The National Communications Support Programme would like to acknowledge the following experts, who provided general comments on the APF structure during the review process:

| Chris Gordon | cg@afriwet.org |
| Alexandre Cabral | Directeur National du projet, Changements Climatiques en Guinée-Bissau |
| Merylyn McKenzie Hedger | Environment Agency, Bristol, UK |
| Javier Gonzales Iwanciw | Ministry of Sustainable Development, Bolivia |
| Rachid Ouali | Ministry of Foreign Affairs, Algiers, Algeria |
| Abdelghani Beloued | Ministère de l'Agriculture et du Developpement Rural, Algiers |
| Sid Ali Ramdane | Ministère de l' Aménagement du Territoire et de l'Environnement, Algiers |
| Yumiko Yasada | UNDP-GEF, Kuala Lumpur, Malaysia |
| Rebecca Carman | UNDP-GEF, New York, United States |
| Yamil Bonduki | UNDP-GEF, New York, United States |
| Qin Dahe | China Meteorological Administration |
| Robert Mendelsohn | Yale School of Forestry and Environmental Studies, New Haven, United States |
| Ana Rosa Moreno | The United States-Mexico Foundation for Science, Mexico City, Mexico |
| Rodel D. Lasco | University of the Philippines, Los Baños |
| Balgis M. E. Osman | Higher Council for Environment & Natural Resources, Sudan |
| Kees Dorland | Institute for Environmental Studies, Netherlands |

The Adaptation Policy Framework went through three major rounds of reviews in November 2002, April 2003 and June 2003. The list of expert reviewers was as follows:

**Antigua and Barbuda**

| Brian Challenger | Ministry of Public Utilities |

**Argentina**

| Gracie laMagrin | Instituto Nacional de Tecnología Agropecuaria |
| Osvaldo Canziani | IPCC WG II Co-Chair |
| Martin de Zurviria | |
| Daniel Bouille | Fundación Bariloche |

**Australia**

| Tas Sakellaris | Australian Greenhouse Office |
| Robert Wasson | Australian National University |
| Chris Mitchell | Commonwealth Scientific & Industrial Research Organisation |
| Peter Whetton | Commonwealth Scientific & Industrial Research Organisation |
| John Handmer | Royal Melbourne Institute of Technology |
| Nick Harvey | University of Adelaide |
| Jon Barnett | University of Melbourne |

**Austria**

| Jill Jäger | Global Environmental Change series |

**Bahamas**

| Philip S. Weech | The Bahamas Environment, Science and Technology Commission |

**Bangladesh**

| Mozaharul Alam | Bangladesh Centre for Advanced Studies |
| Atiq Rahman | Bangladesh Centre for Advanced Studies |
| Nasimul Haque | Bangladesh Centre for Advanced Studies |
| Mizan Khan | Sustainable Development Networking Programme |

**Barbados**

| Neville Trotz | Caribbean Planning for Adaptation to Global Climate Change |

**Bhutan**

| Dechen Tsering | National Environmental Commission |

**Bolivia**

| Javier Gonzales Iwanciw | Nur University – National Climate Change Program |
| Oscar Paz | Programa Nacional de Cambio Climático |

**Brazil**

| Carlos Nobre | Centro de Previsão de Tempo e Estudos Climáticos |
| Roberto Schaeffer | Energy Planning Program |
| Thelma Krug | IPCC Task Force Bureau co-chair |
| José Miguez | Ministerio da Ciencia e Tecnologia |

**Canada**

| Mike Brklacich | Carleton University |
| David Cooper | Convention on Biological Diversity |
| Philip Baker | Canadian International Development Agency |
| Gretchen de Boer | Canadian International Development Agency |
| Pierre Giroux | Canadian International Development Agency |
| Liza LeClerc | Canadian International Development Agency |
| Tana Lowen Stratton | Canadian International Development Agency |
| Patti Edwards | Environment Canada |
| John Drexhage | International Institute of Sustainable Development |
| Pamela Kertland | Natural Resources Canada |
| Hadi Dowlatabadi | Sustainable Development Research Institute |
| Barry Smit | University of Guelph |
| Monirul Mirza | University of Toronto |
| Jim Bruce | Global Change Strategies International |
| John Stewart | |
| Roger Street | Environment Canada |

**Chile**

| | |
|---|---|
| Juan Pedro Solar Searle | Comisión Nacional del Medio Ambiente |
| Alejandro León | Universidad de Chile |

**China**

| | |
|---|---|
| Xu Yinlong | Chinese Academy of Agricultural Sciences |
| M.A. Aimin | State Development Planning Commission |
| Lin Erda | Chinese Academy of Agricultural Sciences |

**Cook Islands**

| | |
|---|---|
| Pasha Carruthers | Pacific Islands Climate Change Assistance Programme |

**Costa Rica**

| | |
|---|---|
| Roberto Villalobos | National Meteorological Institute |
| Ana Rita Chacon | National Meteorological Institute |

**Croatia**

| | |
|---|---|
| Sonja Vidic | Hydro and Meteorology Service |

**Cuba**

| | |
|---|---|
| Avelino Suarez | Cuban Environmental Agency |

**Ecuador**

| | |
|---|---|
| Luis Cáceres | Ministerio del Ambiente |

**Egypt**

| | |
|---|---|
| Ahmed Amin | Information & Decision Support Center |
| Helmi Eid | Soil, Water and Environment Research Institute |
| Mohamed El Raey | University of Alexandria |

**El Salvador**

| | |
|---|---|
| Yvette Aguilar | Environment and Natural Resources Ministry |

**Ethiopia**

| | |
|---|---|
| Abebe Tadege | National Meteorological Services Agency |

**Fiji**

| | |
|---|---|
| Kanayathu C Koshy | Pacific Centre for Environment |
| Mahendra Kumar | University of the South Pacific |

**Finland**

| | |
|---|---|
| Tim Carter | Finnish Environment Institute |

**France**

| | |
|---|---|
| Jan Corfee-Morlot | Organisation for Economic Co-operation and Development |
| Shardul Agrawala | Organisation for Economic Co-operation and Development |
| Monique Mainguet | Université de Reims |

**Gambia**

| | |
|---|---|
| Bubu Jallow | Department of Water Resources |

**Germany**

| | |
|---|---|
| Richard Tol | Hamburg University |
| Anke Herold | Öko-Institut |
| Richard Klein | Potsdam Institute for Climate Impact Research |
| Martin Welp | Potsdam Institute for Climate Impact Research |
| George Manful | United Nations Framework Convention on Climate Change |
| Youssef Nassef | United Nations Framework Convention on Climate Change |
| Martha Perdomo | United Nations Framework Convention on Climate Change |
| Olga Pilisofova | United Nations Framework Convention on Climate Change |
| Graham Sem | United Nations Framework Convention on Climate Change |
| Dennis Tirpak | United Nations Framework Convention on Climate Change |
| Holger Liptow | Deutsche Gesellschaft für Technische Zusammenarbeit |

**Ghana**

| | |
|---|---|
| Chris Gordon | Centre for African Wetlands |

**Guatemala**

| | |
|---|---|
| Carlos Mansilla | Comisión Nacional del Medio Ambiente |

**Honduras**

| | |
|---|---|
| Mirna Marin | Sub-secretaría de Ambiente |

**India**

| | |
|---|---|
| Murari Lal | Indian Institute of Technology |
| Jyoti Parikh | Indira Gandhi Institute of Development Research |
| Rajendra Pauchari | IPCC Chairman |
| Wajih Naqvi | National Institute of Oceanography |
| Sujata Gupta | The Energy and Resources Institute |

**Italy**

| | |
|---|---|
| Gustavo Best | Food and Agriculture Organization |
| Paola Rossi | Università di Bologna |
| Bettina Menne | World Health Organization |

**Japan**

| | |
|---|---|
| Taka  Hiraishi | Institute for Global Environmental Strategies |
| Taka Hiraishi | IPCC Task Force Bureau co-chair |

**Kazakstan**

| | |
|---|---|
| Irina Yesserkepova | Scientific Research Institute of the KAZHYDROMET |

**Kenya**
Richard Odingo          IPCC Vice-Chair
John Nganga             Kenya Meteorological Department
Ravi Sharma             United Nations Environment Programme

**Lebanon**
Samir Safi              Université Libanaise

**Malaysia**
Mastura Mahmud          Universiti Kebangsaan

**Mali**
Mama Konate             Direction Nationale de Météorologie

**Mexico**
Julia Carabias          STAP Chair
Ana Rose Moreno         USA-MEX Foundation for Science
Carlos Gay              Universidad Nacional Autónoma
                        de México
Julia  Martínez         Instituto Nacional de Ecología/
                        SEMARNAT

**Morocco**
Faouzi  Senhaji         Mahgreb

**New Zealand**
Jim Salinger            National Institute of Water &
                        Atmospheric Research Limited
Alistair Woodward       University of Otago
Paul Kench              University of Waikato
John Campbell           University of Waikato
John Hay                University of Waikato
Richard Warrick         University of Waikato

**Nicaragua**
Freddy Picado Trana     Ministerio del Ambiente y los Recursos
                        Naturales
Mario Torres Lezama     Ministerio del Ambiente y los Recursos
                        Naturales

**Niger**
Mohammed Sadeck         African Centre of Meteorological
  Boulahya              Application for Development
Ben Mohamed
  Abdelkrim             Direction de la Météorologie Nationale
**Norway**
Karen O'Brien           Center for International Climate and
                        Environmental Research – Oslo
Harald Dovland          Ministry of the Environment

**Pakistan**
Mahboob Elahi           Ministry of the Environment
Amir Muhammed           National University of Computer &
                        Emerging Sciences

**Panama**
Ligia Castro            Centro del Agua del Trópico Húmedo
                        para América Latina y  el Caribe
Eduardo  Reyes          Autoridad Nacional de Ambiente

**Samoa**
Taito Nakalevu          South Pacific Regional Samoa
                        Environment Programme
Laavasa Malua           Department of Lands, Survey &
                        Environment

**Saudi Arabia**
Taha Zatari             Presidency of Meteorology and
                        Environment

**Senegal**
Mamadou Dansokho        University Cheikh Anta Diop
Youba Sokona            Environmental Development of the
                        Third World
Isabelle Niang Diop     Environmental Development of the
                        Third World

**Slovakia**
Ivan Mojik              Ministry of the Environment

**South Africa**
Lauraine Lotter         Chemical & Allied Industries
                        Association
Ogunlade Davidson       IPCC WG III Co-Chair
Emma Archer             University of Cape Town
Bruce Hewitson          University of Cape Town
Roland Schulze          University of Natal     South
Colleen Vogel           University of Witwatersrand

**Sweden**
Angela Churie
  Kallhauge             Swedish Energy Agency

**Switzerland**
Nicole North            INFRAS
Othmar Schwank          INFRAS
Brett Orlando           International Union for Conservation of
                        Nature and Natural Resources
Annie  Roncerel         United Nations Institute for Training and
                        Research
Carlos Corvalan         World Health Organization
Michael Couglan         World Meteorlogical Organization
Paul Llanso             World Meteorlogical Organization
M.V. K. Sivakumar       World Meteorlogical Organization

**Tanzania**
Hubert Meena            Centre for Energy, Environment, Science
                        & Technology
Richard Muyungi         Division of Environment

**Thailand**
VuteWangwacharakul      Kasetsart University
Liam Salter             World Wildlife Fund

## The Netherlands

| | |
|---|---|
| Bill Hare | Greenpeace |
| Rob Swart | Head of TSU, IPCC WG III |
| Bert Metz | IPCC WG III Co-Chair |
| Madeleen Helmer | Red Cross |
| Kees Dorland | National Institute for Public Health and the Environment |
| Maarten van Aalst | Universiteit Utrecht |
| Roland Rodts | |

## The Philippines

| | |
|---|---|
| Philippine Climate Change Information Center | |
| Rex Cruz | SysTem for Analysis, Research and Training |
| Rodel Lasco | EcoMarket Solutions |

## Togo

| | |
|---|---|
| Ayite-lo Ajavon | Université de Lomé |

## Uganda

| | |
|---|---|
| Stephen Magezi | Department of Meteorology |

## UNDP

| | |
|---|---|
| Susan McDade | Bureau for Development Policy/ Energy and Environment Group |
| Charles McNeill | Bureau for Development Policy/ Energy and Environment Group |
| Minoru Takada | Bureau for Development Policy/ Energy and Environment Group |
| Pascal Girot | Bureau for Development Policy/ SURF – Costa Rica |
| Juha Uitto | Global Environment Facility |
| Arun Kashyap | Bureau for Development Policy/ Energy and Environment Group |
| Martin Krause | Global Environment Facility |
| Walter Baethgen | Uruguay Country Office |

## United Kingdom

| | |
|---|---|
| John Gash | Centre for Ecology and Hydrology |
| John Ingram | Centre for Ecology and Hydrology |
| Caroline Fish | Department for Environment Food and Rural Affairs |
| David Warrilow | Department for Environment Food and Rural Affairs |
| Diana Wilkins | Department for Environment Food and Rural Affairs |
| Penny Bramwell | Department of the Environment, Transport and the Regions |
| Tim Foy | Department for International Development |
| Sam Fankhauser | European Bank for Reconstruction and Development |
| Merylyn McKenzie Hedger | Environment Agency |
| Kate Hampton | Friends of the Earth |

| | |
|---|---|
| Jose Furtado | Imperial College |
| Roy Behnke | Overseas Development Institute |
| Nabeel Hamdi | Oxford Brooks University |
| Dennis Anderson | Scientific and Technical Advisory Panel – Climate Change |
| Nigel Arnell | University of Southampton |
| Peter Jones | University College – London |
| Declan Conway | University of East Anglia |
| Mike Hulme | University of East Anglia |
| Tim Osborn | University of East Anglia |
| Martin Parry | University of East Anglia |
| Emma Tompkins | University of East Anglia |
| Simon Shackley | University of Manchester Institute of Science and Technology |
| Gina Ziervogel | University of Oxford Environmental Change Institute |
| Frans Berkhout | University of Sussex |
| Julia Hertin | University of Sussex |

## United States

| | |
|---|---|
| Adil Najam | Boston University |
| Rawleston Moore | California Postsecondary Agriculture Articulation Collaborative |
| Leonard Nurse | California Postsecondary Agriculture Articulation Collaborative |
| Diana Liverman | Center for Latin American Studies |
| Robin Reid | Consultative Group on International Agricultural Research |
| Peter Thornton | Consultative Group on International Agricultural Research |
| Marc Levy | Center for International Earth Science Information Network, Columbia University |
| Dennis Ojima | Colorado State University |
| Roger Pielke Sr | Colorado State University |
| AntoinetteWannebo | Columbia University |
| Jack Fitzgerald | Department of Energy |
| Bev McIntyre | Department of State |
| Nancy Lewis | East West Center |
| Eileen Shea | East West Center |
| Anne Grambsch | Environmental Protection Agency |
| Michael Slimak | Environmental Protection Agency |
| Yasemin Biro | Global Environment Facility |
| Claudio Volonte | Global Environment Facility |
| William Gutowski | Iowa State University |
| Mohan Munasinghe | IPCC Vice-Chair |
| Susan Solomon | IPCC WG II Co-Chair |
| Jonathon Patz | Johns Hopkins Bloomberg School of Public Health |
| Andrew Githeko | Malaria Research Network, National Library of Medicine |
| Susan Capalbo | Montana State University |
| Ana Iglesias | NASA Goddard Institute for Space Studies |
| Paul Filmer | National Science Foundation |
| Fred Semazzi | North Carolina State University |
| Michael Sale | Oak Ridge National Laboratory |

| | |
|---|---|
| Tom Wilbanks | Oak Ridge National Laboratory |
| Anne Arquit Niederberger | Policy Solutions |
| Otto Doering | Purdue University |
| Joel Smith | Stratus Consulting |
| Neil Leary | SysTem for Analysis, Research and Training |
| Hassan Virji | SysTem for Analysis, Research and Training |
| Tariq Banuri | Tellus Institute |
| Bill Dougherty | Tellus Institute |
| Erika Spanger-Siegfried | Tellus Institute |
| Suzanne Mozer | Union of Concerned Scientists |
| Robert Harriss | University Corporation for Atmospheric Research |
| James Shuttleworth | University of Arizona |
| Soroosh Sorooshian | University of Arizona |
| Emilio Laca | University of California – Davis |
| Paul Freeman | University of Denver |
| Robert Peart | University of Florida |
| Maria Carmen Lemos | University of Michigan |
| Camille Parmesan | University of Texas – Austin |
| Paul Desanker | University of Virginia |
| Dennis Lettenmaier | University of Washington |
| Richard Moss | US Global Change Research Program |
| Ko Barrett | US Agency for International Development |
| John Kimble | US Department of Agriculture |
| Dave Schimmelpfennig | US Department of Agriculture |
| Virginia Burkett | US Geological Survey |
| Patrick Gonzales | US Geological Survey |
| Arthur Horowitz | US Geological Survey |
| Frank Sperling | World Bank |
| Ajay Mathur | World Bank |
| Robert Mendelsohn | Yale University |
| Sally Kane | National Science Foundation |
| Robert Kates | Maine Global Climate Change |

**Uruguay**

| | |
|---|---|
| Cecilia Ramos Mane | Comision Nacional sobre el Cambio Global |

**Uzbekistan**

| | |
|---|---|
| Sergey Myagkov | Main Administration of Hydrometeorology |

**Venezuela**

| | |
|---|---|
| Lelys Bravo de Guenni | Centro de Estadística y Software Matemático, Universidad Simón Bolívar |

**Zambia**

| | |
|---|---|
| Lubinda Aongola | Ministry of Environment and Natural Resources |

**Special Acknowledgements**

| | |
|---|---|
| Maritza Ascencios | UNDP-GEF, New York, United States |
| Yamil Bonduki | UNDP-GEF, New York, United States |
| Rebecca Carman | UNDP-GEF, New York, United States |
| Laurie Douglas | Laurie Douglas Graphic Design, New York, United States |